Springer Proceedings in Mathematics & Statistics

Volume 96

More information about this series at http://www.springer.com/series/10533

Springer Proceedings in Mathematics & Statistics

This book series features volumes composed of select contributions from workshops and conferences in all areas of current research in mathematics and statistics, including OR and optimization. In addition to an overall evaluation of the interest, scientific quality, and timeliness of each proposal at the hands of the publisher, individual contributions are all refereed to the high quality standards of leading journals in the field. Thus, this series provides the research community with well-edited, authoritative reports on developments in the most exciting areas of mathematical and statistical research today.

Denis Ibadula • Willem Veys
Editors

Bridging Algebra, Geometry, and Topology

Springer

Editors
Denis Ibadula
Faculty of Mathematics
and Informatics
Ovidius University
Constanta, Romania

Willem Veys
Department of Mathematics
University of Leuven
Leuven (Heverlee), Belgium

ISSN 2194-1009 ISSN 2194-1017 (electronic)
ISBN 978-3-319-09185-3 ISBN 978-3-319-09186-0 (eBook)
DOI 10.1007/978-3-319-09186-0
Springer Cham Heidelberg New York Dordrecht London

Library of Congress Control Number: 2014950409

Mathematics Subject Classification (2010): 13-XX, 14-XX, 51-XX, 54-XX

Printed on acid-free paper

Springer is part of Springer Science+Business Media (www.springer.com)

Preface

The present volume contains refereed papers which were presented at the International Conference "*Experimental and Theoretical Methods in Algebra, Geometry and Topology*," held in Eforie Nord (near Constanta), Romania, during 20–25 June 2013. The conference was devoted to the 60th anniversary of two distinguished Romanian mathematicians: *Alexandru Dimca* from the Université de Nice-Sophia Antipolis, France and *Ştefan Papadima* from the Institute of Mathematics "Simion Stoilow" of the Romanian Academy.

The conference brought together more than 80 experts in topology, singularities, hyperplane arrangements, combinatorics, algebraic geometry, and commutative algebra from almost all continents of the world. It aimed to strengthen regional and professional networking between high level specialists from different fields of mathematics.

This meeting had a special significance for the community of mathematicians from Romania; it was dedicated to a Romanian mathematician from diaspora and a "local" one, and many other top Romanian mathematicians, living and working abroad, honored the invitation to participate in the conference.

Algebra, geometry, and topology cover a variety of different, but intimately related, research fields in modern mathematics. This book focuses on specific aspects of this interaction.

The selected papers contain original research work and a survey paper. They are intended for a large audience, including researchers and graduate students interested in algebraic geometry, combinatorics, topology, hyperplane arrangements, and commutative algebra. The papers are written by well-known experts from different fields of mathematics, affiliated to universities from all over the word; they cover a broad range of topics and explore the research frontiers of a wide variety of contemporary problems of modern mathematics that are rarely brought together in this manner.

Now, we describe briefly the content of each paper from this volume.

Solutions of a zero-dimensional system of multivariate polynomials over the rationals could be found using the so-called triangular decomposition. Important work on this subject was done by Möller, who has also given an algorithm to compute the triangular decomposition. The purpose of the paper "*Solving via Modular Methods*" by Deeba Afzal, Faira Kanwal Janjua, Gerhard Pfister, and Stefan Steidel is to present a parallel modular algorithm in the line of Moreno Maza–Xie and Li–Moreno Maza with focus on a probabilistic algorithm. Due to the present architecture of the processors, the parallel algorithms are an important research direction in this field.

The paper of Marian Aprodu on "*Lazarsfeld–Mukai Bundles and Applications: II*" is a nice short exposition of the theory of Lazarsfeld–Mukai bundles on surfaces, providing some improvements of known results in the case of rational surfaces as well. The author explains how these bundles can be used in order to bound the dimension of Brill–Noether loci, which is in turn useful in addressing problems on syzygies, like the famous Green conjecture. The paper concludes with a solution of this conjecture for some special curves on special rational surfaces.

Multinets are of interest for their relationship with the fundamental group of a complex hyperplane arrangement complement and the cohomology of one-dimensional local systems. If all multiplicities are equal to 1, then a multinet is a net that is a realization by lines and points of several orthogonal Latin squares. Very few examples of multinets with nontrivial multiplicities are known. In the paper "*Multinets in* $\mathbb{P}^2$," Jeremiah Bartz and Sergey Yuzvinsky present new examples of multinets. These are obtained by using an analogue of nets in $\mathbb{P}^3$ and intersecting them by planes. Prior to the result in this paper, there were almost no systematic constructions of infinite families of multinets. This paper gives a new and natural construction, certain to be of further interest.

The coGalois theory studies the correspondence between subfields of a radical field extension L/K and subgroups of the coGalois group $\text{coG}(L/K) := Tors(L^\times/K^\times)$. In the paper "*A More General Framework for CoGalois Theory*," Şerban Basarab continues his work on abstract co-Galois theory which concerns a continuous action $\Gamma \times A \to A$ of a profinite group on a discrete quasi-cyclic group A. The author investigates Kneser triples and Cogalois triples providing general Kneser and Cogalois criteria. He states problems on the classification of the finite structure, arising naturally from these criteria.

Let P be a product of weighted projective spaces. In his paper "*Connectivity and a Problem of Formal Geometry*," Lucian Bădescu proves an algebraization result for formal-rational functions on certain closed subvarieties X of $P \times P$. First of all, the main theorem of the paper is an improvement of a connectivity result of L. Bădescu and F. Repetto. Secondly, the main result of the paper has two interesting corollaries. The first one extends to arbitrary characteristic and weighted projective spaces a result obtained in characteristic 0 and ordinary projective space by Faltings, and the last one extends also to the weighted case a result of Faltings.

The paper "*Hodge Invariants of Higher-Dimensional Analogues of Kodaira Surfaces*" of Vasile Brînzănescu deals with a class of compact complex manifolds which appear as torus-bundles over an elliptic curve. When they are non-Kählerian, they are called Kodaira manifolds since they are generalizations of the Kodaira surfaces to higher dimension. The main contributions of the paper are the computation of their Hodge numbers and establishing the existence of a holomorphic symplectic structure in the even dimensional case. The author thus brings further examples of holomorphic symplectic manifolds, which were recently intensively studied, although in a different setup.

In "*An Invitation to Quasihomogeneous Rigid Geometric Structures*," Sorin Dumitrescu surveys his results on quasihomogeneous rigid geometric structures of manifolds that encompass more than a decade of work. His results take place in the real or complex setting, sometimes concern general geometric structures, and sometimes concern affine connections or Lorenzian metrics. The problems that have been studied by Dumitrescu concern geometric structures whose isometry pseudogroup has big orbits, and the extent to which this already big orbit is all of the manifold, touching upon many parts of geometry.

The fundamental group of the complement of a singular plane curve, introduced and studied by Zariski as an approach to the deformation classification of surfaces of general type, has been a subject of close interest since then. Usually, this group is very difficult to compute, and it is equally difficult to produce examples of curves with interesting groups. In "*On the Fundamental Groups of Non-generic $\mathbb{R}$-Join-Type Curves*," Christophe Eyral and Mutsuo Oka use the concept of bifurcation graph (a version of Grothendieck's dessins d'enfants, developed by the authors earlier) and show that if the curve is sufficiently generic, then its fundamental group is still the minimal group determined by the multiplicities of the roots.

Motivated by questions in algebraic statistics, there is interest to study the binomial edge ideal of a finite simple graph G. In the paper "*Koszul Binomial Edge Ideals*," Viviana Ene, Jürgen Herzog, and Takayuki Hibi show that if the binomial edge ideal of G defines a Koszul algebra, then G must be chordal and claw free. A converse of this statement is proved for a class of chordal and claw free graphs.

In the paper "*Some Remarks on the Realizability Spaces of (3, 4)-Nets*," Denis Ibadula and Anca Daniela Măcinic prove that in the class of $(3,4)$-nets with double and triple points, lattice isomorphism actually translates into lattice isotopy. Moreover, they disprove the existence of Zariski pairs involving an example of Yoshinaga of a $(3,6)$-net with 48 triple points.

The paper of Ştefan Papadima and Alexander I. Suciu, "*Non-Abelian Resonance: Product and Coproduct Formulas*," is about algebraic aspects of resonance varieties. Originally, the resonance varieties appeared as a natural way to put together information about wedge products of the cohomology of a topological space X with 1-cohomology classes. This is the so-called formal rank 1 case, or, almost equivalently, the "formal abelian" case. In this paper, more general cases are considered by dropping one or both of the assumptions. In other words, the higher rank local systems are considered and possibly non-formal situations. The authors address algebraically, via generalized resonance varieties, the situations arising from the product and the wedge of two topological spaces. They thus address basic foundational questions and obtain concrete results, which will definitely be used in future research in this direction. In particular they describe precisely the rank 2 formal case.

The paper "*Gauss–Lucas and Kuo–Lu Theorems*" of Laurenţiu Păunescu is a nice, short paper which contains elementary observations on the classical Gauss–Lucas Theorem and the Kuo–Lu theorem. An elementary algebraic calculation over the Newton–Puiseux field, only employing its contact order structure, shows that the Kuo–Lu theorem is in fact a Gauss–Lucas type theorem, via a new notion of convexity over the Newton–Puiseux field.

The germ of a plane curve has a minimal desingularization, constructed by blowing up points. The number of blow-ups is an invariant, called blow-up complexity. María Pe Pereira and Patrick Popescu-Pampu, in "*Fibonacci Numbers and Self-Dual Lattice Structures for Plane Branches*," relate the blow-up complexity with the multiplicity and the Milnor number. Fibonacci numbers appear naturally in this study. The proofs are combinatorial, based on Enriques diagrams. The authors construct a partial order and a natural duality on the set of Enriques diagrams with fixed blow-up complexity. The duality is interesting and differs from projective duality in general.

Stanley's conjecture on multigraded modules over the polynomial ring $S = K[x_1, \cdots, x_n]$, where K is a field, has attracted many researchers in the last decade and is still widely open. Some progress has been done by Dorin Popescu in a series of works which treat Stanley's conjecture or weaker forms of it for squarefree multigraded modules of the form I/J where $I \supset J$ are squarefree monomial ideals in S. In the paper "*Four Generated, Squarefree, Monomial Ideals*," Adrian Popescu and Dorin Popescu show that under restricted conditions on the generators of I and J, if the Stanley depth of I/J is $d+1$, then depth $I/J \geq d+1$. Hence Stanley's conjecture holds in this case.

The aim of the paper of J.H.M. Steenbrink, "*Motivic Milnor Fibre for Nondegenerate Function Germs on Toric Singularities*," is to give a combinatorial formula for the motivic Milnor fiber of a nondegenerate function germ on a toroidal variety. The paper provides a nice introduction to the theory of motivic nearby cycles and the use of toroidal methods. The final expression of the given formula contains certain classes of varieties defined by an equation of the form $g = 1$ with g quasi-homogeneous.

In their paper "*The Connected Components of the Space of Alexandrov Surfaces*," Joël Rouyer and Costin Vîlcu study the connected components of the space of (pairwise non-isometric) Alexandrov (two-dimensional) surfaces. Denote by $\mathcal{A}(k)$ the set of all compact Alexandrov surfaces without boundary with curvature bounded below by k, endowed with the topology induced by the Gromov–Hausdorff metric. They determine the connected components of $\mathcal{A}(k)$ and of its closure.

In his paper "*Complements of Hypersurfaces, Variation Maps and Minimal Models of Arrangements*," Mihai Tibăr proves the minimality of the CW-complex structure for complements of hyperplane arrangements in $\mathbb{C}^n$. The proof of this important result, originally due to Randell and Dimca–Papadima, is obtained here using Lefschetz pencils and variation maps within a pencil of hyperplanes. There is a renewed interest in the techniques used in this paper, which are developed by the author in a long series of papers, because of the central role they played in the recent proof of a conjecture of Dimca–Papadima by Huh.

Finally, the paper "*Critical Points of Master Functions and the mKdV Hierarchy of Type* $A_2^{(2)}$" deals with m-parameter families of critical points of the master function associated with the trivial representation of the twisted affine Lie algebra $A_2^{(2)}$. Alexander Varchenko, Tyler Woodruff, and Daniel Wright show that the embedding of the family into the space $\mathcal{M}$ of the Miura opers of type $A_2^{(2)}$ defines a variety which is invariant with respect to all mKdV flows on $\mathcal{M}$, and that this variety is point-wise fixed by all flows of big enough index.

The conference was organized by Ovidius University of Constanta, in cooperation with the Institute of Mathematics "Simion Stoilow" of the Romanian Academy (IMAR) and the Romanian Mathematical Society, and it was a satellite conference of the Joint International Meeting of the American Mathematical Society and the Romanian Mathematical Society, held in 2013 in Alba Iulia, Romania.

The scientific committee of our conference was formed by Dan Burghelea (Ohio State University, USA), Octav Cornea (Centre de Recherches Mathématiques, Canada), Ezra Miller (Duke University, USA), Andras Némethi (Rényi Mathematical Institute, Hungary), Claude Sabbah (Ecole Polytechnique, France), Bernd Sturmfels (University of California, USA), Alexander Suciu (Northeastern University, USA), and Wim Veys (University of Leuven, Belgium).

The organizing committee was formed by Marian Aprodu (IMAR), Wladimir-Georges Boskoff (Ovidius University), Viviana Ene (Ovidius University), Denis Ibadula (Ovidius University), Anca Măcinic (IMAR), Nicolae Manolache (IMAR), and Alexander Suciu (Northeastern University).

The organizers would like to thank the Foundation Compositio Mathematica for providing a significant financial support for organizing this conference. We also gratefully acknowledge the support offered by our sponsors CNRS Franco-Romanian LEA Mathématiques et Modélisation, Institut Universitaire de France, and CNCS Grant PN-II-PCE-2011-3-028, Romanian Mathematical Society.

Special thanks to all contributors for the quality of their papers and to all referees for contributing to the improved scientific level of these proceedings. We would also like to thank Ms. J. Mary Helena, Project manager at Spi Global, Dr. Eve Mayer, Assistant Editor Mathematics from Springer Science and Business Media, and the Springer production team for their patient guidance in the preparation of this volume.

On behalf of all the contributors and the organizers of the conference, we dedicate this volume to the outstanding mathematicians, mentors, colleagues, and friends Alexandru Dimca and Ştefan Papadima, on the occasion of their 60th anniversary!

Constanta, Romania Denis Ibadula
Leuven, Belgium Wim Veys

Contents

Solving via Modular Methods

Deeba Afzal, Faira Kanwal Janjua, Gerhard Pfister, and Stefan Steidel

Abstract In this chapter we present a parallel modular algorithm to compute all solutions with multiplicities of a given zero-dimensional polynomial system of equations over the rationals. In fact, we compute a triangular decomposition using Möller's algorithm (Möller, Appl. Algebra Eng. Commun. Comput. 4:217–230, 1993) of the corresponding ideal in the polynomial ring over the rationals using modular methods, and then apply a solver for univariate polynomials.

Keywords Polynomial solving • Modular solving • Triangular sets

1 Introduction

One possible approach[1] to find the solutions of a zero-dimensional system of multivariate polynomials is the triangular decomposition of the corresponding ideal since triangular systems of polynomials can be solved using a univariate solver recursively. There are already several results in this direction including an implementation in *Maple* [1, 9, 14]. The technique to compute triangular sets has

[1]In SINGULAR [4] this approach is implemented in the library `solve.lib` on the basis of an univariate *Laguerre solver* [16, Sect. 8.9–8.13].

D. Afzal • F.K. Janjua (✉)
Abdus Salam School of Mathematical Sciences, GC University, Lahore, 68-B, New Muslim Town, Lahore 54600, Pakistan
e-mail: deebafzal@gmail.com; fairakanwaljanjua@gmail.com

G. Pfister
Department of Mathematics, University of Kaiserslautern, Erwin-Schrödinger-Str., 67663 Kaiserslautern, Germany
e-mail: pfister@mathematik.uni-kl.de

S. Steidel
Department of Mathematical Methods in Dynamics and Durability, Fraunhofer Institute for Industrial Mathematics ITWM, Fraunhofer-Platz 1, 67663 Kaiserslautern, Germany
e-mail: stefan.steidel@itwm.fraunhofer.de

D. Ibadula, W. Veys (eds.), *Bridging Algebra, Geometry, and Topology*,
Springer Proceedings in Mathematics & Statistics 96,
DOI 10.1007/978-3-319-09186-0_1

been refined in [2, 3, 10], and modularized and parallelized in [11, 15]. We report about a modular and parallel version of the solver in SINGULAR [4] and focus mainly on a probabilistic algorithm to compute the solutions of a polynomial system with multiplicities.

2 Preliminary Technicalities

We recall the definition and some properties of a triangular decomposition. For details we refer to [6, 12, 13].

Let K be a field, $X = \{x_1, \ldots, x_n\}$ a set of variables, $I \subseteq K[X]$ a zero-dimensional ideal, and we fix $>$ to be the lexicographical ordering induced by $x_1 > \ldots > x_n$. If $f \in K[X]$ is a polynomial, then we denote by $\mathrm{LE}(f)$ the leading exponent of f, by $\mathrm{LC}(f)$ the leading coefficient of f, by $\mathrm{LM}(f)$ the leading monomial of f, and by $\mathrm{LT}(f) = \mathrm{LC}(f) \cdot \mathrm{LM}(f)$ the leading term of f.

Definition 1. A set of polynomials $F = \{f_1, \ldots, f_n\} \subseteq K[X]$ is called a *triangular set* if $\mathrm{LT}(f_i) = x_{n-i+1}^{\alpha_i}$ for some $\alpha_i > 0$ and each $i = 1, \ldots, n$.

A list $\mathscr{F} = (F_1, \ldots, F_s)$ of triangular sets is called a *triangular decomposition* of a zero-dimensional ideal $I \subseteq K[X]$ if $\sqrt{I} = \sqrt{\langle F_1 \rangle} \cap \ldots \cap \sqrt{\langle F_s \rangle}$.

Remark 1.

1. A triangular set is a Gröbner basis.
2. A minimal Gröbner basis of a maximal ideal is a triangular set, and the primary decomposition of $\sqrt{I}$ is a triangular decomposition of I.

The following two lemmata are the basis for the algorithm by Möller [12] which avoids primary decomposition.

Lemma 1. *Let* $G = \{g_1, \ldots, g_m\}$ *be a reduced Gröbner basis of the zero-dimensional ideal* $I \subseteq K[X]$ *so that* $\mathrm{LM}(g_1) < \ldots < \mathrm{LM}(g_m)$. *Moreover, for* $i = 1, \ldots, m$, *let* $\alpha_i = \mathrm{LE}(g_i)$ *in* $(K[x_2, \ldots, x_n])\,[x_1]$, *i.e.* $g_i = \sum_{j=0}^{\alpha_i} g'_{ij} x_1^j$ *for suitable* $g'_{ij} \in K[x_2, \ldots, x_n]$. *Then* $G' = \{g'_{1\alpha_1}, \ldots, g'_{m-1\alpha_{m-1}}\}$ *is a Gröbner basis of* $\langle g_1, \ldots, g_{m-1} \rangle : g_m$, *and we have* $\langle G', g_m \rangle = \langle G', G \rangle$.

Proof. The proof can be found in [12, Lemma 7]. □

Lemma 2. *Let* $I \subseteq K[X]$ *be a zero-dimensional ideal, and* $h \in K[X]$. *Then the following hold.*

1. $\sqrt{I} = \sqrt{\langle I, h \rangle} \cap \sqrt{I : h}$.
2. $\dim_K \big(K[X]/I\big) = \dim_K \big(K[X]/\langle I, h \rangle\big) + \dim_K \big(K[X]/(I : h)\big)$.

Proof.

1. The proof is an exercise in [6, Exercise 4.5.3].
2. We consider two exact sequences. The first one

$$0 \longrightarrow (I : h)/I \longrightarrow K[X]/I \xrightarrow{\cdot h} K[X]/I \longrightarrow K[X]/\langle I, h \rangle \longrightarrow 0$$

yields $\dim_K\left((I:h)/I\right) = \dim_K\left(K[X]/\langle I,h\rangle\right)$, and the second one

$$0 \longrightarrow (I:h)/I \longrightarrow K[X]/I \longrightarrow K[X]/(I:h) \longrightarrow 0$$

yields $\dim_K\left(K[X]/I\right) = \dim_K\left((I:h)/I\right) + \dim_K\left(K[X]/(I:h)\right)$. Summarized, we obtain

$$\dim_K\left(K[X]/I\right) = \dim_K\left(K[X]/\langle I,h\rangle\right) + \dim_K\left(K[X]/(I:h)\right). \qquad \square$$

Consequently, for $h_1,\dots,h_l \in K[X]$ and $h_{l+1} = 1$, we are able to apply Lemma 2 inductively and obtain

$$\begin{aligned}\sqrt{I} &= \sqrt{\langle I,h_1\rangle} \cap \sqrt{I:h_1}\\ &= \sqrt{\langle I,h_1,h_2\rangle} \cap \sqrt{\langle I,h_1\rangle : h_2} \cap \sqrt{I:h_1}\\ &= \dots\\ &= \sqrt{\langle I,h_1,\dots,h_l\rangle} \cap \left(\bigcap_{i=1}^{l+1} \sqrt{\langle I,h_1,\dots,h_{i-1}\rangle : h_i}\right).\end{aligned}$$

Together with Lemma 1 we conclude the following corollary.

Corollary 1. *With the assumption of Lemma 1, let $G' \setminus G = \{h_1,\dots,h_l\}$. Then the following hold.*

1. *If $G' \setminus G \neq \emptyset$, then $I \subsetneq \langle G,h_1\rangle$ and $I \subsetneq \langle G\rangle : h_1$.*
2. $\sqrt{I} = \sqrt{\langle G',g_m\rangle} \cap \left(\bigcap_{i=1}^{l+1} \sqrt{\langle G,h_1,\dots,h_{i-1}\rangle : h_i}\right)$ *and, in addition,* $I \subseteq \langle G',g_m\rangle \cap \left(\langle G,h_1,\dots,h_{i-1}\rangle : h_i\right)$.
3. $\dim_K\left(K[X]/I\right) = \sum_{i=1}^{l+1} \dim_K\left(K[X]/(\langle G,h_1,\dots,h_{i-1}\rangle : h_i)\right)$.

With regard to Corollary 1.2, especially $\langle G,h_1,\dots,h_l\rangle = \langle G',g_m\rangle$ with $G' \subseteq K[x_2,\dots,x_n]$ is predestined for induction since it holds $\sqrt{\langle G',g_m\rangle} = \sqrt{\langle F_1',g_m\rangle} \cap \dots \cap \sqrt{\langle F_s',g_m\rangle}$ if $\mathscr{F}' = (F_1',\dots,F_s')$ is a triangular decomposition of G'. Referring to Corollary 1.3, the triangular decomposition obtained by iterating the approach of Corollary 1 respects the multiplicities of the zeros of I. Therefore the zero-sets of different triangular sets are in general not disjoint as the following example shows.

Example 1. Let $G = \{x_2^{10}, x_1x_2^3 + x_2^5, x_1^{11}\} \subseteq \mathbb{Q}[x_1,x_2]$. Then we obtain $G' = \{x_2^{10}, x_2^3\}$, $G' \setminus G = \{x_2^3\}$, and the triangular decomposition $\mathscr{F} = (F_1,F_2)$ of $\langle G\rangle$ with $F_1 = \langle G\rangle : x_2^3 = \langle x_2^7, x_1 + x_2^2\rangle$ and $F_2 = \langle G, x_2^3\rangle = \langle x_2^3, x_1^{11}\rangle$.

Moreover, it holds $\dim_{\mathbb{Q}}\left(\mathbb{Q}[x_1,x_2]/\langle G\rangle\right) = 40$, $\dim_{\mathbb{Q}}\left(\mathbb{Q}[x_1,x_2]/(\langle G\rangle : x_2^3)\right) = 7$, and $\dim_{\mathbb{Q}}\left(\mathbb{Q}[x_1,x_2]/\langle G,x_2^3\rangle\right) = 33$. Note that $\langle G\rangle \subsetneq F_1 \cap F_2$.

Algorithm 1 shows the algorithm by Möller to compute the triangular decomposition of a zero-dimensional ideal.[2]

---

[2]The corresponding procedure is implemented in SINGULAR in the library `triang.lib`.

Algorithm 1 Triangular decomposition ($\texttt{triangM}$)

Input: $I \subseteq K[X]$, a zero-dimensional ideal.
Output: $\mathscr{F} = (F_1, \ldots, F_s)$, a triangular decomposition of $I \subseteq K[X]$ such that $\dim_K\big(K[X]/I\big) = \sum_{i=1}^{s} \dim_K\big(K[X]/\langle F_i \rangle\big)$.

1: compute $G = \{g_1, \ldots, g_m\}$, a reduced Gröbner basis of I with respect to the lexicographical ordering $>$ such that $\mathrm{LM}(g_1) < \ldots < \mathrm{LM}(g_m)$;
2: compute $G' = \{g'_1, \ldots, g'_{m-1}\} \subseteq K[x_2, \ldots, x_n]$ where g'_i is the leading coefficient of g_i in $(K[x_2, \ldots, x_n])[x_1]$;
3: $\mathscr{F}' = \texttt{triangM}(\langle G' \rangle)$;
4: $\mathscr{F} = \{F' \cup \{g_m\} \mid F' \in \mathscr{F}'\}$;
5: **for** $1 \le i \le m-1$ **do**
6:     **if** $g'_i \notin G$ **then**
7:         $\mathscr{F} = \mathscr{F} \cup \texttt{triangM}(\langle G \rangle : g'_i)$;
8:         $G = G \cup \{g'_i\}$;
9: **return** $\mathscr{F}$;

Note that Algorithm 1 is only based on Gröbner basis computations and does not use random elements. Hence, the result is uniquely determined which allows modular computations. In the following we fix Algorithm 1 to compute a triangular decomposition.

Remark 2. Replacing line 7 in Algorithm 1 by $\mathscr{F} = \mathscr{F} \cup \texttt{triangM}\big(\langle G \rangle : g_i'^{\infty}\big)$ we obtain a disjoint triangular decomposition (i.e., $F_i, F_j \in \mathscr{F}$ with $F_i \neq F_j$ implies $\langle F_i \rangle + \langle F_j \rangle = K[X]$). This decomposition does in general not respect the multiplicities (i.e. $\dim_K\big(K[X]/I\big) \neq \sum_{i=1}^{s} \dim_K\big(K[X]/\langle F_i \rangle\big)$).

3 Modular Methods

One possible modular approach for solving a zero-dimensional ideal is to just replace each involved Gröbner basis computation by its corresponding modular algorithm as described in [7]. Particularly, we can replace line 1 in Algorithm 1 by $G = \texttt{modStd}(I)$. In this case it is possible to apply the probabilistic variant since we can easily verify in the end by a simple substitution whether the solutions obtained from the triangular sets are really solutions of the original ideal. Nevertheless, we propose to compute the whole triangular decomposition via modular methods actually. The verification is then similar by just substituting the obtained result into the input polynomials.

We consider the polynomial ring $\mathbb{Q}[X]$, fix a global monomial ordering $>$ on $\mathbb{Q}[X]$, and use the following notation: If $S \subseteq \mathbb{Q}[X]$ is a set of polynomials, then $\mathrm{LM}(S) := \{\mathrm{LM}(f) \mid f \in S\}$ denotes the set of leading monomials of S. If $f \in \mathbb{Q}[X]$ is a polynomial, $I = \langle f_1, \ldots, f_r \rangle \subseteq \mathbb{Q}[X]$ is an ideal, and p is a prime number which does not divide any denominator of the coefficients of $f, f_1, \ldots, f_r$, then we write $f_p := (f \bmod p) \in \mathbb{F}_p[X]$ and $I_p := \big\langle (f_1)_p, \ldots, (f_r)_p \big\rangle \subseteq \mathbb{F}_p[X]$.

In the following, $I = \langle f_1, \dots, f_r \rangle \subseteq \mathbb{Q}[X]$ will be a zero-dimensional ideal. The triangular decomposition algorithm (Algorithm 1) applied to I returns a list of triangular sets $\mathcal{F} = (F_1, \dots, F_s)$ such that $\sqrt{I} = \bigcap_{i=1}^{s} \sqrt{\langle F_i \rangle}$ and $\dim_{\mathbb{Q}}(\mathbb{Q}[X]/I) = \sum_{i=1}^{s} \dim_{\mathbb{Q}}(\mathbb{Q}[X]/\langle F_i \rangle)$.

With respect to modularization, the following lemma obviously holds:

Lemma 3. *With notation as above, let p be a sufficiently general prime number. Then I_p is zero-dimensional, $\big((F_1)_p, \dots, (F_s)_p\big)$ is a list of triangular sets, and $\sqrt{I_p} = \bigcap_{i=1}^{s} \sqrt{\langle (F_i)_p \rangle}$.*

Relying on Lemma 3, the basic idea of the modular triangular decomposition is as follows. First, choose a set $\mathcal{P}$ of prime numbers at random. Second, compute triangular decompositions $\mathcal{F}_p$ of I_p for $p \in \mathcal{P}$. Third, lift the modular triangular sets to triangular sets $\mathcal{F}$ over $\mathbb{Q}[X]$.

The lifting process consists of two steps. First, the set $\mathcal{H} := \{\mathcal{F}_p \mid p \in \mathcal{P}\}$ is lifted to $\mathcal{F}_N$ with $(F_i)_N \subseteq (\mathbb{Z}/N\mathbb{Z})[X]$ and $N := \prod_{p \in \mathcal{P}} p$ by applying the Chinese remainder algorithm to the coefficients of the polynomials occurring in $\mathcal{H}$. Second, we obtain $\mathcal{F}$ with $F_i \subseteq \mathbb{Q}[X]$ by lifting the modular coefficients occurring in $\mathcal{F}_p$ to rational coefficients via the Farey rational map.[3] This map is guaranteed to be bijective provided that $\sqrt{N/2}$ is larger than the moduli of all coefficients of elements in $\mathcal{F}$ with $F_i \subseteq \mathbb{Q}[X]$.

We now define a property of the set of primes $\mathcal{P}$ which guarantees that the lifting process is feasible and correct. This property is essential for the algorithm.

Definition 2. Let $\mathcal{F} = (F_1, \dots, F_s)$ be the triangular decomposition of the ideal I computed by Algorithm 1.

1. A prime number p is called *lucky* for I and $\mathcal{F}$ if $((F_1)_p, \dots, (F_s)_p)$ is a triangular decomposition of I_p. Otherwise p is called *unlucky* for I and $\mathcal{F}$.
2. A set $\mathcal{P}$ of lucky primes for I and $\mathcal{F}$ is called *sufficiently large* for I and $\mathcal{F}$ if

$$\prod_{p \in \mathcal{P}} p \geq \max\{2 \cdot |c|^2 \mid c \text{ coefficient occuring in } \mathcal{F}\}.$$

From a theoretical point of view, the idea of the algorithm is now as follows: Consider a sufficiently large set $\mathcal{P}$ of lucky primes for I and $\mathcal{F}$, compute the triangular decomposition of the I_p, $p \in \mathcal{P}$, via Algorithm 1, and lift the results to the triangular decomposition of I as aforementioned.

From a practical point of view, we face the problem that we do not know in advance whether a prime number p is lucky for I and $\mathcal{F}$.

To handle this problem, we fix a natural number t and an arbitrary set of primes $\mathcal{P}$ of cardinality t. Having computed $\mathcal{H}$, we use the following test to modify $\mathcal{P}$ such that all primes in $\mathcal{P}$ are lucky with high probability:

[3] *Farey fractions* refer to rational reconstruction. A definition of *Farey fractions*, the *Farey rational map*, and remarks on the required bound on the coefficients can be found in [8].

DELETEUNLUCKYPRIMESTRIANG. *We define an equivalence relation on* $(\mathscr{H}, \mathscr{P})$ *by* $(\mathscr{F}_p, p) \sim (\mathscr{F}_q, q) :\Longleftrightarrow \big(\#\mathscr{F}_p = \#\mathscr{F}_q$ *and* $\{\mathrm{LM}(F_p) \mid F_p \in \mathscr{F}_p\} = \{\mathrm{LM}(F_q) \mid F_q \in \mathscr{F}_q\}\big)$. *Then the equivalence class of largest cardinality is stored in* $(\mathscr{H}, \mathscr{P})$*, the others are deleted.*

Since we do not know a priori whether the equivalence class chosen is indeed lucky and whether it is sufficiently large for I and $\mathscr{F}$, we proceed in the following way. We lift the set $\mathscr{H}$ to $\mathscr{F}$ over $\mathbb{Q}[X]$ as described earlier, and test the result with another randomly chosen prime number:

PTESTTRIANG. *We randomly choose a prime number* $p \notin \mathscr{P}$ *such that* p *does not divide the numerator and denominator of any coefficient occurring in a polynomial in* $\{f_1, \ldots, f_r\}$ *or* $\mathscr{F}$. *The test returns true if* $(F_p \mid F \in \mathscr{F})$ *equals the triangular decomposition* $\mathscr{F}_p$ *computed by the fixed Algorithm 1 applied on* I_p*, and false otherwise.*

If PTESTTRIANG returns false, then $\mathscr{P}$ is not sufficiently large for I and $\mathscr{F}$ or the equivalence class of prime numbers chosen was unlucky. In this case, we enlarge the set $\mathscr{P}$ by t new primes and repeat the whole process. On the other hand, if PTESTTRIANG returns true, then we have a triangular decomposition $\mathscr{F}$ of I with high probability. In this case, we compute the solutions $S \subseteq \mathbb{C}^n$ with multiplicities of the triangular sets $F = \{f_1, \ldots, f_n\} \in \mathscr{F}$ as follows. Solve the univariate polynomial $f_1(x_1)$ via a univariate solver counting multiplicities, substitute x_1 in $f_2(x_1, x_2)$ by these solutions of $f_1(x_1)$, solve the corresponding univariate polynomial, and continue inductively this way (call this step SOLVETRIANG). Finally, we verify the result $S \subseteq \mathbb{C}^n$ partially by testing whether the original ideal $I \subseteq \mathbb{Q}[X]$ is contained in every $F \in \mathscr{F}$, and whether the sum of the multiplicities equals the $\mathbb{Q}$-dimension of $\mathbb{Q}[X]/I$ (call this step TESTZERO).

We summarize modular solving in Algorithm 2.[4]

Remark 3. In Algorithm 2, the triangular sets $\mathscr{F}_p$ can be computed in parallel. Furthermore, we can parallelize the final verification whether $I \subseteq \mathbb{Q}[X]$ is contained in every $F \in \mathscr{F}$.

4 Examples and Timings

In this section we provide examples on which we time the algorithm `modSolve` (Algorithm 2) and its parallel version as opposed to the algorithm `solve` (the procedure `solve` is implemented in SINGULAR in the library `solve.lib` and computes all roots of a zero-dimensional input ideal using triangular sets). Timings are conducted by using SINGULAR 3-1-6 on an AMD Opteron 6174 machine with

[4]The corresponding procedures are implemented in SINGULAR in the library `modsolve.lib`.

Algorithm 2 Modular solving (`modSolve`)

Input: $I \subseteq \mathbb{Q}[X]$, a zero-dimensional ideal.
Output: $S \subseteq \mathbb{C}^n$, a set of points in $\mathbb{C}^n$ such that $f(P) = 0$ for all $f \in I$, $P \in S$.

choose $\mathscr{P}$, a list of random primes;
$\mathscr{H} = \emptyset$;
loop
  for $p \in \mathscr{P}$ **do**
    $\mathscr{F}_p =$ `triangM`(I_p), the triangular decomposition of I_p via Algorithm 1;
    $\mathscr{H} = \mathscr{H} \cup \{\mathscr{F}_p\}$;
  $(\mathscr{H}, \mathscr{P}) =$ DELETEUNLUCKYPRIMESTRIANG$(\mathscr{H}, \mathscr{P})$;
  lift $(\mathscr{H}, \mathscr{P})$ to $\mathscr{F}$ over $\mathbb{Q}[X]$ by applying Chinese remainder algorithm and Farey rational map;
  if PTESTTRIANG$(I, \mathscr{F}, \mathscr{P})$ **then**
    $S =$ SOLVETRIANG$(\mathscr{F})$;
    if TESTZERO$(I, \mathscr{F}, S)$ **then**
      return S;
  enlarge $\mathscr{P}$;

48 CPUs, 2.2 GHz, and 128 GB of RAM running the Gentoo Linux operating system. All examples are chosen from The SymbolicData Project [5].

Remark 4. The parallelization of the modular algorithm is attained via multiple processes organized by SINGULAR library code. Consequently, a future aim is to enable parallelization in the kernel via multiple threads.

We choose the following examples—all of them can be found in [5]—to emphasize the superiority of modular solving and especially its parallelization:

Example 2. `Cyclic_7.xml`.

Example 3. `Verschelde_noon6.xml`.

Example 4. `Pfister_1.xml`.

Example 5. `Pfister_2.xml`.

Table 1 summarizes the results where `modSolve`(c) denotes the parallelized version of the algorithm applied on c cores. All timings are given in seconds.

Table 1 Total running times in seconds for computing all roots of the considered examples via `solve`, `modSolve` and its parallelized variant `modSolve`(c) for $c = 10, 20$

Example	`solve`	`modSolve`	`modSolve(10)`	`modSolve(20)`
2	> 18 h	692	217	152
3	517	1,223	522	371
4	526	800	288	165
5	2,250	1,276	323	160

Remark 5. Various experiments reveal that a sensitive choice of $\#\mathscr{P}$, the number of random primes in lines 1 and 13 in Algorithm 2, can decrease the running time enormously. To sum up, it is recommendable to relate c, the number of available cores, to $\#\mathscr{P}$. Particularly, in case of having more than ten cores to ones's disposal it is reasonable to set $c = \#\mathscr{P}$.

References

1. Aubry, P., Moreno Maza, M.: Triangular sets for solving polynomial systems: a comparative implementation of four methods. J. Symb. Comput. **28**, 125–154 (1999)
2. Chen, C., Lemaire, F., Moreno Maza, M., Pan, W., Xie, Y.: Efficient computations of irredundant triangular decompositions with the regular chains library. In: Shi, Y., et al. (eds.) Proceedings of Computer Algebra Systems and Their Applications '07: ICCS 2007, Part II, Lecture Notes in Computer Science, vol. 4488, pp. 268–271 (2007)
3. Dahan, X., Moreno Maza, M., Schost, É., Wu, W., Xie, Y.: Lifting techniques for triangular decompositions. Proceedings of ISSAC '05, Beijing, China, pp. 108–115. ACM Press, New York (2005)
4. Decker, W., Greuel, G.-M., Pfister, G., Schönemann, H.: SINGULAR 3-1-6 — A computer algebra system for polynomial computations. http://www.singular.uni-kl.de (2012)
5. Gräbe, H.-G.: The symbolicdata project — tools and data for testing computer algebra software. http://www.symbolicdata.org (2013)
6. Greuel, G.-M., Pfister, G.: A SINGULAR Introduction to Commutative Algebra, 2nd edn. Springer, London (2007)
7. Idrees, N., Pfister, G., Steidel, S.: Parallelization of modular algorithms. J. Symb. Comput. **46**, 672–684 (2011)
8. Kornerup, P., Gregory, R.T.: Mapping integers and hensel codes onto farey fractions. BIT Numer. Math. **23**(1), 9–20 (1983)
9. Lemaire, F., Moreno Maza, M., Xie, Y.: The RegularChains library in Maple 10. Maplesoft, Canada (2005)
10. Lemaire, F., Moreno Maza, M., Xie, Y.: Making a sophisticated symbolic solver available to different communities of users. Proceedings of Asian Technology Conference in Mathematics '06, Polytechnic University of Hong Kong (2006)
11. Li, X., Moreno Maza, M.: Multithreaded parallel implementation of arithmetic operations modulo a triangular set. Proceedings of Parallel Symbolic Computation '07, London, Canada, pp. 53–59. ACM Press, New York (2007)
12. Möller, H.M.: On decomposing systems of polynomial equations with finitely many solutions. Appl. Algebra Eng. Commun. Comput. **4**, 217–230 (1993)
13. Möller, H.M.: Solving of algebraic equations – an interplay of symbolical and numerical methods. In: Haußmann, W., Jetter, K., Reimer, M. (eds.) Multivariate Approximation, Recent Trends and Results. Mathematical Research, vol. 101, pp. 161–176. Akademie Verlag, Berlin (1997)
14. Moreno Maza, M.: On triangular decompositions of algebraic varieties. Presented at the MEGA 2000 Conference, Bath, UK (2000)

15. Moreno Maza, M., Xie, Y.: Component-level parallelization of triangular decompositions. Proceedings of Parallel Symbolic Computation '07, London, Canada, pp. 69–77. ACM Press, New York (2007)
16. Ralston, A., Rabinowitz, P.: A First Course in Numerical Analysis, 2nd edn. McGraw-Hill, New York (1978)

Lazarsfeld–Mukai Bundles and Applications: II

Marian Aprodu

Dedication to Alexandru Dimca and Ştefan Papadima

Abstract This paper is a continuation of the work by Aprodu (Lazarsfeld-Mukai Bundles and Applications. Commutative Algebra, vol. 1–23. Springer, New York (2013)). We focus on non-$K3$ surfaces providing some improvements of known results.

Keywords Projective curves • Brill-Noether theory • Algebraic surfaces • Syzygies

1 Introduction

The notion of Lazarsfeld–Mukai bundle goes back to the 1980s, when two important problems in algebraic geometry were solved using vector bundle techniques [9, 12]. They were initially defined as vector bundles with particularly nice properties on $K3$ surfaces, and their main applications to date remain within the $K3$ framework. The definition makes sense however in a much larger class of surfaces.

Let S be a surface, C be a smooth curve on S, and A be a g^1_d on C. A natural question related to this setup is the following: can A be lifted to the surface S? The chances for A to be induced by a pencil on S are slim, by the simple fact that we cannot exclude the possibility that $\mathrm{Pic}(S)$ be generated by C itself. This case actually occurs in many situations, for instance on very general $K3$ surfaces. Instead, we can try and lift the pencil $|A|$ to a *different object*, and in doing so we have to decide what kind of *object* would that be. If we recall that an element D in the linear system $|A|$ is a sum of points $x_1+\cdots+x_d$ on the curve C, hence points on the surface S, then it is plausible that any such collection of points on S be cut out by a section in a rank-two vector bundle. Since the points move in a linear system on C,

M. Aprodu (✉)
"Simion Stoilow" Institute of Mathematics of the Romanian Academy, P.O. Box 1-764, RO-014700, Bucharest, Romania
e-mail: Marian.Aprodu@imar.ro

D. Ibadula, W. Veys (eds.), *Bridging Algebra, Geometry, and Topology*, Springer Proceedings in Mathematics & Statistics 96,
DOI 10.1007/978-3-319-09186-0_2

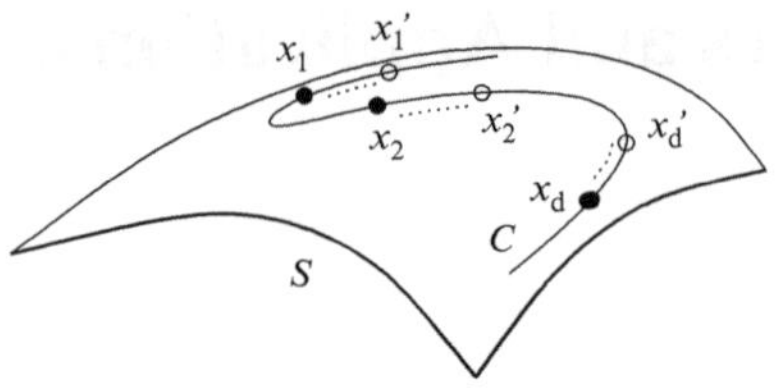

Two divisors in the same linear system on the curve.

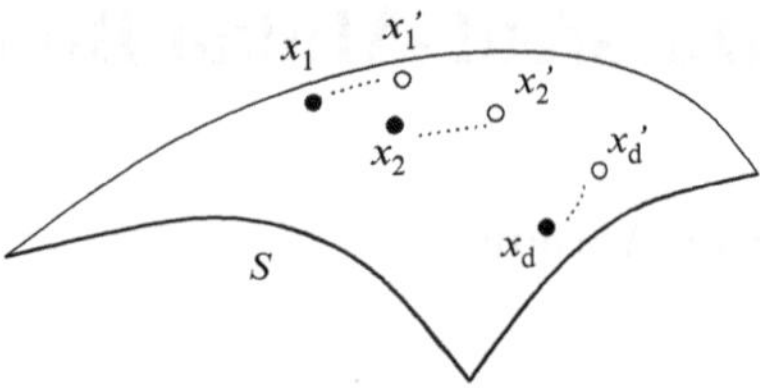

The curve is erased, the points still move.

Fig. 1 Divisors in $|A|$ as moving subschemes of S

the section would also have move in a two-dimensional space of sections over S in this hypothetical bundle. In other words, we want to erase the curve from the picture, keep the moving points and interpret them as elements in a linear family of zero-dimensional subschemes on S, see Fig. 1. If this goal is achieved, then the rank-two bundle in question, which comes equipped with a natural two-dimensional space of global sections, is called a *Lazarsfeld–Mukai* bundle [9, 10, 12]. This strategy works well for regular surfaces [13] and can be extended for higher-dimensional linear systems on C, [9, 10]. The details are explained in Sect. 2.

Lazarsfeld–Mukai bundles proved to be useful in various situations. They appear in the classification of Fano threefolds [12], in classical Brill–Noether problems [9, 10], in higher Brill–Noether theory [6], in syzygy-related questions [4, 15, 16] etc; see [3, 10] for some surveys of this topic.

The specific problem we consider here is the computation of the dimensions of Brill–Noether loci. For curves on regular surfaces, this computation reduces to a dimension calculation of the parameter spaces of Lazarsfeld–Mukai bundles. Beyond the Brill–Noether theoretical interest for this type of problems, the motivation comes from syzygy theory, see Sect. 4 for a more detailed discussion. Our goal is twofold. On the one hand, we review the general theory that is generally focused on the case of $K3$ surfaces. This recollection of facts might be of future use. On the other hand, we slightly improve some results obtained so far in the non-$K3$ case.

The present work is a continuation of [3] and its outline is the following. In Sect. 2 we recall the definition of Lazarsfeld–Mukai bundles on regular surfaces and we prove a general dimension statement, Theorem 1: the Brill–Noether loci have the expected dimension if some suitable vanishing conditions are satisfied. In Sect. 3 we estimate the dimensions of Brill–Noether loci for curves on rational surfaces with an anticanonical bound, Theorem 2. It is an extension of the main result of [11]. In Sect. 4 we discuss some applications to syzygies, based on the main result of [2], Theorem 3. An alternate proof of [1, Theorem 8.1] is given in Example 1.

2 Lazarsfeld–Mukai Bundles on Surfaces with $q = 0$

We follow closely the approach from [10, 13]. Let S be a surface with $h^1(\mathcal{O}_S) = 0$, C be a smooth connected curve of genus g in S and denote $L = \mathcal{O}_S(C)$. The hypothesis $h^1(\mathcal{O}_S) = 0$ is needed for technical reasons and ensures that $T_C|L| =$

$H^0(N_{C|S})$. Let A be a base-point-free complete g^r_d on C and denote by M_A the kernel of

$$\mathrm{ev}_A : H^0(A) \otimes \mathcal{O}_C \to A.$$

The evaluation map lifts to a surjective sheaf morphism $H^0(A) \otimes \mathcal{O}_S \to A$ on S whose kernel $F_{C,A}$ is a vector bundle of rank $(r+1)$. Its dual $E_{C,A} = F^*_{C,A}$ is called a *Lazarsfeld–Mukai bundle*. Dualizing the defining sequence of $E_{C,A}$

$$0 \to F_{C,A} \to H^0(A) \otimes \mathcal{O}_S \to A \to 0 \tag{1}$$

we obtain the defining sequence of $E_{C,A}$

$$0 \to H^0(A)^* \otimes \mathcal{O}_S \to E_{C,A} \to N_{C|S} \otimes A^* \to 0. \tag{2}$$

The following properties of $E_{C,A}$ and $F_{C,A}$ are obtained by direct computation, using the hypotheses $h^1(\mathcal{O}_S) = 0$ and $h^0(C, A) = r + 1$ [10, 11, 13]:

1. $\det(E_{C,A}) = L$,
2. $c_2(E_{C,A}) = d$,
3. $h^0(S, F_{C,A}) = h^1(S, F_{C,A}) = 0$,
4. $\chi(S, F_{C,A}) = h^2(S, F_{C,A}) = (r+1)\chi(\mathcal{O}_S) + g - d - 1$,
5. $h^0(S, E_{C,A}) = r + 1 + h^0(C, N_{C|S} \otimes A^*)$,
6. $E_{C,A}$ is generated off the base locus of $|N_{C|S} \otimes A^*|$ inside C.

Restricting the sequence (1) to the curve C, we obtain a short exact sequence:

$$0 \to N^*_{C|S} \otimes A \to F_{C,A}|_C \to M_A \to 0 \tag{3}$$

which implies, twisting by $K_C \otimes A^*$ and using the adjunction formula,

$$0 \to \omega_S|_C \to F_{C,A} \otimes K_C \otimes A^* \to M_A \otimes K_C \otimes A^* \to 0. \tag{4}$$

Note that $H^0(M_A \otimes K_C \otimes A^*) = \ker(\mu_{0,A})$, where $\mu_{0,A} : H^0(A) \otimes H^0(K_C \otimes A^*) \to H^0(K_C)$ is the Petri map.

Recall [5] that for any r and d, the Brill–Noether loci form a family $\mathcal{W}^r_d(|L|)$ over the open subset of $|L|$ corresponding to smooth curves. The next result uses the hypothesis $h^1(\mathcal{O}_S) = 0$ and follows from the discussion in [13, p. 197] (see also [4, Lemma 2.3]):

Lemma 1. *If $(C, A) \in \mathcal{W}$ is a general pair in an irreducible component of $\mathcal{W}^r_d(|L|)$ dominating over $|L|$, then the coboundary map $H^0(C, M_A \otimes K_C \otimes A^*) \to H^1(C, \omega_S|_C)$ vanishes.*

Lemma 1 exhibits an exact sequence

$$0 \to H^0(C, \omega_S|_C) \to H^0(C, F_{C,A} \otimes K_C \otimes A^*) \to \ker(\mu_{0,A}) \to 0 \tag{5}$$

for a general choice of a pair $(C, A) \in \mathcal{W}$. In particular, $\mathcal{W}$ is a smooth of expected dimension at the point (C, A) if the following equality holds: $h^0(C, \omega_S|_C) = h^0(C, F_{C,A} \otimes K_C \otimes A^*)$. The sequence (5) is useful to estimate the dimension of Brill–Noether loci and, in some situations, smoothness follows from appropriate vanishing conditions:

Theorem 1. *Notation as above. Assume that $h^2(S, L) = 0$. Let (C, A) be a general pair in a dominating component $\mathcal{W}$ such that $h^2(S, F_{C,A} \otimes E_{C,A}) = h^2(S, \mathcal{O}_S)$ and $h^2(S, E_{C,A}) = 0$. Then $\mathcal{W}$ is of dimension $\leq \rho(g, r, d) + \dim|L| + (r+1)h^1(S, E_{C,A})$ at the point (C, A). In particular, if $h^1(S, E_{C,A}) = 0$, then $\mathcal{W}$ is smooth of expected dimension $\rho(g, r, d) + \dim|L|$ at the point (C, A).*

Proof. From the long exact sequence associated with the sequence

$$0 \to \mathcal{O}_S \to L \to N_{C|S} \to 0,$$

applying the vanishing hypothesis $h^2(S, L) = 0$ we obtain an a surjection $H^1(C, N_{C|S}) \to H^2(S, \mathcal{O}_S)$ and hence, since $K_C \cong N_{C|S} \otimes \omega_S|_C$, we have $h^0(C, \omega_S|_C) \geq h^2(S, \mathcal{O}_S)$. From the sequence (5) it follows that $\dim(\ker(\mu_{0,A})) \leq h^0(C, F_{C,A} \otimes K_C \otimes A^*) - h^2(\mathcal{O}_S)$.

Twisting the sequence (2) by $F_{C,A} \otimes \omega_S$, taking global sections in the sequence

$$0 \to H^0(A)^* \otimes F_{C,A} \otimes \omega_S \to F_{C,A} \otimes E_{C,A} \otimes \omega_S \to F_{C,A} \otimes K_C \otimes A^* \to 0 \quad (6)$$

applying Serre duality and using the hypothesis: $h^0(F_{C,A} \otimes \omega_S) = 0$ and $h^0(F_{C,A} \otimes E_{C,A} \otimes \omega_S) = h^2(\mathcal{O}_S)$ we obtain the inequality $h^0(C, F_{C,A} \otimes K_C \otimes A^*) \leq h^2(\mathcal{O}_S) + (r+1)h^1(S, E_{C,A})$.

We obtain $\dim\ker(\mu_{0,A}) \leq (r+1)h^1(S, E_{C,A})$ and hence the conclusion follows. □

Remark 1. The assumption $h^2(S, F_{C,A} \otimes E_{C,A}) = h^2(\mathcal{O}_S)$ is natural. For stable bundles, it is a sufficient condition for the smoothness of the moduli space at the point defined by $E_{C,A}$. Absent this condition, we obtain the weaker estimate

$$\dim_{(C,A)}\mathcal{W} \leq \rho(g, r, d) + \dim|L| + (r+1)h^1(S, E_{C,A}) + (h^2(S, F_{C,A} \otimes E_{C,A}) - h^2(\mathcal{O}_S)).$$

Note that $\mathcal{O}_S$ is a direct summand of $F_{C,A} \otimes E_{C,A}$, its complement is $\mathrm{ad}(E_{C,A})$, the bundle of trace-free endomorphisms, and hence

$$h^2(S, F_{C,A} \otimes E_{C,A}) - h^2(\mathcal{O}_S) = h^2(S, \mathrm{ad}(E_{C,A})).$$

Remark 2. If in addition $p_g(S) = h^2(\mathcal{O}_S) = 0$, then the vanishing of $h^2(S, E_{C,A})$ follows from $h^2(S, F_{C,A} \otimes E_{C,A}) = 0$ and Serre duality in the sequence (6). The condition $h^2(S, L) = h^0(S, L^* \otimes \omega_S) = 0$ is also automatic, as L is effective and hence $h^0(S, L^* \otimes \omega_S) \leq h^0(S, \omega_S)$. Furthermore, the sequence (1) twisted by ω_S shows that $h^1(S, E_{C,A}) = h^0(C, \omega_S \otimes A)$ in this case.

The hypotheses of Theorem 1 are realized in a number of situations, which we enumerate below.

1. *K*3 *surfaces*, [9, 10], see also [4, 13]. In this case, $\omega_S \cong \mathcal{O}_S$ and $N_{C|S} = K_C$. The hypothesis $h^2(S, F_{C,A} \otimes E_{C,A}) = h^2(\mathcal{O}_S) = 1$ is equivalent to simplicity of $E_{C,A}$. The vanishing of $h^1(S, E_{C,A})$ and of $h^2(S, E_{C,A})$ follow from Serre duality and the vanishing of $h^1(S, F_{C,A})$ and of $h^0(S, F_{C,A})$.
2. *Enriques surfaces*, compare to [14]. Assume S is an Enriques surface and consider $X \to S$ the $K3$ universal cover of S. Suppose that for general pair (C, A) in a dominating component $\mathcal{W}$, the associated Lazarsfeld–Mukai bundle $E_{C,A}$ is stable with respect to a given polarization H. Since the property of being a Lazarsfeld–Mukai bundle is an open condition, the main result of [8] (see Theorem on page 88) shows that for a general (C, A), the bundle $E_{C,A}$ is not isomorphic to $E_{C,A} \otimes \omega_S$ and hence $h^2(S, F_{C,A} \otimes E_{C,A}) = h^2(\mathcal{O}_S) = 0$. As noted in Remark 2, the condition $h^2(S, E_{C,A}) = 0$ follows.
3. *Rational surface with an anti-canonical pencil*, [11]. Suppose that $E_{C,A}$ is stable with respect to a given polarization H and $A \not\cong \omega_S^*|_C$. Since ω_S^* is effective, it follows that there are no nonzero morphisms from $E_{C,A} \otimes \omega_S^*$ to $E_{C,A}$, hence $h^2(S, F_{C,A} \otimes E_{C,A}) = h^2(\mathcal{O}_S) = 0$. As pointed out in [11], the existence of an anti-canonical pencil implies $h^1(S, E_{C,A}) = 0$. The vanishing of $h^2(S, E_{C,A}) = h^0(S, F_{C,A} \otimes \omega_S)$ follows from the vanishing of $h^0(S, F_{C,A})$.

3 Lazarsfeld–Mukai Bundles of Rank Two on Rational Surfaces

It was pointed out in Remark 2 that if $q(S) = p_g(S) = 0$, some of the hypotheses of Theorem 1 are superfluous. Under this assumption, since there are fewer conditions, Theorem 1 can be substantially refined. We shall consider the case of rank-two Lazarsfeld–Mukai bundles, i.e., associated with base-point-free complete g^1_d's, on rational surfaces. The main result is the following:

Theorem 2. *Let* S *be a rational surface,* $L \geq 0$ *be an effective line bundle on* S *and denote by* $k \geq 3$ *the maximal gonality of smooth curves on* S *and by* g *their genus. Let* H *be an ample line bundle on* S *and* C *be a smooth curve on* S *of gonality* k*. Suppose that* $\omega_S \cdot H \leq 0$, $-\omega_S \cdot C \geq k$ *and that* C *is of Clifford dimension one. Then, for any integer* d *such that* $k \leq d \leq g - 2k + 2$*, and any component* $\mathcal{W}$ *of* $\mathcal{W}^1_d(|L|)$ *that dominates the linear system* $|L|$*, we have*

$$\dim(\mathcal{W}) \leq \dim|L| + (d - k). \tag{7}$$

Note that of C is an ample divisor, then the polarization H can be chosen to be $\mathcal{O}_S(C)$. If the anticanonical bundle is effective, then the polarization H can be arbitrarily chosen.

Lelli-Chiesa proved this result for rational surfaces S with $h^0(S, \omega_S^*) \geq 2$. Note that this condition implies automatically $-\omega_S \cdot C \geq k$, as an anticanonical pencil will restrict to a pencil on C. For sake of completeness, we present here a full proof covering also the case of surfaces with an anticanonical pencil [11], using the strategy from [4]; the really new case compared to [11] is Subcase I.c.

Proof. We proceed by induction on $d \geq k$. There are several possible cases, according to the behavior of the g^1_d's and of their associated Lazarsfeld–Mukai bundles.

Case I. Assume that for $(C, A) \in \mathcal{W}$ general, A is base-point-free and complete.

Subcase I.a. Assume that for $(C, A) \in \mathcal{W}$ general, $h^1(S, E_{C,A}) = 0$ and $E_{C,A}$ is H-stable. Since $\omega_S^* \cdot H \geq 0$, the stability implies that there is no nonzero morphism form $E_{C,A} \otimes \omega_S^*$ to $E_{C,A}$, i.e. $h^2(S, F_{C,A} \otimes E_{C,A}) = h^0(S, F_{C,A} \otimes E_{C,A} \otimes \omega_S) = 0$. We apply Remark 2 and Theorem 1.

Subcase I.b. Assume that for $(C, A) \in \mathcal{W}$ general, $h^1(S, E_{C,A}) = 0$ and $E_{C,A}$ is not H-stable. If $h^2(S, F_{C,A} \otimes E_{C,A}) = 0$, we apply again Theorem 1, hence we may assume than there is a nonzero morphism from $E_{C,A} \otimes \omega_S$ to $E_{C,A}$. The bundle $E_{C,A}$ has a maximal destabilizing subsheaf M, which induces an extension

$$0 \to M \to E_{C,A} \to N \otimes \mathcal{I}_\xi \to 0, \tag{8}$$

with $M \cdot H \geq N \cdot H$, where ξ is a zero-dimensional locally complete intersection subscheme of length $\ell = \ell(\xi) = M \cdot N - d$. Note that if $E_{C,A} \not\cong M \oplus N$ then we have (compare to [4] Lemma 3.4):

$$\dim \mathrm{Hom}(E_{C,A}, N_{C|S} \otimes A^*) = h^0(S, F_{C,A} \otimes E_{C,A}) = \dim \mathrm{Hom}(N \otimes \mathcal{I}_\xi, M) + 1. \tag{9}$$

We suppose that $E_{C,A}$ is indecomposable. Fix N and ℓ and denote by $\mathcal{P}_{N,\ell}$ the parameter space of bundles E with Chern classes $c_1(E) = L$ and $c_2(E) = d$ given by extensions of type (8), and by $\mathcal{G}_{N,\ell}$ the Grassmann bundle over $\mathcal{P}_{N,\ell}$ whose fiber over E is $G(2, H^0(E))$. If we assume that $\mathcal{P}_{N,\ell}$ contains Lazarsfeld–Mukai bundles corresponding to $\mathcal{W}$, then we have a rational map

$$\pi_{N,\ell} : \mathcal{G}_{N,\ell} \to \mathcal{W}$$

whose fiber over $E_{C,A}$ is the projectivization of $\mathrm{Hom}(E_{C,A}, N_{C|S} \otimes A^*)$. Since $\mathrm{Pic}(S)$ is discrete and the Lazarsfeld–Mukai condition is open, it follows that for a given N and ℓ, the map $\pi_{N,\ell}$ is dominant. Hence, using (9), it suffices to prove to prove that

$$\dim \mathcal{G}_{N,\ell} - \dim \mathrm{Hom}(N \otimes \mathcal{I}_\xi, M) \leq \dim |L| + (d - k). \tag{10}$$

Similarly to [4, Lemma 3.10] and [11, Lemma 4.1], we obtain the inequality

$$M \cdot N + \omega_S \cdot N + 2 \geq k. \tag{11}$$

We have

$$\dim \mathcal{G}_{N,\ell} \leq \dim G(2, H^0(E_{C,A})) + \dim S^{[\ell]} + (\dim \mathrm{Ext}^1(N \otimes \mathcal{I}_\xi, M) - 1),$$

and hence, using Serre duality and observing that $M \cdot H \geq N \cdot H \geq (N + \omega_S) \cdot H \geq 0$ implies $\mathrm{Ext}^2(N \otimes \mathcal{I}_\xi, M) = \mathrm{Hom}(M, N \otimes \mathcal{I}_\xi \otimes \omega_S) = 0$, we obtain the estimate:

$$\begin{aligned}\dim \mathcal{G}_{N,\ell} - \dim \mathrm{Hom}(N \otimes \mathcal{I}_\xi, M) \leq \dim G(2, H^0(E_{C,A}))\\ +2\ell - \chi(S, M^* \otimes N \otimes \omega_S \otimes \mathcal{I}_\xi) - 1.\end{aligned}$$

By the assumption $h^1(S, E_{C,A}) = 0$, we have $h^0(C, \omega_S \otimes A) = 0$, Remark 2. It follows that $h^0(C, N_{C|S} \otimes A^*) = g - d - 1 - \omega_S \cdot C$. Since $h^0(S, E_{C,A}) = 2 + h^0(C, N_{C|S} \otimes A^*)$ it implies

$$\dim G(2, H^0(E_{C,A})) = 2(g - d - 1 - \omega_S \cdot C)$$

The conclude the proof of inequality (10) we compute by the Riemann–Roch theorem (compare to [11])

$$\chi(S, M^* \otimes N \otimes \omega_S \otimes \mathcal{I}_\xi) = g - 2N \cdot M - \omega_S \cdot M - \ell.$$

and use (11) and the inequality:

$$h^0(S, L) \geq g - \omega_S \cdot C = \chi(S, L);$$

note that $h^2(S, L) = 0$, since L is effective on a surface with $p_g = 0$.

Subcase I.c. Assume that for all $(C, A) \in \mathcal{W}$, $h^1(S, E_{C,A}) > 0$. From Remark 2, we are in the situation $h^0(C, \omega_S \otimes A) > 0$, in particular, $A \in \{\omega_S^*|_C\} + W^0_{d+\omega_S \cdot C}(C)$. Since $\omega_S \cdot C \leq -k$, it follows that A moves in a family of dimension $\leq d - k$.

Case II. Assume that for any $(C, A) \in \mathcal{W}$, A has base-points. Then we apply the inductive argument and reduce to the previous case. Note that this case cannot occur for $d = k$. □

4 Syzygies of Curves

In recent curve theory a lot of effort has been put into understanding the relations between syzygies of canonical curves (algebraic objects) and the existence of special linear series (geometric objects). The interest in clarifying these deep relationships between algebraic and geometric properties is high, as failure of vanishing of syzygies produces interesting determinantal cycles on various moduli spaces, and the canonical case is the most natural and basic situation. The precise relationship is predicted by *Green's conjecture*: the ideal of a non-hyperelliptic curve is generated by quadrics, and the Clifford dimension controls the number of steps up to which the syzygies are linear. In the language of Koszul cohomology using duality [7], it amounts to the following relation

$$K_{p,1}(C, K_C) = 0 \text{ for all } p \geq g - c - 1,$$

for any curve C of genus g and Clifford index c; for the precise definitions of the objects involved in the statement, we refer to [7]. Green's conjecture is known to be true for general curves, [15, 16], and moreover the dimension computations of Brill–Noether loci, in particular conditions similar to (7), are related to syzygies of canonical curves. This relationship is explained in [2, 3] and has been used in [4] for curves on $K3$ surfaces. In our case, Green's conjecture is satisfied for *general* curves, in the linear system $|L|$, which verify the hypotheses of Theorem 2. However, under stronger hypotheses, we can prove Green's conjecture for *every* curve in the corresponding linear system:

Theorem 3. *Under the assumptions of Theorem 2, suppose moreover that*

$$g - k \geq h^0(S, \omega_S^{\otimes 2}(C)).$$

Then any smooth curve in $|L|$ is of gonality k, Clifford dimension one, and satisfies Green's conjecture.

Proof. The long exact sequence associated with

$$0 \to \omega_S \to \omega_S \otimes L \to K_C \to 0$$

shows that the restriction morphism $H^0(S, \omega_S \otimes L) \to H^0(C, K_C)$ is an isomorphism and provides us with a long exact sequence on Koszul cohomology [7, (1.d.4)]

$$0 = K_{p,1}(S, -C, \omega_S \otimes L) \to K_{p,1}(S, \omega_S \otimes L) \to K_{p,1}(C, K_C) \\ \to K_{p-1,2}(S, -C, \omega_S \otimes L) \to \cdots$$

Green's vanishing Theorem [7] (3.a.1) implies

$$K_{p-1,2}(S, -C, \omega_S \otimes L) = 0$$

for $p \geq h^0(S, \omega_S^{\otimes 2}(C)) + 1$, in particular, for any C and any $p \geq g - k + 1$ we obtain an isomorphism

$$K_{p,1}(S, \omega_S \otimes L) \xrightarrow{\sim} K_{p,1}(C, K_C).$$

Since Green's conjecture is valid for general curves $C \in |L|$ which have gonality k and Clifford dimension one, we infer that $K_{p,1}(S, \omega_S \otimes L) = 0$ for any $p \geq g-k+1$. In particular, $K_{p,1}(C, K_C) = 0$ for any $p \geq g-k+1$ and *any* smooth curve C. From the Green–Lazarsfeld non-vanishing Theorem [7, Appendix], it follows that *any* smooth curve C must have gonality k and Clifford dimension one. □

Example 1 (Smooth Curves on Hirzebruch Surfaces). The hypotheses of Theorem 3 are realized for curves on Hirzebruch surfaces $S = \Sigma_e$ with $e \geq 2$, hence we obtain an alternate proof of the results of [1].

Indeed, denote by C_0 the minimal section and by F the class of a fiber, and let $C \equiv kC_0 + mF$ with $m \geq ke$ by a smooth curve on S. The gonality of C is k and the genus of C is computed by the formula

$$g = (k-1)\left(m - 1 - \frac{ke}{2}\right).$$

The first condition $-\omega_S \cdot C \geq k$, from Theorem 2, is easily verified, as $(2C_0+(e+2)F)\cdot(kC_0+mF) = -ke+2m+2k$. The second condition $g-k \geq h^0(S, \omega_S^{\otimes 2}(C))$, from Theorem 3, is verified by direct computation, using the vanishing of h^1 of $\omega_S^{\otimes 2}(C) \equiv (k-4)C_0 + (m-2e-4)F$ and applying the Riemann–Roch Theorem.

Acknowledgements This work was partly supported by a Humboldt Fellowship and by the CNCS/UEFISCDI grant PN-II-ID-PCE-2011-3-0288 contract no. 132/05.10.2011. I would like to thank Humboldt Universität for hospitality and to G. Farkas for useful discussions on the topic.

References

1. Aprodu, M.: On the vanishing of higher syzygies of curves. Math. Z. **241**(1), 1–15 (2002)
2. Aprodu, M.: Remarks on syzygies of d-gonal curves. Math. Res. Lett. **12**(2–3), 387–400 (2005)
3. Aprodu, M.: Lazarsfeld-Mukai Bundles and Applications. Commutative Algebra, vol. 1–23. Springer, New York (2013)
4. Aprodu, M., Farkas, G.: Green's conjecture for curves on arbitrary K3 surfaces. Compos. Math. **147**(3), 839–851 (2011)
5. Arbarello, E., Cornalba, M.: Su una congetura di Petri. Comment. Math. Helv. **56**(1), 1–38 (1981)

6. Farkas, G., Ortega, A.: Higher rank Brill-Noether theory on sections of K3 surfaces. Int. J. Math. **23**(7), 1250075, 18 pp (2012)
7. Green, M.: Koszul cohomology and the geometry of projective varietie. J. Differ. Geom. **19**(1), 125–171 (1984)
8. Kim, H.: Moduli spaces of stable vector bundles on Enriques surfaces. Nagoya Math. J. **150**, 85–94 (1998)
9. Lazarsfeld, R.: Brill–Noether–Petri without degenerations. J. Differ. Geom. **23**, 299–307 (1986)
10. Lazarsfeld, R.: A sampling of vector bundle techniques in the study of linear series. Lectures on Riemann surfaces (Trieste, 1987), pp. 500–559. World Scientific, Teaneck (1989)
11. Lelli-Chiesa, M.: Green's Conjecture for curves on rational surfaces with an anticanonical pencil. Math. Z. (2013). doi: 10.1007/s00209-013-1164-7
12. Mukai, S.: Biregular classification of Fano 3-folds and Fano manifolds of coindex 3. Proc. Natl. Acad. Sci. USA **86**, 3000–3002 (1989)
13. Pareschi, G.: A proof of Lazarsfeld's theorem on curves on K3 surfaces. J. Algebraic Geom. **4**(1), 195–200 (1995)
14. Rasmussen, N.H., Zhou, S.: Pencils of small degree on curves on unnodal Enriques surfaces. arXiv:1307.5526
15. Voisin, C.: Green's generic syzygy conjecture for curves of even genus lying on a K3 surface. J. Eur. Math. Soc. (JEMS) **4**(4), 363–404 (2002)
16. Voisin, C.: Green's canonical syzygy conjecture for generic curves of odd genus. Compos. Math. **141**(5), 1163–1190 (2005)

Multinets in $\mathbb{P}^2$

Jeremiah Bartz and Sergey Yuzvinsky

Abstract Multinets are certain configurations of lines and points with multiplicities in the complex plane $\mathbb{P}^2$. They are used in the studies of resonance and characteristic varieties of complex hyperplane arrangement complements and cohomology of Milnor fibers. If all multiplicities equal 1 then a multinet is a net that is a realization by lines and points of several orthogonal latin squares. Very few examples of multinets with nontrivial multiplicities are known. In this paper, we present new examples of multinets. These are obtained by using an analogue of nets in $\mathbb{P}^3$ and intersecting them by planes.

Keywords Nets • Multinets • Line arrangements • Pencils of curves

1 Introduction

Multinets are certain configurations of lines and points with multiplicities in the complex projective plane $\mathbb{P}^2$. More exactly they are multi-arrangements of projective lines partitioned in three blocks with some extra properties (see Sect. 2). They appeared in [3, 6] in the study of resonance and characteristic varieties of the complement of a complex hyperplane arrangement. More recently, multinets have been used to study the cohomology of Milnor fibers such as in [2].

Although the notion of multinets has been around since 2006, very few examples of multinets with nontrivial multiplicities are known. In the paper, we recall some definitions, describe a new method to obtain multinets, and give quite a few new examples. For that we consider analogues of nets in $\mathbb{P}^3$ and intersect them by planes.

The paper is organized as follows. In Sect. 2, we recall basic definitions and properties of multinets. In Sect. 3, we give the general idea of constructing multinets.

J. Bartz
Department of Mathematics, Francis Marion University, Florence, SC 29506, USA
e-mail: jbartz@fmarion.edu

S. Yuzvinsky (✉)
Department of Mathematics, University of Oregon, Eugene, OR 97403, USA
e-mail: yuz@uoregon.edu

D. Ibadula, W. Veys (eds.), *Bridging Algebra, Geometry, and Topology*,
Springer Proceedings in Mathematics & Statistics 96,
DOI 10.1007/978-3-319-09186-0_3

Section 4 contains the main part of the paper. We systematically go over different cases of constructed multinets. The main parameters are the numbers of lines and points of various multiplicities. All cases are classified except multinets with all lines having multiplicity 1 and points having multiplicities 1 and 2. For that case we have only examples and a uniform upper bound on the number of points of multiplicity 2. In Sect. 5, we discuss briefly the combinatorics inside blocks. The conclusion is that this combinatorics in all our examples is defined by multinet structure, i.e., the combinatorics between blocks. Finally some open questions and conjectures are collected in Sect. 6.

2 Preliminaries

2.1 Pencils of Curves and Multinets in $\mathbb{P}^2$

There are several equivalent ways to define multinets in $\mathbb{P}^2$. We introduce them here using pencils of plane curves. A *pencil of plane curves* is a line in the projective space of homogeneous polynomials from $\mathbb{C}[x_1, x_2, x_3]$ of some fixed degree d. Any two distinct curves of the same degree generate a pencil, and conversely a pencil is determined by any two of its curves C_1, C_2. An arbitrary curve C in the pencil (called a *fiber*) is $C = aC_1 + bC_2$ where $[a : b] \in \mathbb{P}^1$. Every two fibers in a pencil intersect in the same set of points $\mathscr{X} = C_1 \cap C_2$, called the *base* of the pencil. If fibers do not have a common component (called a *fixed component*), then the base is a finite set of points.

A curve of the form $\prod_{i=1}^{q} \alpha_i^{m_i}$, where α_i are distinct linear forms and $m_i \in \mathbb{Z}_{>0}$ for $1 \leq i \leq q$, is called *completely reducible*. Such a curve is called *reduced* if $m_i = 1$ for each i. We are interested in connected pencils of plane curves without fixed components and at least three completely reducible fibers. By connectivity here we mean the nonexistence of a reduced fiber whose distinct components intersect only at $\mathscr{X}$. For brevity we say that such a pencil is *of Ceva type*.

Definition 1. The union of all completely reducible fibers (with a fixed partition into fibers, also called *blocks*) of a Ceva type pencil of degree d is called a (k, d)*-multinet* where k is the number of the blocks. The base $\mathscr{X}$ of the pencil is determined by the multinet structure and called the *base* of the multinet.

If the intersection of each two fibers is transversal, i.e., $|\mathscr{X}| = d^2$ and hence all blocks are reduced, then the multinet is called a *net*. If all blocks are reduced but $|\mathscr{X}| < d^2$, then we call the multinet *proper* and *light*. If there are non-reduced blocks we call the multinet *proper* and *heavy*.

From the viewpoint of projective geometry, a (k, d)-multinet is a multi-arrangement $\mathscr{A}$ of lines in $\mathbb{P}^2$ provided with multiplicities $m(\ell) \in \mathbb{Z}_{>0}$ ($\ell \in \mathscr{A}$) and partitioned into k blocks $\mathscr{A}_1, \ldots, \mathscr{A}_k$ ($k \geq 3$) subject to the following two condition.

(i) Let $\mathcal{X}$ be the set of the intersections of lines from different blocks. For each point $P \in \mathcal{X}$, the number

$$m(P) = \sum_{\ell \in \mathcal{A}_i, P \in \ell} m(\ell)$$

is independent on i. This number is called the *multiplicity* of P.

A net can be defined as a multinet with $m(\ell) = m(P) = 1$ for all $\ell \in \mathcal{A}$ and $P \in \mathcal{X}$. A proper light multinet has $m(\ell) = 1$ for $\ell \in \mathcal{A}$ but for some $P \in \mathcal{X}$ we have $m(P) \neq 1$.

(ii) For every two lines ℓ and ℓ' from the same block, there exists a sequence of lines from that block $\ell = \ell_0, \ell_1, \ldots, \ell_r = \ell'$ such that $\ell_{i-1} \cap \ell_i \notin \mathcal{X}$ for $1 \leq i \leq r$.

Thus, multinets can be defined purely combinatorially using an incidence relation. Note that the multiplicity $m(\ell)$ for each $\ell \in \mathcal{A}$ equals the multiplicity of its corresponding linear factor in the completely reducible fibers of the Ceva pencil.

2.2 Properties of Multinets and Examples That Have Been Known

There are several important properties of multinets.

Proposition 1. *Let $\mathcal{A}$ be a (k, d)-multinet. Then:*

(1) $\sum_{\ell \in \mathcal{A}_i} m(\ell) = d$, independent of i;
(2) $\sum_{\ell \in \mathcal{A}} m(\ell) = dk$;
(3) $\sum_{P \in \mathcal{X}} m(P)^2 = d^2$ (Bézout's theorem);
(4) $\sum_{P \in \mathcal{X} \cap \ell} m(P) = d$ for every $\ell \in \mathcal{A}$;
(5) There are no multinets with $k \geq 5$;
(6) All multinets with $k = 4$ are nets.

The first four numerical equalities are easy and proved in [3]. The last two properties are harder to establish and proved in [11, 13]. We now recall several examples of proper multinets that have been known since [3].

Example 1. A $(k, 1)$-net consists of k lines intersecting all at one point with each block consisting of one line. This case corresponds to a so-called local resonance component. It is considered to be trivial and we will often tacitly assume that $d > 1$.

Example 2. For each $n \geq 1$, a $(3, 2n)$-multinet is given by the pencil generated by polynomials $x^n(y^n - z^n)$ and $y^n(x^n - z^n)$ with the third completely reducible fiber being $z^n(x^n - y^n)$. These are the projectivizations of the reflection arrangements for the full monomial groups $G(n, 1, 3)$ (see [7]). For $n = 1$, it gives the only (up to projective isomorphism) $(3,2)$-net of Coxeter type A_3; for $n = 2$, it is the $(3,4)$-multinet of Coxeter type B_3. These multinets are heavy when $n > 1$.

Example 3. The cubics xyz and $x^3 + y^3 + z^3$ generate a Ceva pencil with four completely reducible fibers. They give the $(4,3)$-net known as the Hesse configuration. It is the only known multinet with four blocks. A long-standing conjecture is that the Hasse configuration is the unique 4-net.

2.3 Constructions of Nets

From a combinatorial viewpoint, (k,d)-nets are realizations of $k-2$ pairwise orthogonal Latin squares of size d (after identifying all blocks). If $k = 3$, the Latin square gives a multiplication table of a quasi-group. Thus one can view such a net as a representation of a Latin square or a quasi-group (see [12]).

If this quasi-group is a group, the representation can be reconstructed using the complex torus $(\mathbb{C}^*)^2$. The list of groups that represent a net have been recently completed by Korchmaros et al. in [4, 5] confirming a conjecture by Yuzvinsky in [14]. They also discovered new ways to construct the respective nets. Examples of nets representing quasi-groups which are not groups were constructed first by Stipins in [10].

3 Construction of Multinets

3.1 Multinets in Higher Dimensions

It is possible to generalize the notion of multinets to $\mathbb{P}^r$ $(r > 2)$ using pencils of homogeneous polynomials of $r+1$ variables. It is known that no multinet exists for $r \geq 5$ and every multinet in $\mathbb{P}^3$ or $\mathbb{P}^4$ would be a net with three blocks (see [8]).

The only known nets in $\mathbb{P}^r$ for $r > 2$ are the $(3,2n)$-nets in $\mathbb{P}^3$ given for each $n \in \mathbb{Z}_{>0}$ by the defining polynomial

$$Q_n = [(x_0^n - x_1^n)(x_2^n - x_3^n)][(x_0^n - x_2^n)(x_1^n - x_3^n)][(x_0^n - x_3^n)(x_1^n - x_2^n)]$$

where the brackets determine the blocks.

This is the collection of all (projectivizations of) reflection hyperplanes of the finite complex reflection group known as the monomial group $G(n,n,4)$ (see [7]). For $n = 2$ it is the Coxeter group of type D_4.

Each block of Q_n is partitioned in two *half-blocks* (determined by parentheses) of degree n each. Notice that all the planes of a half-block intersect at one line, called the *base of the half-block*. For instance the base of the leftmost half-block is given by the system $x_0 = 0, x_1 = 0$.

3.2 Construction of Multinets

Unlike for nets, there have been no known systematic ways to construct proper multinets, i.e., multinets which are not nets. Here we suggest a way that has produced a variety of new examples.

Intersect Q_n with a plane H that does not belong to Q_n. The resulting multiarrangement in H is denoted by $\mathscr{A}^H$ and referred to as the *arrangement induced by* Q_n. The pencil in $\mathbb{P}^3$ corresponding to Q_n induces a pencil in $\mathbb{P}^2$ with three completely reducible fibers. It may happen that the pencil has a fixed component. In this case, we cancel the fixed components obtaining a smaller arrangement $\mathscr{A}_0^H$ with a multinet structure. Abusing the notation slightly we will call $\mathscr{A}^H$ (if there is no fixed component) or $\mathscr{A}_0^H$, provided with the partitions into fibers of the induced pencil, the *induced multinet*.

In the rest of the paper, the following convention is applied. We use a homogeneous coordinate system $[x_0 : x_1 : x_2 : x_3]$ in $\mathbb{P}^3$. If needed we can change the coordinates using symmetries of Q_n. In particular, in examples below we can always assume that the plane H does not contain the point $[1 : 0 : 0 : 0]$ (by permuting coordinates if needed). Then intersecting with H amounts to substituting x_0 by a linear combination of other coordinates from the equation of H. We also can change the coordinates by multiplying them by any nth roots of unity.

4 Multinets Induced by Q_n

4.1 General Position and Survey of Possibilities

If H is in general position, by which we mean that it does not contain any elements of the intersection lattice L of Q_n, then $\mathscr{A}^H$ is a $(3, 2n)$-net realizing the dihedral group of order $2n$ (see [12]). In particular, $\mathscr{A}^H$ contains the $(3, n)$-net realizing the cyclic group C_n as a subarrangement. For instance for $n = 3$, $\mathscr{A}^H$ realizes the dihedral group of order 6.

If H contains some elements of L, then $\mathscr{A}^H$ falls in one or more of the following classes. In the first four cases we assume that $\mathscr{A}^H$ does not have fixed components.

1. H contains the base of a half-block; equivalently $\mathscr{A}^H$ contains a line of multiplicity n.
2. H contains the intersection of two planes from different half-blocks of a block; equivalently $\mathscr{A}^H$ contains a line of multiplicity 2.

   Now we assume that H does not contain any line from L whence $\mathscr{A}^H$ is light.
3. H contains the intersection of two half-block bases; equivalently $\mathscr{A}^H$ has a point of multiplicity n.

4. H contains the intersection of six hyperplanes, one from each half-block; equivalently $\mathcal{A}^H$ has a point of multiplicity 2.
5. H belongs to both classes 3 and 4; equivalently $\mathcal{A}^H$ has fixed components (see a proof in 4.7).

All these cases are discussed separately below. When we discuss points of multiplicity 2 in case 4 we always assume that $n > 2$ in order not to confuse them with points of multiplicity n. When we discuss them in the presence of fixed components in case 5, we assume $n > 3$ in order to have $2 \neq n-1$ or assume $n > 4$ if there are two fixed components.

4.2 Heavy Induced Multinets (Lines of Multiplicity n)

H contains the base of precisely one half-block if and only if $\mathcal{A}^H$ is a proper heavy multinet with only one line of multiplicity n. For instance, if H is given by $x_0 = cx_1$ ($c \neq 0$ nor it is a root of unity of degree n) containing the base of the block $x_2^n - x_3^n$, then the blocks of $\mathcal{A}^H$ are determined by the polynomials:

$$x_1^n(x_2^n - x_3^n),\ (c^n x_1^n - x_2^n)(x_2^n - x_3^n),\ (c^n x_1^n - x_3^n)(x_1^n - x_2^n).$$

If H contains the bases of two half-blocks, then these half-blocks are from different blocks (since the bases of the half-blocks from the same block do not intersect in $\mathbb{P}^3$) and H contains also the base of a half-block from the third block. Such H is a coordinate plane, for example, $x_0 = 0$. Then every block of $\mathcal{A}^H$ contains exactly one line of multiplicity n and the underlying arrangement is the reflection arrangements for the full monomial groups $G(n, 1, 3)$ mentioned before. For instance, if H is given by $x_0 = 0$, then $\mathcal{A}^H$ has blocks:

$$x_1^n(x_2^n - x_3^n),\ x_2^n(x_1^n - x_3^n),\ x_3^n(x_1^n - x_2^n).$$

4.3 Heavy Induced Multinets (Lines of Multiplicity 2)

$\mathcal{A}^H$ has a line of multiplicity 2 if and only if H contains the intersection of precisely two planes of Q_n from different half-blocks of a block. If H is generic with that condition, then $\mathcal{A}^H$ has precisely one line of multiplicity 2. Also H could contain several such lines from different blocks which produces up to three lines of multiplicity 2 in $\mathcal{A}^H$ if n is even, and up to two such lines if n is odd.

For instance, suppose H is given by $a(x_0 - x_1) - (x_2 - x_3) = 0$ where $a \neq 0$. Then $\mathcal{A}^H$ has the factor $(x_2 - x_3)^2$ in one block. If a is not a root of unity of degree n, then all other factors have multiplicity 1. If $a = 1$ and n is odd, then $\mathcal{A}^H$ contains the additional factor $(x_1 - x_3)^2$ in another block. If $a = 1$ and n is even,

$\mathscr{A}^H$ contains the additional factors $(x_1 - x_3)^2$ and $(x_1 + x_2)^2$ in the second and third block, respectively. The multiplicities of all other factors are equal to 1 in each of these situations. A similar effect is produced when a is another root of unity of degree n.

Finally it is easy to see that $\mathscr{A}^H$ cannot have both: a line of multiplicity n and a line of multiplicity 2. Indeed by Sect. 4.2 the former forces H to have equation of the form $Ax_i - Bx_j = 0$ for some distinct i and j. The latter forces H to contain a line of intersection of precisely two planes of Q_n from different half-blocks of the same block. This block must coincide with the block containing $x_i^n - x_j^n$. Furthermore H must coincide with a plane of this block (given by $x_i - \zeta x_j = 0$ where ζ is a root of unity of degree n). That is, H is a plane of $\mathscr{A}$ and not an allowable choice of H (see Sect. 3.2).

4.4 *Light Induced Multinets (Points of Multiplicity n)*

In the rest of this section, by a 'point' and 'line' we mean respectively a point or a line of $\mathbb{P}^3$ *from the intersection lattice of* Q_n. If any of these lie on H, then they have multiplicity coming from the multinet $\mathscr{A}^H$ or $\mathscr{A}_0^H$.

Recall that a multinet is light if it is proper and all of its lines have multiplicity 1. $\mathscr{A}^H$ is a light multinet with a point of multiplicity n if and only if the plane H contains the point of intersection of two bases of half-blocks (they must be from different blocks).

We can make this more concrete. Every one of six bases of half-blocks can be given by the equations $x_i = x_j = 0$ where $\{i, j\} \subset \{0, 1, 2, 3\}$. Also two non-disjoint bases intersect at one of the points having one coordinate 1 and others 0. Thus the induced multinet is light with a point of multiplicity n if and only if H contains precisely one of these points. For instance, if H is given by $Ax_0 + Bx_1 + Cx_2 = 0$ with generic coefficients, then the induced multinet is light with the only one multiple point $[0 : 0 : 0 : 1]$ of multiplicity n.

Light $\mathscr{A}^H$ cannot have more than one point of multiplicity n. If $\mathscr{A}^H$ has two points of multiplicity n, then H contains two distinct points with one coordinate 1 and others 0. It follows that H is given by $Ax_i + Bx_j = 0$ for some distinct i and j. Thus H contains the base of a half-block and $\mathscr{A}^H$ is heavy (see Sect. 4.2).

4.5 *Light Induced Multinets (Points of Multiplicity 2)*

Since a point P of multiplicity 2 in a light multinet has exactly two lines from each block intersecting at it, the plane H must contain the point of intersection of six planes, two from each block of Q_n. Moreover planes in every one of these pairs must come from different half-blocks since otherwise H would contain the base of a block and the induced multinet would be heavy. Conversely if H contains the intersection of such planes, say P, then P has multiplicity 2 in $\mathscr{A}^H$. If H is generic otherwise, $\mathscr{A}^H$ is light and contains precisely one point of multiplicity 2.

For instance, if H is given by $Ax_0 + Bx_1 + Cx_2 + Dx_3 = 0$ with $A+B+C+D = 0$ (whence passing through $P = [1:1:1:1]$) and generic otherwise, P has multiplicity 2 in $\mathscr{A}^H$ while the other points and all lines have multiplicity 1.

Let us notice that for a point P' of intersection of only four planes, two from one block and two from another, there are always two planes from the third block passing through P'.

For future use we characterize these points of intersection more explicitly. Let four planes be given by

$$x_0 - \zeta^a x_1 = 0,\ x_2 - \zeta^b x_3 = 0,\ x_0 - \zeta^c x_2 = 0,\ x_1 - \zeta^d x_3 = 0$$

where ζ is primitive root of unity and its exponents are arbitrary from a cyclic group C_n (in additive notation). For the intersection to exist, the equality $a + d = b + c$ is needed and then the intersection is

$$[\zeta^{a+d} : \zeta^d : \zeta^b : 1].$$

Conversely, let $P' \in H$ where $P' = [\zeta^a : \zeta^b : \zeta^c : 1]$ with ζ, a, b, c as above. Clearly P' lies in the four planes:

$$x_0 - \zeta^a x_3 = 0,\ x_1 - \zeta^b x_3 = 0,\ x_1 - \zeta^{b-c} x_2 = 0,\ x_0 - \zeta^{a-c} x_2 = 0$$

that proves the converse.

In particular this proves the following Lemma.

Lemma 1. *If a point P of $\mathbb{P}^3$ has multiplicity 2 on some plane H ($H \notin Q_n$) passing through P, then it has homogeneous coordinates that are roots of unity of degree n. If P has such coordinates, then it has multiplicity 2 on H for every plane H ($H \notin Q_n$) passing through P.*

4.6 Light Induced Multinets (Several Points of Multiplicity 2)

A plane H can have several points described in the previous subsection whence $\mathscr{A}^H$ can have several points of multiplicity 2. A partial classification of induced light multinets with double points for $n \leq 6$ is given in [1].

First we consider light induced multinets without points of multiplicity n. The current maximal number of points of multiplicity 2 known for light multinets is 8.

Example 4. Take $n = 8$ and fix a primitive root of unity ζ of degree 8. Let H be given by

$$x_0 - (\zeta + 1)x_1 - \zeta^3 x_2 + (\zeta^3 + \zeta)x_3 = 0.$$

Then $\mathscr{A}^H$ is light, has no fixed components, and has 8 points of multiplicity 2 (with all other points having multiplicity 1). These 8 points are as follows:

$$\begin{array}{ll} [1:1:1:1] & [\zeta^5:\zeta^2:\zeta^3:1] \\ [\zeta^2:\zeta:1:1] & [\zeta^5:\zeta^3:\zeta^5:1] \\ [\zeta^2:\zeta^2:\zeta^6:1] & [\zeta^7:1:\zeta:1] \\ [\zeta^4:\zeta^3:\zeta^6:1] & [\zeta^7:\zeta:\zeta^3:1]. \end{array}$$

Each of these points lies on H and on exactly six hyperplanes of Q_8 (one from each half-block). For instance, $[\zeta^2:\zeta:1:1]$ lies on the six hyperplanes

$$\begin{array}{lll} x_0-\zeta x_1 & x_0-\zeta^2 x_2 & x_0-\zeta^2 x_3 \\ x_2-x_3 & x_1-\zeta x_3 & x_1-\zeta x_2. \end{array}$$

On the other hand, the following surprising result holds.

Theorem 2. *Let $n > 3$ and $\mathscr{A}^H$ a light induced multinet without points of multiplicity n. Then the number of points of multiplicity 2 in it is less than 2^{96}(independently of n).*

Proof. We fix H and without any loss of generality assume that it is given by $Ax_0 + Bx_1 + Cx_2 - x_3 = 0$. If $\mathscr{A}^H$ has a point of multiplicity 2 then by Lemma 1

$$A\zeta^a + B\zeta^b + C\zeta^c = 1 \tag{1}$$

for some primitive root of unity ζ of degree n and $a, b, c \in C_n$.

Now we prove that the relation (1) of roots of unity is non-degenerate, that is, no proper partial sum of the left-hand side is 0.

If for instance, $A\zeta^a = 0$, i.e., $A = 0$ then H contains the point $[1:0:0:0]$ that has multiplicity n in $\mathscr{A}^H$ which contradicts a condition of the theorem. Furthermore suppose $A\zeta^a + B\zeta^b = 0$, i.e., $[A:B] = [\zeta^b : -\zeta^a]$ and $C = \zeta^{-c}$. Then the equation for H becomes

$$D\zeta^b x_0 - D\zeta^a x_1 + \zeta^{-c} x_2 - x_3 = D\zeta^b(x_0 - \zeta^{a-b} x_1) + \zeta^{-c}(x_2 - \zeta^c x_3) = 0.$$

The equation shows that H contains the line of intersection of 2 planes (from one block): $x_0 - \zeta^{b-a} x_1 = 0$ and $x_2 - \zeta^c x_3 = 0$. The intersection of these planes with H gives in $\mathscr{A}^H$ a line of multiplicity 2 whence $\mathscr{A}^H$ is heavy.

We finish the proof applying the result from [9] (for $k = 3$) which says that the number of non-degenerate solutions in roots of unity of a given equation $\sum_{i=1}^k A_i x_i = 1$ with complex coefficients is bounded from above by $2^{4(k+1)!}$.

4.7 Fixed Components

$\mathscr{A}^H$ has a fixed component if and only if H contains the intersection of two planes from different blocks. If we fix such a component, then by permuting coordinates and scaling them by a root of unity we can assume that the intersecting planes are $x_0 - x_2 = 0$ and $x_1 - x_2 = 0$ whence H is given by

$$Ax_0 + Bx_1 + Cx_2 = 0 \tag{2}$$

with $A + B + C = 0$. It is clear that $ABC \neq 0$ since otherwise H would be a plane from Q_n.

Now we show that it suffices to consider light multinets.

Lemma 2. *If $\mathscr{A}^H$ has a fixed component, then the multiplicity of every line in H equals 1.*

Proof. The result follows immediately by comparing (2) with conditions for heavy induced multinets from Sects. 4.2 and 4.3.

Using this lemma we consider only light multinets for the remainder of this subsection. Now we rephrase the existence of fixed components in terms of multiplicity of points.

Proposition 3. *Assume that $\mathscr{A}^H$ does not have lines of multiplicity greater than 1. Then the following statements hold:*

(i) H has both a point of multiplicity n and a point of multiplicity 2 if and only if $\mathscr{A}^H$ has a fixed component;
(ii) Each pair of points of multiplicities n and 2 has a unique H passing through the pair such that $\mathscr{A}^H$ has a fixed component.

Proof. (i) Suppose H contains points P_1 of multiplicity n and P_2 of multiplicity 2. Using a permutation of coordinates we can assume that $P_1 = [0 : 0 : 0 : 1]$. Also using Lemma 1 we can assume that $P_2 = [\zeta^a : \zeta^b : 1 : \zeta^c]$ where ζ is a primitive root of unity of degree n while $a, b, c \in C_n$. Thus an equation of H can be reduced to

$$Ax_0 + Bx_1 + Cx_2 = 0$$

with $C = -A\zeta^a - B\zeta^b$. We can write the equation also as

$$A(x_0 - \zeta^a x_2) = -B(x_1 - \zeta^b x_2). \tag{3}$$

Note that in the last equation $A, B \neq 0$. Indeed if only one of them is 0, then H coincides with a plane of Q_n which is not an allowable choice for H (see Sect. 3.2). If both vanish, then H is given by $x_2 = 0$ which gives a line in $\mathscr{A}^H$ of multiplicity n which is also not allowed by the condition of the lemma.

Thus we see that two blocks of $\mathscr{A}^H$ have $x_0 - \zeta^a x_2$ as a common component whence all three of them do. This implies that $\mathscr{A}^H$ has a fixed component.

In order to prove the converse suppose $\mathscr{A}^H$ has a fixed component whence we can assume that H is given by (2). Then H contains the points $[0:0:0:1]$ and $[1:1:1:1]$ which have multiplicity n and 2, respectively.

(ii) Part (i) of the proof shows also that H is uniquely determined by the multiple points P_i $(i = 1, 2)$. For instance, if coordinates of the points are chosen as in part (i) and H is given by $Ax_0 + Bx_1 + Cx_2 + Dx_3 = 0$ we obtain $A : B = (x_0 - \zeta^a x_2) : (-x_1 + \zeta^b x_2)$, $C = -A\zeta^a - B\zeta^b$, and $D = 0$.

The next result about fixed components is more subtle and requires some preparation.

Proposition 4. *Let $U \in \mathbb{C}$, $U \neq 0, 1$. The equation*

$$U = \frac{1-\xi}{1-\eta} \tag{4}$$

has at most one solution $(\xi, \eta) \in \mathbb{C}^2$ such that $|\xi| = |\eta| = 1$.

Proof. Suppose that there are two solutions of (4): (ξ, η) and (ξ_1, η_1) and their respective complex arguments are $\alpha, \beta, \alpha_1, \beta_1 \in (0, 2\pi)$. It is clear that for any number $z = e^{i\gamma}$ we have

$$1 - z = 2\sin\left(\frac{\gamma}{2}\right) e^{i\left(\frac{\gamma-\pi}{2}\right)}.$$

Hence Eq. (4) implies

$$\alpha - \beta = \alpha_1 - \beta_1 \tag{5}$$

$$\sin\left(\frac{\alpha}{2}\right)\sin\left(\frac{\beta_1}{2}\right) = \sin\left(\frac{\alpha_1}{2}\right)\sin\left(\frac{\beta}{2}\right) \tag{6}$$

$$\cos\left(\frac{\alpha}{2}\right)\cos\left(\frac{\beta_1}{2}\right) = \cos\left(\frac{\alpha_1}{2}\right)\cos\left(\frac{\beta}{2}\right) \tag{7}$$

where Eq. (5) is the equality of the arguments, Eq. (6) comes from the equality of the moduli, and Eq. (7) follows from (5, 6). Also, (5, 6, 7) imply the following equations for cotangents

$$\cot\left(\frac{\alpha}{2}\right)\cot\left(\frac{\beta_1}{2}\right) = \cot\left(\frac{\alpha_1}{2}\right)\cot\left(\frac{\beta}{2}\right) \tag{8}$$

$$\cot\left(\frac{\alpha}{2}\right) - \cot\left(\frac{\beta}{2}\right) = \cot\left(\frac{\alpha_1}{2}\right) - \cot\left(\frac{\beta_1}{2}\right). \tag{9}$$

If $\cot\left(\frac{\beta}{2}\right) = 0$, i.e., $\frac{\beta}{2} = \frac{\pi}{2}$, then (8) implies either $\cot\left(\frac{\alpha}{2}\right) = 0$ whence $\xi = \eta$ or $\cot\left(\frac{\beta_1}{2}\right) = 0$ whence $\eta = \eta_1$. The first situation is not possible by the hypotheses of the proposition. The second and (4) show $(\xi_1, \eta_1) = (\xi, \eta)$.

If $\cot\left(\frac{\beta}{2}\right) \neq 0$, resolve Eq. (8) for $\cot\left(\frac{\alpha_1}{2}\right)$ and plug it into Eq. (9). Since $\frac{\alpha}{2}, \frac{\beta}{2} \in (0, \pi)$ and $\alpha \neq \beta$, we can cancel the factor $\cot\left(\frac{\alpha}{2}\right) - \cot\left(\frac{\beta}{2}\right)$ to obtain $\cot\left(\frac{\beta_1}{2}\right) = \cot\left(\frac{\beta}{2}\right)$ whence $\frac{\beta_1}{2} = \frac{\beta}{2}$. The equality $(\xi_1, \eta_1) = (\xi, \eta)$ again follows.

Remark 1. When the first version of the paper appeared on arXiv we received a comment from Joe Buhler with a more elegant proof (of the previous proposition) whose main idea he attributed to Richard Stong. With their permission, we exhibit this proof below.

When z ranges over the pointed unit circle without 1 in the complex plane the set of $\{1 - z\}$ is the pointed unit circle S without 0 centered at 1. The proposition is about the multiplicative relation of the form

$$ab = cd$$

where $a, b, c, d \in S$, and (say) a is distinct from both c and d. By taking inverses, this is equivalent to the same relation in the set S^{-1} of the inverses of elements of S, which is the set of complex numbers with the real part $\frac{1}{2}$. But if

$$(1/2 + ir)(1/2 + is) = (1/2 + it)(1/2 + iu)$$

with r, s, t, u real then equating real and imaginary parts one deduces that the sums and products of distinct elements inside the sets $\{r, s\}$ and $\{t, u\}$ are equal. Thus these sets coincide.

Theorem 5. *Suppose $\mathscr{A}^H$ does not have lines of multiplicity greater than 1. If it has a fixed component F, then there is a (unique) point of multiplicity n in H and precisely two sets of n points of multiplicity 2 each. Moreover the point of multiplicity n lies in F together with one set of points of multiplicity 2; the other set lies in $H \setminus F$.*

Proof. Since $\mathscr{A}^H$ has a fixed component we can use (2) and assume that F is given by the system $x_0 = x_1 = x_2$ and H is given by the equation $A(x_0 - x_2) + B(x_1 - x_2) = 0$ with $AB(A + B) \neq 0$. This immediately implies that multiplicity n point $P_0 = [0 : 0 : 0 : 1] \in F \subset H$. The set of points of multiplicity 2 in F is $P_c = [1 : 1 : 1 : \zeta^c]$ for $c \in C_n$ and they are clearly the only points in F with roots of unity as coordinates.

Now suppose a point $[\zeta^a : \zeta^b : 1 : \zeta^c] \in H \setminus F$. In particular $a \neq b$ nor $a, b \neq 0$. Then we have

$$\frac{1 - \zeta^a}{1 - \zeta^b} = U$$

where $U = -\frac{B}{A}$. Applying Proposition 4 we obtain that a and b are determined uniquely. This gives precisely n points in $H \setminus F$ since c runs through the group C_n.

Corollary 1. *No induced arrangements $\mathscr{A}^H$ can have more than two fixed components.*

Proof. Suppose $\mathscr{A}^H$ has a fixed components. By Theorem 5 there are at most $2n$ double points and precisely n of them lie in one fixed component. By Proposition 3 (ii) these sets of n points lying in distinct components are disjoint.

If $\mathscr{A}^H$ has a fixed component, it must be canceled in order to obtain a multinet. For instance, let H is given by $A(x_0 - x_1) + B(x_0 - x_2) = 0$, with $AB(A+B) \neq 0$. Upon intersecting with H we substitute $\frac{Ax_1+Bx_2}{A+B}$ for x_0 and cancel $x_1 - x_2$. We may obtain a multinet $\mathscr{A}_0^H$ such that $P_0 = [0 : 0 : 1]$ acquires multiplicity $n - 1$. For instance, this is the case for H given by $3x_0 - 2x_1 - x_2 = 0$.

Choosing H more carefully we can get $\mathscr{A}^H$ having another fixed component. For instance, if H is given by $x_0 = (\zeta + 1)x_1 - \zeta x_2$ where ζ is a primitive root of unity of degree n, then $\mathscr{A}^H$ has two common factors $x_1 - x_2$ and $x_1 - \zeta x_2$ which results in a light $(3, 2n - 2)$-multinet upon cancellation. The point P_0 of $\mathscr{A}_0^H$ has multiplicity $n - 2$ and all other points have multiplicity 1.

According to Proposition 3 (i), every H such that $\mathscr{A}^H$ has a fixed component must have a point of multiplicity n and points of multiplicity 2 that all lies on fixed components. Thus $\mathscr{A}_0^H$ does not have any points of multiplicity 2.

4.8 Summary of Properties of Induced Multinets from Q_n

The multinets induced from Q_n possess the following properties (we suppose $n > 3$).

1. The multiplicity of lines takes only values $1, 2$, and n. There may be 1 or 3 lines of multiplicity n. There may be one line of multiplicity 2. Also, there may be three such lines (if n is even) or two (if n is odd).
2. The multiplicity of points takes values from the list $\{1, 2, n - 2, n - 1, n\}$.
3. A light multinet has at most one point of multiplicity n or at most one point of multiplicity $n - 1$ or at most one point of multiplicity $n - 2$. These three cases are disjoint and each does not allow any other point with multiplicity larger than 1.
4. A light multinet can have several points of multiplicity 2 if it does not have points of multiplicity greater than 2. The number of these points is bounded independently of n by 2^{96}.

5 Combinatorics Inside Blocks

In this section we discuss the possibilities for combinatorics of lines and points inside a block of an induced multinet.

First suppose that $\mathscr{A}^H$ is light, does not have a fixed component, and H is given by the equation $Ax_0+Bx_1+Cx_2+Dx_3 = 0$. Because of the conditions on $\mathscr{A}^H$, the plane H does not contain the base line of any of the half-blocks. On the other hand, it has one point of intersection with every base line. For instance, for the half-block $x_0^n - x_1^n$ the point is $[0 : 0 : -D : C]$. Thus in $\mathscr{A}^H$ every half-block consists of lines intersecting all at one point (i.e., forming a pencil of dimension 1). For generic H the base point of the pencil is not in $\mathscr{X}$ whence these points for different half-blocks are distinct. If however H is passing through the intersection of two base lines, then that point is in $\mathscr{X}$. For instance if H is given by $Ax_0 + Bx_1 + Cx_2 = 0$, then the point $[0 : 0 : 0 : 1]$ in it is the base point of the pencils in three half-blocks from three different blocks.

If $\mathscr{A}^H$ has fixed components, then the same claim holds for $\mathscr{A}_0^H$ except the amount of lines in each of three half-blocks involved in the fixed components decreases either to $n-1$ or $n-2$.

Now suppose that $\mathscr{A}^H$ is heavy. If it has lines of multiplicity n (one or three), then the respective half-block contains precisely one of these lines instead of a pencil. If $\mathscr{A}^H$ has lines of multiplicity 2, then one of the lines in the respective block connect the base vertices of two half-blocks (i.e., the intersections of the half-block bases with H). Recall that $\mathscr{A}^H$ cannot have lines of both multiplicities: n and 2 (see Sect. 4.3).

Summing up the discussion in this section we conclude that the multiplicities of lines of an induced multinet determine the combinatorics inside blocks.

6 Conjectures and Open Problems

For multinets there are more open questions than answers. Here are some of the former.

Problem 1. Make the upper bound in Theorem 2 smaller. (We conjecture that it can be significantly decreased.)

Problem 2. All induced multinets can be obtained from nets by deformation (moving the plane H). Prove or disprove the conjecture from [8] that all multinets have this property.

Problem 3. Are there nets in $\mathbb{P}^3$ other than Q_n?

Problem 4. There are nets which are not induced by Q_n. Examples include every $(3, 2k+1)$-net for $k \in \mathbb{Z}_{>0}$. The light multinet in Fig. 2 of [3] is also not induced from Q_n (a proof should include that it is not induced from Q_6 after a cancellation).

Are there heavy multinets not induced from Q_n?

References

1. Bartz, J.: Multinets in $\mathbb{P}^2$ and $\mathbb{P}^3$. Ph.D. thesis, University of Oregon (2013)
2. Denham, G., Suciu, A.: Multinets, parallel connections, and Milnor fibrations of arrangements. arXiv:1209.3414v2
3. Falk, M., Yuzvinsky, S.: Multinets, ressonance varieties, and pencils of plane curves. Compositio Math. **143**, 1069–1088 (2007)
4. Korchmaros, G., Nagy, G.P., Pace, N.: k-nets embedded in a projective plane over a field. arXiv:1306.5779
5. Korchmaros, G., Nagy, G., Pace, N.: 3-nets realizing a group in a projective plane. arXiv:1104.4439v3
6. Libgober, A., Yuzvinsky, S.: Cohomology of the Orlik-Solomon algebras and local systems. Compositio Math. **121**, 337–361 (2000)
7. Orlik, P., Terao, H.: Arrangements of Hyperplanes. Springer, New York (1992)
8. Pereira, J.V., Yuzvinsky, S.: Completely reducible hypersurfaces in a pencil. Adv. Math. **219**(2), 672–688 (2008)
9. Schlickewei, H.P.: Equations in roots of unity. Acta Arith. **LXXVI**(2), 99–108 (1996)
10. Stipins, J.: Old and new examples of k-nets in $\mathbb{P}^2$. math.AG/0701046. Arxiv
11. Stipins, J.: On finite k-nets in the complex projective plane. Ph.D. thesis, The University of Michigan (2007)
12. Yuzvinsky, S.: Realization of finite Abelian groups by nets in $\mathbb{P}^2$. Compositio Math. **140**, 1614–1624 (2004)
13. Yuzvinsky, S.: A new bound on the number of special fibers in a pencil of curves. Proc. Am. Math. Soc. **137**, 1641–1648 (2009)
14. Yuzvinsky, S.: Resonance varieties of arrangement complements. Adv. Stud. Pure Math. **62**, 553–570 (2012)

A More General Framework for CoGalois Theory

Şerban A. Basarab†

Abstract The coGalois theory studies the correspondence between subfields of a radical field extension L/K and subgroups of the coGalois group $\mathrm{coG}(L/K)$-the torsion of the quotient group $L^\times/K^\times$. Its abstract version concerns a continuous action of a profinite group Γ on a discrete quasi-cyclic group A, establishing a Galois connection between closed subgroups of Γ and subgroups of the group $Z^1(\Gamma, A)$ of continuous 1-cocycles. More generally, we study in the present work triples $(\Gamma, \mathfrak{G}, \eta)$, where Γ is a profinite group, $\mathfrak{G}$ is a profinite operator Γ-group, and $\eta : \Gamma \longrightarrow \mathfrak{G}$ is a continuous 1-cocycle such that $\eta(\Gamma)$ topologically generates $\mathfrak{G}$. To any such triple one assigns a natural coGalois connection between closed subgroups of Γ and closed Γ-invariant normal subgroups of $\mathfrak{G}$. In the abelian context, examples concern the coGalois theory of separable radical extensions, an additive analogue of it based on Witt calculus and higher Artin-Schreier theory, and an extension of the abstract cyclotomic framework to Galois algebras. Kneser triples and coGalois triples are investigated, and general Kneser and coGalois criteria are provided. Problems on the classification of certain finite algebraic structures arising naturally from these criteria are stated and partial solutions are given.

Keywords Galois connection • Profinite group • CoGalois group • Kneser criterion • Generating cocycle • Spectral space • Radical extension • Operator group

2000 *Mathematics Subject Classification*: 20E15, 20E18, 12G05, 12F10, 06A15, 06E15.

†This work of Basarab is one of his last submitted papers before he suddenly passed away in July 2014.

D. Ibadula, W. Veys (eds.), *Bridging Algebra, Geometry, and Topology*,
Springer Proceedings in Mathematics & Statistics 96,
DOI 10.1007/978-3-319-09186-0_4

1 Introduction

The so-called coGalois theory is a more or less recent development [1, 4, 5, 8, 9, 17, 30] of the study of finite radical extensions carried out, in chronological order, by H. Hasse 1930 (see the second edition [18] of his mimeographed lectures on Class Field Theory), A. Besicovitch [14], L.J. Mordell [26], C.L. Siegel [38], M. Kneser [21] and A. Schinzel [34], among others. Roughly speaking, the coGalois theory is an extension of the *Kummer theory*, being somewhat dual to the very classical *Galois theory*. For details and more references the reader may consult the monograph [2].

More precisely, given a radical field extension L/K, the main goal of the coGalois theory is to investigate the relation between the intermediate fields of the extension L/K and the subgroups of the torsion subgroup of the multiplicative factor group $L^{\times}/K^{\times}$, called the *coGalois group* of the extension L/K, denoted $\mathrm{coG}(L/K)$. Assuming that the extension L/K is Galois with $\Gamma := \mathrm{Gal}(L/K)$, $\mathrm{coG}(L/K)$ is canonically isomorphic, via Hilbert's Theorem 90, to the group $Z^1(\Gamma, \mu(L))$ of all continues 1-cocycles (crossed homomorphisms) of the profinite group Γ with coefficients in the group $\mu(L)$ of all roots of unity in L [9]. Notice that the multiplicative group $\mu(L)$ is quasi-cyclic, so it is isomorphic (in a noncanonical way) to a subgroup of the additive group $\mathbb{Q}/\mathbb{Z}$.

By analogy with Neukirch's *Abstract Galois Theory* within his *Abstract Class Field Theory* [27], an abstract, group theoretic framework of the coGalois theory is developed in the papers [3, 11, 12], where basic concepts of the field theoretic coGalois theory, as well as their main properties, are generalized to arbitrary continuous actions of profinite groups on discrete quasi-cyclic groups.

Such a continuous action $\Gamma \times A \longrightarrow A$, where Γ is a profinite group and the discrete quasi-cyclic group A is identified with a subgroup of $\mathbb{Q}/\mathbb{Z}$, establishes via the evaluation map $\Gamma \times Z^1(\Gamma, A) \longrightarrow A$, $(\gamma, \alpha) \mapsto \alpha(\gamma)$, a Galois connection between the lattice $\mathbb{L}(\Gamma)$ of all closed subgroups of Γ and the lattice $\mathbb{L}(Z^1(\Gamma, A))$ of all subgroups of the torsion abelian group $Z^1(\Gamma, A)$ of continuous 1-cocycles.

On the other hand, the continuous action of Γ on A endows the dual group $Z^1(\Gamma, A)^{\vee} = \mathrm{Hom}(Z^1(\Gamma, A), \mathbb{Q}/\mathbb{Z}) = \mathrm{Hom}(Z^1(\Gamma, A), A)$ with a natural structure of profinite Γ-module, related to Γ through a continuous cocycle $\eta : \Gamma \longrightarrow Z^1(\Gamma, A)^{\vee}$ with the property that the abelian profinite group $Z^1(\Gamma, A)^{\vee}$ is topologically generated by its closed subset $\eta(\Gamma)$. The continuous cocycle η plays a key role in [3, 11, 12] for the study of several interesting classes of subgroups of Γ and $Z^1(\Gamma, A)$ induced by the Galois connection above (*radical, hereditarily radical, Kneser, hereditarily Kneser, coGalois* and *strongly coGalois*).

The major role played by the continuous cocycle η above in what we may call *cyclotomic abstract coGalois theory* is the motivation for a more general approach of coGalois theory with potential new applications. The present paper, based on the unpublished preprint [13], is organized in six sections (2–7). Some basic notions used throughout the paper (profinite and spectral spaces, profinite groups and operator groups, subgroup lattices, Galois and coGalois connections) are briefly explained in Sect. 2. Section 3 introduces a general abstract framework for coGalois theory having as main object of investigation the triples $(\Gamma, \mathfrak{G}, \eta)$, where Γ is a profinite group, $\mathfrak{G}$ is a profinite operator Γ-group, and $\eta : \Gamma \longrightarrow \mathfrak{G}$ is a continuous 1-cocycle with the property that $\eta(\Gamma)$ topologically generates the profinite group $\mathfrak{G}$; such a cocycle η is called *generating cocycle*. To any such triple one assigns a natural coGalois connection between the lattice $\mathbb{L}(\Gamma)$ of all closed subgroups of Γ and the modular lattice $\mathbb{L}(\mathfrak{G})$ of all closed Γ-invariant normal subgroups of $\mathfrak{G}$, called *ideals* of the profinite Γ-group $\mathfrak{G}$. The main properties of this coGalois connection are collected in Proposition 3.6.

The general framework presented in Sect. 3 is applied in Sect. 4 to an abelian context which extends as most as possible the framework of cyclotomic coGalois theory by considering continuous actions of profinite groups Γ on arbitrary discrete torsion abelian groups A, not necessarily quasi-cyclic. To any such Γ-module A one assigns a profinite Γ-module $\mathfrak{G}$ and a generating cocycle $\eta : \Gamma \longrightarrow \mathfrak{G}$ such that the Galois connection between the lattice $\mathbb{L}(\Gamma)$ and the lattice $\mathbb{L}(Z^1(\Gamma, A))$ of all subgroups of the torsion abelian group $Z^1(\Gamma, A)$ is obtained by composing the coGalois connection between $\mathbb{L}(\Gamma)$ and $\mathbb{L}(\mathfrak{G})$, introduced in Sect. 3, with a natural Galois connection between the lattices $\mathbb{L}(\mathfrak{G})$ and $\mathbb{L}(Z^1(\Gamma, A))$ (Propositions 4.3, 4.6, 4.8, Corollary 4.9). Section 4 ends with four relevant examples: the first two examples are concerned with the coGalois theory of separable radical extensions and its abstract cyclotomic version; the third example is devoted to an additive analogue of the coGalois theory of separable radical extensions, based on Witt calculus and higher Artin-Schreier theory, while the fourth example is an extension of the cyclotomic context to Galois algebras.

Some of the main notions and results of the cyclotomic abstract coGalois theory are extended in Sects. 5 and 6 to the general framework introduced in Sect. 3. Here two remarkable types of triples $(\Gamma, \mathfrak{G}, \eta)$ are investigated: the *Kneser triples*, where the cocycle $\eta : \Gamma \longrightarrow \mathfrak{G}$ is surjective, and the *coGalois triples*, i.e., Kneser triples for which the associated coGalois connection is perfect. Given a triple $(\Gamma, \mathfrak{G}, \eta)$, where η is a generating cocycle, the main properties of the space $\mathbb{K}(\mathfrak{G})$ ($\mathbb{CG}(\mathfrak{G})$), consisting of the ideals $\mathbf{a}$ of the profinite Γ-group $\mathfrak{G}$ for which the induced triple $(\Gamma, \mathfrak{G}/\mathbf{a}, \eta_{\mathbf{a}} : \Gamma \longrightarrow \mathfrak{G}/\mathbf{a})$ is Kneser (coGalois), are collected in Propositions 5.19, 6.4. A special attention is paid in Sect. 5.1.2 to a procedure for obtaining bijective cocycles by deformation of a profinite group via a continuous action on itself. General *Kneser* and *coGalois criteria* are provided by Propositions 5.22, 6.6, and two remarkable classes of finite algebraic structures (*minimal non-Kneser* and *Kneser minimal non-coGalois triples*) arising naturally from these general criteria are introduced. The open Problems 5.24, 6.8 are concerned with the classification of these finite algebraic structures.

Partial answers to Problem 5.24 are given in Sect. 7. In particular, the special case when $\Gamma, \mathfrak{G}$ are abelian is completely solved under an additional assumption on the local subring of the endomorphism ring of the abelian p-group $\mathfrak{G}$, generated by Γ (Propositions 7.7, 7.8). As an immediate consequence, we find again [3, Lemma 1.18, Theorem 1.20], the abstract version of the classical *Kneser criterion for separable radical extensions* [21], [2, Theorem 11.1.5].

2 Preliminaries

2.1 *Profinite and Spectral Spaces*

A *profinite space*, also called *Stone* or *boolean space*, is a compact Hausdorff and totally disconnected topological space. A *spectral space*, also called *coherent* or *quasi-boolean space*, is a compact T_0 topological space which admits a base of compact sets for its topology; equivalently, a topological space X is spectral if the family of compact open sets is closed under finite intersections (in particular, X itself is compact) and forms a base for the topology of X, and X is *sober*, i.e., every irreducible closed subset of X is the closure of a unique point of X.

The spectral spaces form a category **SPECS** having as morphisms the so-called coherent maps, i.e., the maps $f : X \longrightarrow Y$ for which $f^{-1}(V)$ is a compact open subset of the spectral space X provided V is a compact open subset of the spectral space Y. In particular, the coherent maps are continuous, so **SPECS** is a non-full subcategory of the category **TOP** of topological spaces. We denote by **PFS** the full subcategory of **TOP** whose objects are the profinite spaces. Obviously, **PFS** is also a full subcategory of **SPECS**.

By *Stone Representation Theorem*, **PFS** and **SPECS** are duals to the categories of boolean algebras and (bounded) distributive lattices respectively. For more details concerning the topological spaces above and their dual structures the reader may consult [10, 19, 20, 31].

2.2 *Profinite Groups*

A *profinite group* is a group object in **PFS**, i.e., a compact totally disconnected topological group; equivalently, a topological group Γ is profinite if the identity element 1 of Γ admits a fundamental system $\mathscr{U}$ of open neighborhoods U such that U is a normal subgroup of Γ, and $\Gamma = \varprojlim_{U \in \mathscr{U}} \Gamma/U$, the inverse limit of the inverse system of discrete finite groups $\{\Gamma/U \mid U \in \mathscr{U}\}$. For details on profinite groups see [28, 32, 35].

Let $\mathscr{C}$ be a *variety* of finite groups, i.e., a nonempty class of finite groups, closed under subgroups, quotients, and finite direct products. The profinite groups Γ, with $\Gamma/U \in \mathscr{C}$ for all open normal subgroups $U \subseteq \Gamma$, called *pro-$\mathscr{C}$ groups*, form a full subcategory **P**$\mathscr{C}$**G** of the category **PFG** of profinite groups, with continuous homomorphisms. In particular, for any subgroup $A \subseteq \mathbb{Q}/\mathbb{Z}$, let $\mathbf{FAb}_A$ denote the variety consisting of those finite abelian groups G for which $\frac{1}{\exp(G)}\mathbb{Z}/\mathbb{Z} \subseteq A$. For instance, for $A = \mathbb{Q}/\mathbb{Z}$, $(\mathbb{Q}/\mathbb{Z})(p) \cong \mathbb{Q}_p/\mathbb{Z}_p$, $\frac{1}{n}\mathbb{Z}/\mathbb{Z}$, we obtain the varieties of finite abelian groups, of finite abelian p-groups for any prime number p, and of finite abelian groups of exponent dividing $n \geq 1$. Moreover any variety of finite abelian groups is of the form above. Setting $A^\vee := \mathrm{Hom}(A, \mathbb{Q}/\mathbb{Z})$, the Pontryagin dual of the discrete quasi-cyclic group A, with the canonical structure of profinite ring, the pro-$\mathbf{FAb}_A$ groups are identified with the profinite $A^\vee$-modules, the duals of discrete (torsion) $A^\vee$-modules.

2.3 The Lattice of Closed Subgroups of a Profinite Group

For any profinite group Γ, we denote by $\mathbb{L}(\Gamma)$ the poset with respect to inclusion of all closed subgroups of Γ; moreover $\mathbb{L}(\Gamma)$ is a bounded lattice with naturally defined operations $\wedge = \cap$ and $\vee$. In addition, $\mathbb{L}(\Gamma)$ becomes a spectral space as a closed subspace of the spectral space of all closed subsets of the underlying profinite space of Γ. The *spectral topology* τ_s on $\mathbb{L}(\Gamma)$ is defined by the base of compact open sets $\mathbb{L}(\Delta)$ for Δ ranging over all open subgroups of Γ. Note that for any $\Lambda \in \mathbb{L}(\Gamma)$, the closure of the one-point set $\{\Lambda\}$ is $\overline{\{\Lambda\}} = \mathbb{L}(\Gamma \mid \Lambda)$, the set of all closed subgroups of Γ lying over Λ. Thus the spectral space $\mathbb{L}(\Gamma)$ is irreducible with the generic point $\{1\}$, while Γ is its unique closed point. Since the poset $\mathbb{L}(\Gamma)$ is the inverse limit of the inverse system of finite posets $\mathbb{L}(\Gamma/\Delta)$ for Δ ranging over $\mathscr{N}(\Gamma)$, the set of all open normal subgroups of Γ, with natural order-preserving connecting maps, the topology τ_s is exactly the inverse limit of the T_0 topologies induced by the partial order given by inclusion on the finite sets $\mathbb{L}(\Gamma/\Delta)$.

The *boolean (profinite) completion* τ_b of the spectral topology τ_s on $\mathbb{L}(\Gamma)$, also called the *patch topology*, is the topology with the base of clopen sets $\mathscr{V}_{\Delta,\Delta'} = \{\Lambda \in \mathbb{L}(\Gamma) \mid \Lambda\Delta = \Delta'\}$ for all pairs (Δ, Δ'), with $\Delta \in \mathscr{N}(\Gamma)$ and $\Delta' \in \mathbb{L}(\Gamma|\Delta)$. The profinite space above is the inverse limit of the discrete finite spaces $\mathbb{L}(\Gamma/\Delta)$ for Δ ranging over $\mathscr{N}(\Gamma)$. A subset $\mathscr{U}$ of $\mathbb{L}(\Gamma)$ is τ_s-open if and only if $\mathscr{U}$ is both τ_b-open and a *lower* subset of $\mathbb{L}(\Gamma)$; the later condition means that $\Lambda \in \mathscr{U}$ and $\Lambda' \in \mathbb{L}(\Lambda)$ imply $\Lambda' \in \mathscr{U}$.

Remark 2.1. One checks easily that the canonical action of the profinite group Γ on the spectral space $\mathbb{L}(\Gamma)$, $(\gamma, \Lambda) \mapsto \gamma\Lambda\gamma^{-1}$, and the join operation $(\Lambda_1, \Lambda_2) \mapsto \Lambda_1 \vee \Lambda_2$ are coherent maps, in particular, continuous, while the meet operation $(\Lambda_1, \Lambda_2) \mapsto \Lambda_1 \cap \Lambda_2$ is continuous, not necessarily coherent.

2.4 Profinite Operator Groups

Let Γ be a profinite group. By a *profinite (operator)* Γ*-group* we understand a profinite group $\mathfrak{G}$ together with a continuous action by automorphisms $\Gamma \times \mathfrak{G} \longrightarrow \mathfrak{G}$, $(\gamma, g) \mapsto \gamma g$; equivalently, the profinite group $\mathfrak{G}$ possesses a system of neighborhoods of the identity consisting of open Γ-invariant normal subgroups. Denote by $\mathrm{Fix}_\Gamma(\mathfrak{G})$ the kernel of the action, a closed normal subgroup of Γ. The abelian profinite Γ-groups are usually called *profinite* Γ*-modules*. For any nontrivial subgroup $A \subseteq \mathbb{Q}/\mathbb{Z}$, the pro-$\mathbf{FAb}_A$ Γ-groups are identified with the profinite (left) $A^\vee[[\Gamma]]$-modules.

We denote by **PFOG** the category having as objects the pairs $(\Gamma, \mathfrak{G})$, where Γ is a profinite group and $\mathfrak{G}$ is a profinite Γ-group. The morphisms $(\Gamma, \mathfrak{G}) \longrightarrow (\Gamma', \mathfrak{G}')$ are pairs $(\varphi : \Gamma \longrightarrow \Gamma', \psi : \mathfrak{G} \longrightarrow \mathfrak{G}')$ of continuous homomorphisms such that $\psi(\gamma g) = \varphi(\gamma)\psi(g)$ for all $\gamma \in \Gamma, g \in \mathfrak{G}$. The composition law in **PFOG** is naturally defined. On the other hand, we consider the category **SEPI** of *splitting epimorphisms* having as objects the tuples $(\Gamma, \mathfrak{E}, p, s)$ consisting of profinite groups $\Gamma, \mathfrak{E}$, an epimorphism $p : \mathfrak{E} \longrightarrow \Gamma$, and a continuous homomorphic section $s : \Gamma \longrightarrow \mathfrak{E}$ of p. As morphisms $(\Gamma, \mathfrak{E}, p, s) \longrightarrow (\Gamma', \mathfrak{E}', p', s')$ in **SEPI**, we take the pairs $(\varphi : \Gamma \longrightarrow \Gamma', \psi : \mathfrak{E} \longrightarrow \mathfrak{E}')$ of profinite group morphisms satisfying $\varphi \circ p = p' \circ \psi$ and $\psi \circ s = s' \circ \varphi$, with the natural composition law.

Lemma 2.2. *The categories* **PFOG** *and* **SEPI** *are equivalent.*

Proof. Let F : **PFOG** $\longrightarrow$ **SEPI** denote the covariant functor induced by the map assigning to any profinite operator group $(\Gamma, \mathfrak{G})$ the object $(\Gamma, \mathfrak{E}, p, s)$ of **SEPI**, where $\mathfrak{E}$ is the semidirect product $\mathfrak{G} \rtimes \Gamma$ induced by the continuous action of Γ on $\mathfrak{G}$, $p : \mathfrak{E} \longrightarrow \Gamma$ is the natural projection with $\mathrm{Ker}\,(p) = \mathfrak{G}$, and $s : \Gamma \longrightarrow \mathfrak{E}$ is the canonical homomorphic section of p which identifies Γ with a closed subgroup of $\mathfrak{E}$ satisfying $\mathfrak{G} \cdot \Gamma = \mathfrak{E}$, $\mathfrak{G} \cap \Gamma = \{1\}$. One checks easily that the functor F is faithfully full and essentially surjective, and hence it yields an equivalence of categories. Note that the inverse equivalence **SEPI** $\longrightarrow$ **PFOG** is induced by the map assigning to an object $(\Gamma, \mathfrak{E}, p, s)$ of **SEPI** the profinite Γ-group $\mathfrak{G} = \mathrm{Ker}\,(p)$ with the action of Γ defined by $\gamma g := s(\gamma)\, g\, s(\gamma)^{-1}$ for $\gamma \in \Gamma, g \in \mathfrak{G}$.

Note that the equivalent categories **PFOG** and **SEPI** have inverse limits. Moreover the equivalence above is naturally extended to an equivalence of suitable categories of bundles containing **PFOG** and **SEPI** as reflective subcategories [13].

2.4.1 The Lattice of Ideals of a Profinite Operator Group

For any profinite Γ-group $\mathfrak{G}$, we denote by $\mathbb{L}(\mathfrak{G})$ the poset with respect to inclusion of all Γ-invariant closed normal subgroups of $\mathfrak{G}$, called *ideals* of the profinite Γ-group $\mathfrak{G}$. The ideals of $\mathfrak{G}$ are exactly those closed normal subgroups of the semidirect product $\mathfrak{G} \rtimes \Gamma$ which are contained in $\mathfrak{G}$. $\mathbb{L}(\mathfrak{G})$ is a bounded *modular* lattice, dual to the lattice of all quotients of the profinite Γ-group $\mathfrak{G}$; an ideal $\mathbf{a} \in \mathbb{L}(\mathfrak{G})$ is open if and only if the quotient Γ-group $\mathfrak{G}/\mathbf{a}$ is finite.

In addition, $\mathbb{L}(\mathfrak{G})$ is equipped with a spectral topology defined by the base of compact open sets $\mathbb{L}(\mathbf{a}) = \{\mathbf{b} \in \mathbb{L}(\mathfrak{G}) \mid \mathbf{b} \subseteq \mathbf{a}\}$ for $\mathbf{a}$ ranging over the open ideals of $\mathfrak{G}$. The join operation $(\mathbf{a}, \mathbf{b}) \mapsto \mathbf{a} \vee \mathbf{b} := \mathbf{a} \cdot \mathbf{b}$ is coherent, while the meet operation $(\mathbf{a}, \mathbf{b}) \mapsto \mathbf{a} \cap \mathbf{b}$ is continuous, not necessarily coherent.

2.5 Galois and CoGalois Connections

The notions of Galois and coGalois connections are remarkable cases of the more general concept of *adjunction* from category theory.

According to Ore [29], a *Galois connection* is a system (X, Y, φ, ψ), where X, Y are posets, and $\varphi : X \longrightarrow Y$, $\psi : Y \longrightarrow X$ are *order-reversing* maps satisfying $x \leq \psi(\varphi(x))$, $y \leq \varphi(\psi(y))$ for all $x \in X, y \in Y$. It follows that the maps φ and ψ are *quasi-inverse* to one another, i.e., $\varphi \circ \psi \circ \varphi = \varphi$, $\psi \circ \varphi \circ \psi = \psi$, and the order-preserving endomaps $\psi \circ \varphi : X \longrightarrow X$, $\varphi \circ \psi : Y \longrightarrow Y$ are *closure operators* with $X_c := \{x \in X \mid \psi(\varphi(x)) = x\} = \psi(Y)$, $Y_c := \{y \in Y \mid \varphi(\psi(y)) = y\} = \varphi(X)$ as sets of closed points. The Galois connection (X, Y, φ, ψ) is *perfect* if $X = X_c$, $Y = Y_c$, i.e., the maps $\varphi : X \longrightarrow Y$, $\psi : Y \longrightarrow X$ are anti-isomorphisms inverse to one another.

By duality, a *coGalois connection* is a system (X, Y, φ, ψ), where X, Y are posets, and $\varphi : X \longrightarrow Y, \psi : Y \longrightarrow X$ are *order-preserving* maps satisfying $x \leq \psi(\varphi(x))$, $\varphi(\psi(y)) \leq y$ for all $x \in X, y \in Y$. The coGalois connection (X, Y, φ, ψ) is *perfect* if $\varphi : X \longrightarrow Y, \psi : Y \longrightarrow X$ are *isomorphisms* inverse to one another.

2.5.1 Standard Galois Connections

Let L/K be an arbitrary field extension. According to [22], [37, Sect. 6.3: A generalization of Galois Theory], $\Gamma := \mathrm{Gal}(L/K) = \{\gamma \in \mathrm{Aut}(L) : \sigma|_K = 1_K\}$ has a natural structure of *totally disconnected (Hausdorff) topological group* with respect to the weakest topology for which the action of Γ on the discrete field L is continuous; the subgroups $\mathrm{Gal}(L/F)$, where $K \subseteq F \subseteq L$ ranges over the finitely generated field extensions of K, form a system of open neighborhoods of the identity 1_L. For algebraic extensions L/K, Γ is compact, whence profinite, and the topology above is usually called *Krull topology*. Note that Γ is locally compact if and only if $\mathrm{tr.deg}\, L/K < \infty$.

Consider the bounded lattices (with respect to inclusion) $\mathcal{L}(L/K)$ of all subfields of L containing K, $\mathcal{L}(\Gamma)$ of all subgroups of Γ, and $\mathbb{L}(\Gamma)$ of all closed subgroups of Γ; note that $\mathbb{L}(\Gamma)$ is a sub-semilattice of $\mathcal{L}(\Gamma)$ with respect to the meet operation (intersection), while its retract $\mathcal{L}(\Gamma) \longrightarrow \mathbb{L}(\Gamma)$, $\Lambda \mapsto \overline{\Lambda}$, is a morphism of semilattices with respect to the join operations. These lattices are naturally related through the order-reversing maps

$$\mathcal{D} : \mathcal{L}(\Gamma) \longrightarrow \mathcal{L}(L/K),\ \Lambda \mapsto L^{\Lambda} := \{x \in L \mid \forall \gamma \in \Lambda, \gamma(x) = x\}$$

(*Dedekind connection*), and

$$\mathcal{K} : \mathcal{L}(L/K) \longrightarrow \mathcal{L}(\Gamma), F \mapsto \mathrm{Gal}(L/F)$$

(*Krull connection*).

$\mathcal{D}$ factorizes through the canonical surjection $\mathcal{L}(\Gamma) \to \mathbb{L}(\Gamma)$, $\Lambda \mapsto \overline{\Lambda}$, and $\mathcal{K}$ factorizes through the embedding $\mathbb{L}(\Gamma) \hookrightarrow \mathcal{L}(\Gamma)$.

The systems $(\mathcal{L}(L/K), \mathcal{L}(\Gamma), \mathcal{K}, \mathcal{D})$ and $(\mathcal{L}(L/K), \mathbb{L}(\Gamma), \mathcal{K}, \mathcal{D})$ are both Galois connections. The first one is perfect if and only if L/K is a *finite Galois extension* (solution of *Steinitz's problem* [39]), while the latter one is perfect if and only if L/K is an *algebraic* (not necessarily finite) *Galois extension* [22]. In the first case, the Galois group Γ is finite of order $[L : K]$, while in the latter case, Γ is a profinite group isomorphic to $\varprojlim \mathrm{Gal}(F/K)$, F ranging over the finite normal extensions of K contained in L.

Remarks 2.3. (1) According to Barbilian [7], a field extension L/K is *Dedekindian* (i.e., $\forall F \in \mathscr{L}(L/K)$, $\mathscr{D}(\mathscr{K}(F)) = F^{\mathrm{Gal}(L/F)} = F$) if and only if for all $F \in \mathscr{L}(L/K)$, the relative algebraic closure of F in L is Galois over K. In particular, assuming char $K > 0$, L/K is Dedekindian if and only if L/K is an algebraic Galois extension.

(2) Extending Barbilian's paper [7], Krull investigates in [23] the field extensions L/K, called locally normal by Barbilian, and simply *normal* by Krull, having the property that for all $F \in \mathscr{L}(L/K)$, the relative algebraic closure of F in L is normal over the base field K. He shows that an arbitrary field extension L/K is normal if and only if for every *Steinitz decomposition* $F \in \mathscr{L}(L/K)$ (i.e., F purely transcendental over K, and L/F algebraic), the algebraic extension L/F is normal (necessarily infinite); in particular, the purely transcendental extensions are not normal. Using this general notion of normality, Barbilian's main result reads as follows: *L/K is Dedekindian if and only if L/K is normal and separable*. The following open problem is raised by Krull in [23]: *Do there exist transcendental normal extensions which are not algebraically closed?* [1]

(3) Though the Galois algebraic extensions play a key role in the most applications of the field arithmetic, there exist also significant contexts in which the Galois groups of transcendental extensions are essential tools of investigation. As a relevant example, we mention the automorphism group of the field of modular functions, rational over the maximal abelian extension of $\mathbb{Q}$ [37, Ch. 6]. Another situation, treated in [33], concerns finitely generated extensions F/K of transcendence degree 1 over an algebraically closed field K of characteristic 0. In this case, the *absolute Galois group* $\mathrm{Gal}(\tilde{F}/F)$ is a free profinite group, and hence, its structure tells nothing about the field F. However, considering two finitely generated extensions F_1, F_2 of K, contained into an algebraically closed field L of transcendence degree 1 over K, the author proves that F_1 and F_2 are isomorphic over K, provided there is a continuous and open automorphism of $\mathrm{Gal}(L/K)$ inducing by restriction an isomorphism $\mathrm{Gal}(L/F_1) \cong \mathrm{Gal}(L/F_2)$.

[1]Unfortunately, I couldn't find in literature some references to this fundamental problem stated by Krull in 1953. I found only a paper in Arch. Math. **61** (1993), 238–240, with the title *Gibt es nichttriviale absolut-normale Körpererweiterungen?*, but, strangely enough, this paper contains only some of Barbilian and Krull's notions and results without to mention their papers in bibliography !

3 An Abstract Framework for CoGalois Theory

3.1 Generating Cocycles

For any profinite Γ-group $\mathfrak{G}$, let $Z^1(\Gamma, \mathfrak{G})$ denote the set of all continuous 1-cocycles (crossed homomorphisms) of Γ with coefficients in $\mathfrak{G}$, i.e., the continuous maps $\eta : \Gamma \longrightarrow \mathfrak{G}$ satisfying $\eta(\sigma\tau) = \eta(\sigma) \cdot \sigma\eta(\tau)$ for all $\sigma, \tau \in \Gamma$; in particular, $\eta(\gamma^{-1}) = \gamma^{-1}\eta(\gamma)^{-1}$ for all $\gamma \in \Gamma$, and $\eta(1) = 1$. The set $Z^1(\Gamma, \mathfrak{G})$ contains the trivial cocycle $\gamma \mapsto 1$ as a privileged element, so it is an object of the category **PS** of pointed sets with naturally defined morphisms. $Z^1(\Gamma, \mathfrak{G})$ becomes an abelian group whenever $\mathfrak{G}$ is a profinite Γ-module.

Note that $\mathrm{Ker}\,(\eta) := \eta^{-1}(1)$ is a closed subgroup of Γ for every $\eta \in Z^1(\Gamma, \mathfrak{G})$. Set $Z^1(\Gamma \mid \Lambda, \mathfrak{G}) := \{\eta \in Z^1(\Gamma, \mathfrak{G}) \mid \Lambda \subseteq \mathrm{Ker}\,(\eta)\}$, where Λ is a closed subgroup of Γ.

We denote by $\mathscr{Z}^1$ the *category of* 1*-cocycles* whose objects are the triples $(\Gamma, \mathfrak{G}, \eta)$ consisting of a profinite group Γ, a profinite Γ-group $\mathfrak{G}$ and a cocycle $\eta \in Z^1(\Gamma, \mathfrak{G})$. As morphisms $(\Gamma, \mathfrak{G}, \eta) \longrightarrow (\Gamma', \mathfrak{G}', \eta')$ we take those morphisms $(\varphi : \Gamma \longrightarrow \Gamma', \psi : \mathfrak{G} \longrightarrow \mathfrak{G}')$ in the category **PFOG** of profinite operator groups which in addition are compatible with the cocycles η and η', i.e., $\psi \circ \eta = \eta' \circ \varphi$. The composition law in $\mathscr{Z}^1$ is induced from the category **PFOG**.

Definition 3.1. $\eta \in Z^1(\Gamma, \mathfrak{G})$ is said to be a *generating cocycle* (for short *g-cocycle*) if the profinite group $\mathfrak{G}$ is topologically generated by its closed subset $\eta(\Gamma)$.

For any g-cocycle $\eta \in Z^1(\Gamma, \mathfrak{G})$, we obtain

$$\mathrm{Fix}_{\,\Gamma}(\mathfrak{G}) = \mathrm{Fix}_{\,\Gamma}(\eta(\Gamma)) = \{\gamma \in \Gamma \mid \forall \sigma \in \Gamma,\ \eta(\gamma\sigma) = \eta(\gamma)\eta(\sigma)\},$$

while the core of $\Delta := \mathrm{Ker}(\eta)$ is $\tilde{\Delta} := \bigcap_{\gamma \in \Gamma} \gamma\Delta\gamma^{-1} = \mathrm{Fix}_{\,\Gamma}(\mathfrak{G}) \cap \Delta$.

We denote by $\mathscr{G}\mathscr{Z}^1$ the full subcategory of $\mathscr{Z}^1$ whose objects are the g-cocycles.

On the other hand, consider the category of *pairs of homomorphic sections*, denoted $\mathscr{P}\mathscr{H}\mathscr{S}$, having as objects the tuples $(\Gamma, \mathfrak{E}, p, s_1, s_2)$ consisting of two profinite groups $\Gamma, \mathfrak{E}$, a splitting epimorphism $p : \mathfrak{E} \longrightarrow \Gamma$, and a pair (s_1, s_2) of homomorphic continuous sections of p. A morphism $(\Gamma, \mathfrak{E}, p, s_1, s_2) \longrightarrow (\Gamma', \mathfrak{E}', p', s_1', s_2')$ is a pair of continuous homomorphisms $(\varphi : \Gamma \longrightarrow \Gamma', \psi : \mathfrak{E} \longrightarrow \mathfrak{E}')$ satisfying $\varphi \circ p = p' \circ \psi$, $\psi \circ s_i = s_i' \circ \varphi$, $i = 1, 2$. The composition law in $\mathscr{P}\mathscr{H}\mathscr{S}$ is naturally defined. We denote by $\mathscr{G}\mathscr{P}\mathscr{H}\mathscr{S}$ the full subcategory of those objects $(\Gamma, \mathfrak{E}, p, s_1, s_2)$ of $\mathscr{P}\mathscr{H}\mathscr{S}$ for which $\mathfrak{E}$ is topologically generated by the union $s_1(\Gamma) \cup s_2(\Gamma)$.

Lemma 3.2. *The categories $\mathscr{Z}^1$ and $\mathscr{P}\mathscr{H}\mathscr{S}$, as well as the corresponding full subcategories $\mathscr{G}\mathscr{Z}^1$ and $\mathscr{G}\mathscr{P}\mathscr{H}\mathscr{S}$, are equivalent.*

Proof. We have to extend the equivalence of the categories **PFOG** and **SEPI** provided by Lemma 2.2. To a cocycle $\eta \in Z^1(\Gamma, \mathfrak{G})$ we first assign the quadruple

$(\Gamma, \mathfrak{E}, p, s_1)$ associated with the profinite Γ-group $\mathfrak{G}$, where $\mathfrak{E}$ is the semidirect product $\mathfrak{G} \rtimes \Gamma$ induced by the action of Γ on $\mathfrak{G}$, $p : \mathfrak{E} \longrightarrow \Gamma$ is the natural projection and $s_1 : \Gamma \longrightarrow \mathfrak{E}$ is the canonical homomorphic section to p identifying Γ with a closed subgroup of $\mathfrak{E}$ satisfying $\mathfrak{G} \cdot \Gamma = \mathfrak{E}$, $\mathfrak{G} \cap \Gamma = \{1\}$. Next we extend the quadruple above by adding the homomorphic section $s_2 : \Gamma \longrightarrow \mathfrak{E}$ to p induced by the cocycle $\eta : s_2(\gamma) = \eta(\gamma)s_1(\gamma)$ for all $\gamma \in \Gamma$.

To obtain the inverse equivalence, we assign to an object $(\Gamma, \mathfrak{E}, p, s_1, s_2)$ of $\mathscr{PHS}$ the profinite Γ-group $\mathfrak{G} = \mathrm{Ker}(p)$, with the action of Γ defined by $\gamma g := s_1(\gamma)\, g\, s_1(\gamma)^{-1}$ for $\gamma \in \Gamma, g \in \mathfrak{G}$, and the cocycle $\eta \in Z^1(\Gamma, \mathfrak{G})$ induced by the homomorphic section $s_2 : \eta(\gamma) = s_2(\gamma)s_1(\gamma)^{-1}$ for $\gamma \in \Gamma$.

Note that the equivalent categories $\mathscr{Z}^1$ and $\mathscr{PHS}$, as well as their full subcategories $\mathscr{GZ}^1$ and $\mathscr{GPHS}$, have inverse limits and free products of bundles [13].

3.1.1 Universal Cocycles

Given a profinite group Γ and a closed subgroup $\Delta \subseteq \Gamma$, there exists uniquely (up to isomorphism) a pair $(\Omega_{\Gamma,\Delta}, \omega_{\Gamma,\Delta})$ consisting of a profinite Γ-group $\Omega_{\Gamma,\Delta}$ and a g-cocycle $\omega_{\Gamma,\Delta} : \Gamma \longrightarrow \Omega_{\Gamma,\Delta}$, with $\mathrm{Fix}_\Gamma(\Omega_{\Gamma,\Delta}) = \tilde{\Delta} = \bigcap_{\gamma\in\Gamma} \gamma\Delta\gamma^{-1}$, $\mathrm{Ker}\,(\omega_{\Gamma,\Delta}) = \Delta$, such that for all profinite Γ-groups $\mathfrak{G}$, the map

$$\mathrm{Hom}_\Gamma(\Omega_{\Gamma,\Delta}, \mathfrak{G}) \longrightarrow Z^1(\Gamma \mid \Delta, \mathfrak{G}),\ \varphi \mapsto \varphi \circ \omega_{\Gamma,\Delta}$$

is a functorial bijection [13]. To construct the *universal pair* $(\Omega_{\Gamma,\Delta}, \omega_{\Gamma,\Delta})$, we consider the (generalized) free profinite group $(\Omega_X, \omega_X : X \longrightarrow \Omega_X)$ generated by the pointed profinite space $X := (\Gamma/\Delta\,;\,\Delta)$. Since for any $\sigma \in \Gamma$, the continuous injective map $X \longrightarrow \Omega_X,\ \tau\Delta \mapsto \omega_X(\sigma\Delta)^{-1}\,\omega_X(\sigma\tau\Delta)$ extends uniquely to an automorphism, Ω_X becomes a profinite Γ-group denoted $\Omega_{\Gamma,\Delta}$, with $\mathrm{Fix}_\Gamma(\Omega_{\Gamma,\Delta}) = \tilde{\Delta}$, while the map $\omega_{\Gamma,\Delta} : \Gamma \longrightarrow \Omega_{\Gamma,\Delta},\ \gamma \mapsto \omega_X(\gamma\Delta)$, is a g-cocycle with kernel Δ as desired. In particular, for $\Delta = 1$, the action of Γ on $\Omega_\Gamma := \Omega_{\Gamma,1}$ is faithful, and the g-cocycle $\omega_\Gamma := \omega_{\Gamma,1}$ is injective. As a profinite group, Ω_Γ is free of rank

$$\kappa = \begin{cases} |\Gamma| - 1 & \text{if } \Gamma \text{ is finite,} \\ \max(\aleph_0, \mathrm{rank}\,(\Gamma)) & \text{if } \Gamma \text{ is infinite.} \end{cases}$$

Remarks 3.3. (1) The pairs $(\Omega_{\Gamma,\Delta}, \omega_{\Gamma,\Delta})$ are also universal objects for certain *embedding problems* for profinite operator groups [13].

(2) Let $\mathscr{C}$ be a variety of finite groups containing nontrivial groups. For any profinite group Γ and any closed subgroup $\Delta \subseteq \Gamma$, let $\Omega^{\mathscr{C}}_{\Gamma,\Delta}$ be the maximal pro-$\mathscr{C}$ quotient of $\Omega_{\Gamma,\Delta}$, with the induced action of Γ, and denote by $\omega^{\mathscr{C}}_{\Gamma,\Delta} : \Gamma \longrightarrow \Omega^{\mathscr{C}}_{\Gamma,\Delta}$ the g-cocycle induced by $\omega_{\Gamma,\Delta}$. Since the variety $\mathscr{C}$ contains nontrivial groups, it follows from the construction of $\Omega_{\Gamma,\Delta}$ [13]

that $\mathrm{Ker}\,(\omega^{\mathscr{C}}_{\Gamma,\Delta}) = \Delta$, in particular, $\omega^{\mathscr{C}}_{\Gamma} : \Gamma \longrightarrow \Omega^{\mathscr{C}}_{\Gamma}$ is injective. The g-cocycles $\eta : \Gamma \longrightarrow \mathfrak{G}$, with $\Delta \subseteq \mathrm{Ker}\,(\eta)$ and $\mathfrak{G}$ a pro-$\mathscr{C}$ Γ-group, are, up to isomorphism, in 1–1 correspondence with the quotients of the pro-$\mathscr{C}$ Γ-group $\Omega^{\mathscr{C}}_{\Gamma,\Delta}$.

In particular, taking $\mathscr{C} = \mathbf{FAb}_A$, where A is an arbitrary nontrivial subgroup of $\mathbb{Q}/\mathbb{Z}$, the pro-$\mathscr{C}$ Γ-groups are identified with the profinite (left) $A^{\vee}[[\Gamma]]$-modules, and for any such module $\mathfrak{G}$, every continuous 1-cocycle $\eta : \Gamma \longrightarrow \mathfrak{G}$ extends uniquely to a derivation of the complete group algebra $A^{\vee}[[\Gamma]]$ into $\mathfrak{G}$, i.e., to a continuous $A^{\vee}$-linear map $D : A^{\vee}[[\Gamma]] \longrightarrow \mathfrak{G}$ satisfying $D(fg) = D(f) \cdot \varepsilon(g) + f \cdot D(g)$ for all $f, g \in A^{\vee}[[\Gamma]]$, where $\varepsilon : A^{\vee}[[\Gamma]] \longrightarrow A^{\vee}, \gamma \in \Gamma \mapsto 1$, is the augmentation map. $\Omega^{\mathscr{C}}_{\Gamma}$ becomes the module $\Omega_{A^{\vee}[[\Gamma]]}$ of (noncommutative) differential forms of $A^{\vee}[[\Gamma]]$, the injective universal cocycle $\omega^{\mathscr{C}}_{\Gamma} : \Gamma \longrightarrow \Omega^{\mathscr{C}}_{\Gamma}$ extends to the universal derivation $d : A^{\vee}[[\Gamma]] \longrightarrow \Omega_{A^{\vee}[[\Gamma]]}, f \mapsto df$, and the map $d\gamma \mapsto \gamma - 1, \gamma \in \Gamma$, extends to a canonical isomorphism of profinite $A^{\vee}[[\Gamma]]$-modules from $\Omega_{A^{\vee}[[\Gamma]]}$ onto the augmentation ideal I of $A^{\vee}[[\Gamma]]$, the kernel of the augmentation map ε. For any closed subgroup Δ of Γ, the canonical isomorphism above induces an isomorphism of profinite $A^{\vee}[[\Gamma]]$-modules from $\Omega^{\mathscr{C}}_{\Gamma,\Delta}$ onto the quotient of the augmentation ideal I by the left closed ideal J_{Δ} of $A^{\vee}[[\Gamma]]$ generated by $\{\delta - 1 \mid \delta \in \Delta\}$.

3.2 *The CoGalois Connection Associated with a Generating Cocycle*

Fix an object $(\Gamma, \mathfrak{G}, \eta)$ of the category $\mathscr{G}\mathscr{Z}^1$; thus Γ is a profinite group acting continuously on the profinite group $\mathfrak{G}$, and $\eta : \Gamma \longrightarrow \mathfrak{G}$ is a g-cocycle, so the profinite group $\mathfrak{G}$ is topologically generated by $\eta(\Gamma)$.

Set $\Delta = \mathrm{Ker}(\eta) := \eta^{-1}(1)$, $\Delta' = \mathrm{Fix}_{\Gamma}(\mathfrak{G})$, and $\tilde{\Delta} = \bigcap_{\gamma \in \Gamma} \gamma \Delta \gamma^{-1} = \Delta \cap \Delta'$.

Definition 3.4. The triple $(\Gamma, \mathfrak{G}, \eta)$ is *normalized* (we say also that the g-cocycle η is *normalized*) if the closed normal subgroup $\tilde{\Delta}$ is trivial. In particular, if Γ is abelian, then η is normalized if and only if η is injective.

Given a g-cocycle $\eta : \Gamma \longrightarrow \mathfrak{G}$, its *normalization* is obtained by replacing the profinite group Γ and the cocycle $\eta : \Gamma \longrightarrow \mathfrak{G}$ with the quotient $\Gamma' := \Gamma/\tilde{\Delta}$ and the cocycle $\eta' : \Gamma' \longrightarrow \mathfrak{G}$ induced by η, with $\mathrm{Ker}(\eta') = \Delta/\tilde{\Delta}$, respectively.

Thus we may assume from the beginning that the triple $(\Gamma, \mathfrak{G}, \eta)$ is normalized. We associate to $(\Gamma, \mathfrak{G}, \eta)$ two bounded lattices related through natural maps induced by the g-cocycle $\eta : \Gamma \longrightarrow \mathfrak{G}$: the lattice $\mathbb{L}(\Gamma \mid \Delta)$ of all closed subgroups of Γ lying over Δ and the modular lattice $\mathbb{L}(\mathfrak{G})$ of all ideals of the profinite Γ-group $\mathfrak{G}$, dual to the lattice of all quotients of $\mathfrak{G}$. According to Sects. 2.3 and 2.4.1, $\mathbb{L}(\Gamma \mid \Delta)$ and $\mathbb{L}(\mathfrak{G})$ are also irreducible spectral spaces with generic points Δ and $\{1\}$ respectively, coherent join operations and continuous meet operations.

The posets $\mathbb{L}(\Gamma \mid \Delta)$ and $\mathbb{L}(\mathfrak{G})$ are naturally related through the following canonical order-preserving maps induced by the cocycle $\eta : \Gamma \longrightarrow \mathfrak{G}$

$$\mathscr{I} : \mathbb{L}(\Gamma \mid \Delta) \longrightarrow \mathbb{L}(\mathfrak{G}), \Lambda \mapsto \mathscr{I}(\Lambda) := \text{the ideal generated by } \eta(\Lambda),$$

and

$$\mathscr{S} : \mathbb{L}(\mathfrak{G}) \longrightarrow \mathbb{L}(\Gamma \mid \Delta), \mathbf{a} \mapsto \mathscr{S}(\mathbf{a}) := \eta^{-1}(\mathbf{a}).$$

For any subset $X \subseteq \Gamma$, let $\Lambda \in \mathbb{L}(\Gamma \mid \Delta)$ be the closed subgroup generated by the union $X \cup \Delta$; it follows that $\mathscr{I}(X) = \mathscr{I}(\Lambda)$, where $\mathscr{I}(X)$ is the ideal of the profinite Γ-group $\mathfrak{G}$ generated by $\eta(X)$. Note that $\mathfrak{G}/\mathscr{I}(\Lambda)$ is the maximal quotient $\mathfrak{H}$ of the profinite Γ-group $\mathfrak{G}$ for which the cocycle obtained by composing $\eta : \Gamma \longrightarrow \mathfrak{G}$ with the natural projection $\mathfrak{G} \longrightarrow \mathfrak{H}$ vanishes on Λ, and, for any $\mathbf{a} \in \mathbb{L}(\mathfrak{G})$, $\mathscr{S}(\mathbf{a}) = \mathrm{Ker}(\eta_{\mathbf{a}})$, where $\eta_{\mathbf{a}} : \Gamma \longrightarrow \mathfrak{G}/\mathbf{a}$ is the cocycle obtained by composing $\eta : \Gamma \longrightarrow \mathfrak{G}$ with the natural projection $\mathfrak{G} \longrightarrow \mathfrak{G}/\mathbf{a}$.

Remark 3.5. To give an alternative description of the operators $\mathscr{I}$ and $\mathscr{S}$ above, set $\mathfrak{E} := \mathfrak{G} \rtimes \Gamma$, and let $\Gamma_i \cong \Gamma, i = 1, 2$, be the complements of the closed normal subgroup $\mathfrak{G}$ of $\mathfrak{E}$ induced by the canonical section $s_1 : \Gamma \longrightarrow \mathfrak{E}$ and the homomorphic section $s_2 : \Gamma \longrightarrow \mathfrak{E}$ determined by the cocycle $\eta : \Gamma \longrightarrow \mathfrak{G}$, with $\Delta = \mathrm{Ker}(\eta)$, respectively. As η is by assumption a g-cocycle, the profinite group $\mathfrak{E}$ is topologically generated by the union $\Gamma_1 \cup \Gamma_2$. For any $\Lambda \in \mathbb{L}(\Gamma \mid \Delta)$, let $\Lambda_i \subseteq \Gamma_i, i = 1, 2$, denote the image of Λ through the homomorphic sections above. As $\Delta \subseteq \Lambda$, we obtain $\Lambda_1 \cap \Lambda_2 = \Gamma_1 \cap \Gamma_2 = s_1(\Delta) = s_2(\Delta) \cong \Delta$. Denote by $\tilde{\mathfrak{G}}$ the intersection of $\mathfrak{G}$ with the closed subgroup $\tilde{\Lambda}$ of $\mathfrak{E}$ generated by the union $\Lambda_1 \cup \Lambda_2$, and note that $\tilde{\mathfrak{G}}$ is topologically generated by $\eta(\Lambda)$, while Λ_1 and Λ_2 are complements of $\tilde{\mathfrak{G}}$ in $\tilde{\Lambda}$. It follows that $\mathscr{I}(\Lambda)$ is the smallest closed normal subgroup of $\mathfrak{E}$ containing $\tilde{\mathfrak{G}}$. On the other hand, given $\mathbf{a} \in \mathbb{L}(\mathfrak{G})$, let $\mathfrak{H} := \mathfrak{G}/\mathbf{a}$, $\mathfrak{E}' := \mathfrak{H} \rtimes \Gamma$ and $\Gamma_i' \cong \Gamma, i = 1, 2$, be the complements of $\mathfrak{H}$ in $\mathfrak{E}'$ induced by the canonical section $s_1' : \Gamma \longrightarrow \mathfrak{E}'$ and the homomorphic section s_2' determined by the cocycle $\eta' : \Gamma \longrightarrow \mathfrak{H}$, obtained by composing the cocycle $\eta : \Gamma \longrightarrow \mathfrak{G}$ with the projection $\mathfrak{G} \longrightarrow \mathfrak{H}$, respectively. It follows that $\mathscr{S}(\mathbf{a}) \cong \Gamma_1' \cap \Gamma_2'$ is the image of $\Gamma_1' \cap \Gamma_2'$ through the projection $\mathfrak{E}' \longrightarrow \Gamma$.

The next result collects together the main properties of the operators $\mathscr{I}$ and $\mathscr{S}$ as defined above.

Proposition 3.6. *Let $\eta \in Z^1(\Gamma, \mathfrak{G})$ be a g-cocycle, $\Delta :=$ Ker (η). The following assertions hold.*

(1) *For all $\Lambda \in \mathbb{L}(\Gamma \mid \Delta), \mathbf{a} \in \mathbb{L}(\mathfrak{G})$, $\Lambda \subseteq \mathscr{S}(\mathscr{I}(\Lambda))$, $\mathscr{I}(\mathscr{S}(\mathbf{a})) \subseteq \mathbf{a}$, so the pair of operators $(\mathscr{I}, \mathscr{S})$ establishes a coGalois connection between the posets $\mathbb{L}(\Gamma \mid \Delta)$ and $\mathbb{L}(\mathfrak{G})$.*

(2) *For all families* $(\mathbf{a}_i)_{i\in I}$ *and* $(\Lambda_i)_{i\in I}$, *with* $\mathbf{a}_i \in \mathbb{L}(\mathfrak{G})$, $\Lambda_i \in \mathbb{L}(\Gamma \mid \Delta)$,

$$\mathscr{S}(\bigcap_{i\in I}\mathbf{a}_i) = \bigcap_{i\in I}\mathscr{S}(\mathbf{a}_i) \ \ and \ \ \mathscr{J}(\bigvee_{i\in I}\Lambda_i) = \bigvee_{i\in I}\mathscr{J}(\Lambda_i),$$

i.e., $\mathscr{S}$ *and* $\mathscr{J}$ *are complete semi-lattice morphisms with respect to* $\cap$ *and* $\vee$ *respectively.*

(3) *The map* $\mathscr{J} : \mathbb{L}(\Gamma \mid \Delta) \longrightarrow \mathbb{L}(\mathfrak{G})$ *is coherent, in particular continuous.*

(4) *The map* $\mathscr{S} : \mathbb{L}(\mathfrak{G}) \longrightarrow \mathbb{L}(\Gamma \mid \Delta)$ *is continuous.*

Proof. The assertions (1) and (2) are obvious.

(3) Let $\mathbf{b}$ be an open ideal of $\mathfrak{G}$. As the quotient Γ-group $\mathfrak{G}/\mathbf{b}$ is finite and the map $\Gamma/\mathscr{S}(\mathbf{b}) \longrightarrow \mathfrak{G}/\mathbf{b}$ induced by the cocycle $\eta : \Gamma \longrightarrow \mathfrak{G}$ is injective, it follows that $\mathscr{S}(\mathbf{b})$ is an open subgroup of Γ lying over Δ. Consequently, the inverse image

$$\{\Lambda \in \mathbb{L}(\Gamma \mid \Delta) \mid \mathscr{J}(\Lambda) \subseteq \mathbf{b}\} = \{\Lambda \in \mathbb{L}(\Gamma \mid \Delta) \mid \Lambda \subseteq \mathscr{S}(\mathbf{b})\} = \mathbb{L}(\mathscr{S}(\mathbf{b}) \mid \Delta)$$

of the basic compact open set $\mathbb{L}(\mathbf{b})$ of the spectral space $\mathbb{L}(\mathfrak{G})$ through the map $\mathscr{J}$ is open and compact as desired.

(4) For any open subgroup $\Lambda \in \mathbb{L}(\Gamma \mid \Delta)$, let $\mathscr{W} := \{\mathbf{b} \in \mathbb{L}(\mathfrak{G}) \mid \mathscr{S}(\mathbf{b}) \subseteq \Lambda\}$ denote the inverse image through the map $\mathscr{S}$ of the basic open set $\mathbb{L}(\Lambda \mid \Delta)$ of the spectral space $\mathbb{L}(\Gamma \mid \Delta)$. As $\mathscr{S}(\{1\}) = \Delta \subseteq \Lambda$, $\mathscr{W}$ is nonempty. For any $\mathbf{b} \in \mathscr{W}$, denote by $\mathbb{L}(\mathfrak{G} \mid \mathbf{b})_o$ the poset of all *open* ideals of $\mathfrak{G}$ lying over $\mathbf{b}$, so $\mathbf{b} = \bigcap_{\mathbf{a}\in\mathbb{L}(\mathfrak{G}\mid\mathbf{b})_o}\mathbf{a}$. By (2) and by the compactness of Γ it follows that there exists $\mathbf{a} \in \mathbb{L}(\mathfrak{G} \mid \mathbf{b})_o$ such that $\mathscr{S}(\mathbf{b}) \subseteq \mathscr{S}(\mathbf{a}) \subseteq \Lambda$. Consequently, the nonempty set $\mathscr{W}_{\max}$ of all maximal members of $\mathscr{W}$ with respect to inclusion consists of open ideals of the profinite Γ-group $\mathfrak{G}$, and hence $\mathscr{W} = \bigcup_{\mathbf{c}\in\mathscr{W}_{\max}}\mathbb{L}(\mathbf{c})$ is open as a union of basic open sets of the spectral space $\mathbb{L}(\mathfrak{G})$.

Remark 3.7. By contrast with the map $\mathscr{J}$ which is always coherent, the continuous map $\mathscr{S}$ is not coherent in general. For instance, let $\Gamma = (\hat{\mathbb{Z}}, +)$ and $\mathfrak{G} = \prod_{p\in\mathscr{P}'}(\mathbb{Z}/p\mathbb{Z}, +)$, where $\mathscr{P}'$ is the set of the odd prime numbers p for which the order $f_p \mid (p-1)$ of $2 \bmod p$ is even. Consider the continuous action $\Gamma \times \mathfrak{G} \longrightarrow \mathfrak{G}$, $(\gamma, g) \mapsto 2^{\gamma}g$ and the coboundary $\eta : \Gamma \longrightarrow \mathfrak{G}, \gamma \mapsto 2^{\gamma} - 1$; since η sends the topological generator 1 of Γ to a topological generator of $\mathfrak{G}$, η is a g-cocycle. Note that $\Delta := \mathrm{Fix}_{\Gamma}(\mathfrak{G}) = \mathrm{Ker}\,(\eta) = \bigcap_{p\in\mathscr{P}'} f_p\hat{\mathbb{Z}}$. For the open subgroup $\Lambda := 2\hat{\mathbb{Z}} \in \mathbb{L}(\Gamma \mid \Delta)$, we obtain, with the notation from Proposition 3.6,(4), $\mathscr{W}_{\max} = \{\prod_{p\in\mathscr{P}'\setminus\{l\}}\mathbb{Z}/p\mathbb{Z} \mid l \in \mathscr{P}'\}$, the set of all maximal open subgroups of $\mathfrak{G}$, whence the open set $\mathscr{W}$ is not compact since the set $\mathscr{P}'$ is infinite (any odd prime number $p \not\equiv \pm 1 \bmod 8$ belongs to $\mathscr{P}'$).

Thus, given g-cocycles $\eta : \Gamma \longrightarrow \mathfrak{G}$, it is natural to look for suitable closed subspaces X of the spectral space $\mathbb{L}(\mathfrak{G})$ for which the restriction $\mathscr{S}|_X : X \longrightarrow \mathbb{L}(\Gamma \mid \Delta)$ becomes coherent.

Other useful properties of g-cocycles are collected in the next lemma whose proof is straightforward.

Lemma 3.8. *Let $\eta \in Z^1(\Gamma, \mathfrak{G})$ be a g-cocycle, $\Delta :=$ Ker (η), $\Delta' :=$ Fix$_\Gamma(\mathfrak{G})$, $\tilde{\Delta} := \bigcap_{\gamma\in\Gamma} \gamma\Delta\gamma^{-1} = \Delta \cap \Delta'$, $\Delta'' := \{\sigma \in \Gamma \mid \forall\gamma \in \Gamma, \gamma\,\eta(\sigma) = \eta(\gamma\sigma\gamma^{-1})\}$, $\overline{\Delta} := \Delta' \cap \Delta''$. Then, the following assertions hold.*

(1) *Δ'' is the maximal closed normal subgroup $\Lambda \subseteq \Gamma$ for which the restrictiom map $\eta\,|_\Lambda : \Lambda \longrightarrow \mathfrak{G}$ is Γ-equivariant; in particular, $\overline{\Delta}$ is a closed normal subgroup of Γ containing $\tilde{\Delta}$.*
(2) *If $\mathfrak{G}$ is abelian, then $\Delta' \subseteq \Delta''$, so $\overline{\Delta} = \Delta'$.*
(3) *$\eta(\overline{\Delta}) = \mathscr{J}(\overline{\Delta})$ is contained in the center $C(\mathfrak{G})$ of $\mathfrak{G}$.*
(4) *$\overline{\Delta} = \Delta' \cap \mathscr{S}(C(\mathfrak{G}))$.*
(5) *η induces by restriction an isomorphism $\overline{\Delta}/\tilde{\Delta} \cong \eta(\overline{\Delta})$ of profinite Γ-modules.*

4 Continuous Actions on Discrete Abelian Groups

In this section we apply the general framework provided by Sect. 3.2 to the case of continuous actions of profinite groups on discrete abelian groups, including as particular cases among others the framework of the classical coGalois theory of separable radical extensions as well as its abstract version developed in [3, 11, 12].

4.1 *The CoGalois Group of a Discrete Module*

Let Γ be a profinite group, E a discrete Γ-module, and $A := t(E)$ the torsion group of the abelian group E with the induced action of Γ. In the following we extend to arbitrary discrete Γ-modules E the notion of *coGalois group* coG(L/K) of a field extension L/K, introduced by Greither and Harrison [17], which plays a major role in the *coGalois theory of radical extensions*.

Setting $H^0(\Gamma, E) = E^\Gamma = \{x \in E \mid \gamma x = x \text{ for all } \gamma \in \Gamma\}$, we denote

$$\text{Rad}(E \mid E^\Gamma) := \{x \in E \mid nx \in E^\Gamma \text{ for some } n \in \mathbb{N} \setminus \{0\}\}.$$

Rad$(E \mid E^\Gamma)$ is a Γ-submodule of E, Rad$(E \mid E^\Gamma)^\Gamma = E^\Gamma$, and $t(\text{Rad}(E \mid E^\Gamma)) = t(E) = A$. By the *coGalois group* of the discrete Γ-module E, denoted by coG(E), we understand the quotient group Rad$(E \mid E^\Gamma)/E^\Gamma$, which is nothing else than the torsion group $t(E/E^\Gamma)$ of the quotient group E/E^Γ. Note that coG(E) is a discrete Γ-module, and coG$(A) = A/A^\Gamma$ is a Γ-submodule of coG(E). Since coG$(E) =$ coG(Rad$(E \mid E^\Gamma)$), in order to study the coGalois group of a discrete Γ-module E we may assume without loss that $E =$ Rad$(E \mid E^\Gamma)$, so coG$(E) = E/E^\Gamma$.

The profinite group Γ and the discrete abelian group $E = \mathrm{Rad}(E \mid E^{\Gamma})$ are naturally related through the map $\Gamma \times E \longrightarrow A$, $(\gamma, x) \mapsto \gamma x - x$, which induces a map $\Gamma/\mathrm{Fix}_{\Gamma}(E) \times \mathrm{coG}(E) \longrightarrow A$ relating the profinite quotient group $\Gamma/\mathrm{Fix}_{\Gamma}(E)$ and the discrete abelian torsion group $\mathrm{coG}(E)$, so we may also assume that $\mathrm{Fix}_{\Gamma}(E)$ is trivial, i.e., the action of Γ on E is faithful.

Consider the homomorphism $\theta : E \longrightarrow Z^1(\Gamma, A)$ defined by $\theta(x)(\gamma) = \gamma x - x$ for $x \in E, \gamma \in \Gamma$, where $Z^1(\Gamma, A)$ is the discrete torsion abelian group of all continuous 1-cocycles of Γ with coefficients in A. As $\mathrm{Ker}\,(\theta) = E^{\Gamma}$, $\mathrm{coG}(E) \cong \theta(E)$ is identified with a subgroup of $Z^1(\Gamma, A)$, while $\mathrm{coG}(A) \cong \theta(A) = B^1(\Gamma, A)$, the subgroup of 1-coboundaries $f_a : \Gamma \longrightarrow A, \gamma \mapsto \gamma a - a$, for $a \in A$. Consequently, the quotient group $\mathrm{coG}(E)/\mathrm{coG}(A)$ is identified with a subgroup of $H^1(\Gamma, A) := Z^1(\Gamma, A)/B^1(\Gamma, A)$.

Using the exact sequence of cohomology groups in low dimensions associated with the short exact sequence of discrete Γ-modules

$$0 \longrightarrow A \longrightarrow E \longrightarrow E/A \longrightarrow 0,$$

we obtain

Lemma 4.1. *Let E be an arbitrary discrete Γ-module, with $\mathrm{Rad}(E \mid E^{\Gamma})$ may be properly contained in E, and $A := t(E)$. Then the following assertions are equivalent.*

(1) $\mathrm{coG}(E) \cong Z^1(\Gamma, A)$.
(2) $\mathrm{coG}(E)/\mathrm{coG}(A) \cong H^1(\Gamma, A)$.
(3) $H^1(\Gamma, E) = 0$.
(4) $H^1(\Gamma, \mathrm{Rad}(E \mid E^{\Gamma})) = 0$.

Remark 4.2. Consider the canonical continuous action of Γ on $Z^1(\Gamma, A)$ defined by

$$(\sigma\alpha)(\gamma) = \sigma\alpha(\sigma^{-1}\gamma\sigma) = \alpha(\gamma) + (\gamma\alpha(\sigma) - \alpha(\sigma)) \text{ for } \alpha \in Z^1(\Gamma, A), \sigma, \gamma \in \Gamma,$$

i.e., $\sigma\alpha = \alpha + f_{\alpha(\sigma)}$. Thus the homomorphism $\theta : E \longrightarrow Z^1(\Gamma, A)$ is Γ-equivariant, and hence $\mathrm{coG}(E)$ is identified with a Γ-submodule of $Z^1(\Gamma, A)$. In particular, if $\mathrm{coG}(A) = 0$, i.e., $A^{\Gamma} = A$, then $\theta(\sigma x) = \theta(x)$ for all $\sigma \in \Gamma, x \in E$, and $\mathrm{coG}(E)$ is identified with a subgroup of the torsion abelian group $\mathrm{Hom}(\Gamma, A)$ of continuous homomorphisms from Γ to A.

4.2 *The Galois Connection Associated with a Torsion Module*

Let Γ be a profinite group acting continuously on a discrete torsion abelian group A. Consider the *evaluation map* $\Gamma \times Z^1(\Gamma, A) \longrightarrow A$, $(\gamma, \alpha) \mapsto \alpha(\gamma)$, relating the profinite group Γ and the discrete torsion Γ-module $Z^1(\Gamma, A)$, with the action of Γ defined as in Remark 4.2.

Consider also the lattice $\mathbb{L}(\Gamma)$, its modular sublattice $\mathbb{L}_n(\Gamma)$ of all closed normal subgroups of Γ, the modular lattice $\mathbb{L}(Z^1(\Gamma, A))$ of all subgroups of $Z^1(\Gamma, A)$, and its sublattice $\mathbb{L}_\Gamma(Z^1(\Gamma, A))$ of all Γ-submodules of $Z^1(\Gamma, A)$. Note that $\mathbb{L}(Z^1(\Gamma, A))$ is also an irreducible spectral space with the basic compact sets $\mathbb{L}(Z^1(\Gamma, A) \mid F) := \{G \in \mathbb{L}(Z^1(\Gamma, A)) \mid F \subseteq G\}$ for F ranging over all finite subgroups of $Z^1(\Gamma, A)$, generic point $Z^1(\Gamma, A)$, and the unique closed point $\{0\}$. The action

$$\Gamma \times \mathbb{L}(Z^1(\Gamma, A)) \longrightarrow \mathbb{L}(Z^1(\Gamma, A)),\ (\gamma, G) \mapsto \gamma G = \{\gamma\alpha \mid \alpha \in G\}$$

and the meet operation

$$\mathbb{L}(Z^1(\Gamma, A)) \times \mathbb{L}(Z^1(\Gamma, A)) \longrightarrow \mathbb{L}(Z^1(\Gamma, A)),\ (G_1, G_2) \mapsto G_1 \cap G_2$$

are coherent maps, while the join operation

$$\mathbb{L}(Z^1(\Gamma, A)) \times \mathbb{L}(Z^1(\Gamma, A)) \longrightarrow \mathbb{L}(Z^1(\Gamma, A)),\ (G_1, G_2) \mapsto G_1 + G_2$$

is a continuous map, not necessarily coherent.

The posets $\mathbb{L}(\Gamma)$ and $\mathbb{L}(Z^1(\Gamma, A))$ are related through the canonical order-reversing maps

$$\mathbb{L}(\Gamma) \longrightarrow \mathbb{L}(Z^1(\Gamma, A)),\ \Lambda \mapsto \Lambda^\perp := Z^1(\Gamma \mid \Lambda, A) = \{\alpha \in Z^1(\Gamma, A) \mid \alpha|_\Lambda = 0\}$$

and

$$\mathbb{L}(Z^1(\Gamma, A)) \longrightarrow \mathbb{L}(\Gamma),\ G \mapsto G^\perp := \bigcap_{\alpha \in G} \alpha^\perp,$$

where $\alpha^\perp := \mathrm{Ker}\,(\alpha) = \{\gamma \in \Gamma \mid \alpha(\gamma) = 0\}$ is an open subgroup of Γ for all $\alpha \in Z^1(\Gamma, A)$.

For $\Lambda \in \mathbb{L}(\Gamma)$, let $\mathrm{res}^\Gamma_\Lambda : Z^1(\Gamma, A) \longrightarrow Z^1(\Lambda, A),\ \alpha \mapsto \alpha|_\Lambda$, be the restriction homomorphism. It follows that $\Lambda^\perp = \mathrm{Ker}\,(\mathrm{res}^\Gamma_\Lambda)$ and $(\mathrm{res}^\Gamma_\Lambda(G))^\perp = G^\perp \cap \Lambda$ for all $\Lambda \in \mathbb{L}(\Gamma), G \in \mathbb{L}(Z^1(\Gamma, A))$.

In the following we assume without loss that the closed normal subgroup $Z^1(\Gamma, A)^\perp$ of Γ is trivial, and hence the closed normal subgroup $B^1(\Gamma, A)^\perp = \mathrm{Fix}_\Gamma(A)$ is abelian since $\alpha(\sigma\tau) = \alpha(\sigma) + \alpha(\tau) = \alpha(\tau\sigma)$ for $\alpha \in Z^1(\Gamma, A)$, $\sigma, \tau \in \mathrm{Fix}_\Gamma(A)$, while the profinite quotient group $\Gamma/\mathrm{Fix}_\Gamma(A)$ is identified with a closed subgroup of the totally disconnected topological group $\mathrm{Aut}(A)$ for which the subgroups $\mathrm{Aut}(A \mid F) := \{\varphi \in \mathrm{Aut}(A) \mid \varphi|_F = 1_F\}$ for F ranging over the finite subgroups of A form a fundamental system of open neighborhoods of 1_A.

The next result is an analogue of Proposition 3.6, and the proof is similar.

Proposition 4.3. *The following assertions hold.*

(1) *The pair of order-reversing maps*

$$\mathbb{L}(\Gamma) \longrightarrow \mathbb{L}(Z^1(\Gamma, A)),\ \Lambda \mapsto \Lambda^\perp,\ \mathbb{L}(Z^1(\Gamma, A)) \longrightarrow \mathbb{L}(\Gamma),\ G \mapsto G^\perp,$$

establishes a Galois connection between the posets $\mathbb{L}(\Gamma)$ *and* $\mathbb{L}(Z^1(\Gamma, A))$, *i.e.,* $X \subseteq X^{\perp\perp}$ *for any element* X *of* $\mathbb{L}(\Gamma)$ *or* $\mathbb{L}(Z^1(\Gamma, A))$.

(2) *The map* $\Lambda \mapsto \Lambda^\perp$ *is a coherent complete-semi-lattice morphism* $(\mathbb{L}(\Gamma), \vee) \longrightarrow (\mathbb{L}(Z^1(\Gamma, A)), \wedge)$ *satisfying* $(\sigma\Lambda\sigma^{-1})^\perp = \sigma \cdot \Lambda^\perp$ *for all* $\sigma \in \Gamma$, $\Lambda \in \mathbb{L}(\Gamma)$.

(3) *The map* $G \mapsto G^\perp$ *is a continuous, not necessarily coherent, complete-semi-lattice morphism* $(\mathbb{L}(Z^1(\Gamma, A)), \vee) \longrightarrow (\mathbb{L}(\Gamma), \wedge)$ *satisfying* $(\sigma \cdot G)^\perp = \sigma \cdot G^\perp \cdot \sigma^{-1}$ *for all* $\sigma \in \Gamma$, $G \in \mathbb{L}(Z^1(\Gamma, A))$.

(4) *The maps above induce by restriction a Galois connection between the posets* $\mathbb{L}_n(\Gamma)$ *and* $\mathbb{L}_\Gamma(Z^1(\Gamma, A))$.

Remark 4.4. Let E be a discrete Γ-module such that $t(E) = A$, and assume without loss that $\mathrm{coG}(E) := t(E/E^\Gamma) = E/E^\Gamma$ and $\mathrm{Fix}_\Gamma(E)$ is trivial. According to Remark 4.2, $\mathrm{coG}(E)$ is identified with a Γ-submodule of $Z^1(\Gamma, A)$, so $Z^1(\Gamma, A)^\perp \subseteq \mathrm{coG}(E)^\perp = \mathrm{Fix}_\Gamma(E) = \{1\}$, and hence $Z^1(\Gamma, A)^\perp$ is trivial. The assertions of Proposition 4.3 remain valid for the order-reversing maps

$$\mathbb{L}(\Gamma) \longrightarrow \mathbb{L}(\mathrm{coG}(E)),\ \Lambda \mapsto \Lambda^\perp \cap \mathrm{coG}(E),\ \mathbb{L}(\mathrm{coG}(E)) \longrightarrow \mathbb{L}(\Gamma),\ G \mapsto G^\perp.$$

4.3 The Associated Profinite Module and Generating Cocycle

Consider the same data as in Sect. 4.2. We construct a profinite Γ-module $\mathfrak{G}$, a sort of dual of the discrete torsion Γ-module $Z^1(\Gamma, A)$ with respect to the discrete torsion Γ-module A, and a natural continuous g-cocycle $\eta : \Gamma \longrightarrow \mathfrak{G}$. To this end, we consider the abelian group $\mathfrak{H} := \mathrm{Hom}(Z^1(\Gamma, A), A)$, and for any subgroup $G \subseteq Z^1(\Gamma, A)$, we denote by r_G the restriction homomorphism $\mathfrak{H} \longrightarrow \mathrm{Hom}(G, A)$, $\varphi \mapsto \varphi|_G$, with kernel $\mathfrak{H}_G$. $\mathfrak{H}$ becomes a totally disconnected (Hausdorff) topological group with the compact-open topology, for which the subgroups $\mathfrak{H}_F$, with F ranging over the *finite* subgroups of $Z^1(\Gamma, A)$, serve as a fundamental system of open neighborhoods of the null homomorphism. The profinite group Γ acts continuously on $\mathfrak{H}$ according to the rule $(\gamma\varphi)(\alpha) := \gamma\varphi(\alpha)$ for $\gamma \in \Gamma$, $\varphi \in \mathfrak{H}$, $\alpha \in Z^1(\Gamma, A)$, and the canonical map $\eta : \Gamma \longrightarrow \mathfrak{H}$, $\gamma \mapsto \eta_\gamma$, defined by $\eta_\gamma(\alpha) := \alpha(\gamma)$ for $\gamma \in \Gamma, \alpha \in Z^1(\Gamma, A)$, is a continuous 1-cocycle; note that with the notation from Sect. 4.2, $\eta^{-1}(\mathfrak{H}_F) = F^\perp = \bigcap_{\alpha \in F} \alpha^\perp$ is an open subgroup of Γ provided F is a finite subgroup of $Z^1(\Gamma, A)$. As $\mathrm{Ker}\,(\eta) = Z^1(\Gamma, A)^\perp$ is a closed normal subgroup of Γ, we may assume without loss that the cocycle η is injective.

Next let us consider the subgroup $\langle\eta(\Gamma)\rangle \subseteq \mathfrak{H}$ generated by $\eta(\Gamma)$; for any finite subgroup $F \subseteq Z^1(\Gamma, A)$, let $\mathfrak{G}^F := r_F(\langle\eta(\Gamma)\rangle) = \langle\eta_\gamma|_F : \gamma \in \Gamma\rangle \cong \langle\eta(\Gamma)\rangle/(\langle\eta(\Gamma)\rangle \cap \mathfrak{H}_F)$. Since any $\alpha \in Z^1(\Gamma, A)$ is a continuous map, and hence locally constant, with values in the torsion abelian group A, we deduce that $\mathfrak{G}^F \subseteq \mathrm{Hom}(F, A)$ is a finite abelian group for all finite subgroups $F \subseteq Z^1(\Gamma, A)$. Since the induced topology on the subgroup $\langle\eta(\Gamma)\rangle \subseteq \mathfrak{H}$ is determined by the collection of the subgroups $\langle\eta(\Gamma)\rangle \cap \mathfrak{H}_F$ of finite index filtered from below, it follows that the closure $\mathfrak{G} := \overline{\langle\eta(\Gamma)\rangle}$ is a profinite abelian group isomorphic with $\varprojlim_F \mathfrak{G}^F$, topologically generated by $\eta(\Gamma)$. Moreover $\mathfrak{G}$ is a profinite Γ-module with the induced action of Γ, and $\eta : \Gamma \longrightarrow \mathfrak{G}$ is an injective g-cocycle.

Remarks 4.5. (1) Using Remark 3.3 (2), the profinite Γ-module $\mathfrak{G}$ is the quotient of $\Omega_\Gamma^{\mathrm{ab}}$, identified with the augmentation ideal of $\hat{\mathbb{Z}}[[\Gamma]]$, by the closed submodule

$$\bigcap\nolimits_{\varphi \in \mathrm{Hom}_\Gamma(\Omega_\Gamma^{\mathrm{ab}}, A)} \mathrm{Ker}\,(\varphi),$$

while the g-cocycle $\eta : \Gamma \longrightarrow \mathfrak{G}$ is induced by $\omega_\Gamma^{\mathrm{ab}} : \Gamma \longrightarrow \Omega_\Gamma^{\mathrm{ab}}$. See also Proposition 4.7.

(2) Aside from the continuous action $\Gamma \times \mathfrak{G} \longrightarrow \mathfrak{G}$, $(\sigma, \varphi) \mapsto \sigma\varphi$, considered above, there is another continuous action of the profinite group Γ on the profinite abelian group $\mathfrak{G} \subseteq \mathrm{Hom}(Z^1(\Gamma, A), A)$ induced by the continuous actions of Γ on the discrete torsion abelian groups $Z^1(\Gamma, A)$ and A, defined by

$$({}^\sigma\varphi)(\alpha) := \sigma \cdot \varphi(\sigma^{-1}\alpha) = (\sigma\varphi)(\alpha) + (\sigma\varphi)(f_{\alpha(\sigma^{-1})})$$

for $\sigma \in \Gamma, \varphi \in \mathfrak{G}, \alpha \in Z^1(\Gamma, A)$. In particular, taking $\varphi = \eta_\gamma$ for $\gamma \in \Gamma$, we obtain ${}^\sigma\eta_\gamma = \eta_{\sigma\gamma\sigma^{-1}}$ for all $\sigma \in \Gamma$, whence the injective continuous map $\eta : \Gamma \longrightarrow \mathfrak{G}, \gamma \mapsto \eta_\gamma$, is Γ-equivariant with respect to the action of Γ on itself by inner automorphisms and the afore defined action $\Gamma \times \mathfrak{G} \longrightarrow \mathfrak{G}$, $(\sigma, \varphi) \mapsto {}^\sigma\varphi$.

We denote by $\mathbb{L}(\mathfrak{G})$ the modular lattice of all closed Γ-submodules of $\mathfrak{G}$ (with respect to the action $\Gamma \times \mathfrak{G} \longrightarrow \mathfrak{G}$, $(\sigma, \varphi) \mapsto \sigma\varphi$), and by $\mathbb{L}_\Gamma(\mathfrak{G})$ its sublattice consisting of those Γ-submodules which are also invariant under the action $\Gamma \times \mathfrak{G} \longrightarrow \mathfrak{G}$, $(\sigma, \varphi) \mapsto {}^\sigma\varphi$.

According to Sect. 2.4.1, $\mathbb{L}(\mathfrak{G})$ is also an irreducible spectral space with basic compact open sets $\mathbb{L}(\mathbf{a}) := \{\mathbf{b} \in \mathbb{L}(\mathfrak{G}) \mid \mathbf{b} \subseteq \mathbf{a}\}$ for all open $\mathbf{a} \in \mathbb{L}(\mathfrak{G})$, generic point $\{0\}$, unique closed point $\mathfrak{G}$, coherent join operation ($+$), and continuous meet operation ($\cap$).

Applying Proposition 3.6 to the triple $(\Gamma, \mathfrak{G}, \eta)$ above, we obtain

Proposition 4.6. *The following assertions hold.*

(1) *The order-preserving maps*

$$\mathscr{J} : \mathbb{L}(\Gamma) \longrightarrow \mathbb{L}(\mathfrak{G}),\ \Lambda \mapsto \textit{the closed } \Gamma-\textit{submodule generated by } \eta(\Lambda),$$

$$\mathscr{S} : \mathbb{L}(\mathfrak{G}) \longrightarrow \mathbb{L}(\Gamma),\ \mathbf{a} \mapsto \eta^{-1}(\mathbf{a}),$$

establishes a coGalois connection between the posets $\mathbb{L}(\Gamma)$ *and* $\mathbb{L}(\mathfrak{G})$, *i.e.,* $\Lambda \subseteq \mathscr{S}(\mathscr{J}(\Lambda))$ *and* $\mathscr{J}(\mathscr{S}(\mathbf{a})) \subseteq \mathbf{a}$ *for all* $\Lambda \in \mathbb{L}(\Gamma)$, $\mathbf{a} \in \mathbb{L}(\mathfrak{G})$.

(2) *The coherent map* $\Lambda \mapsto \mathscr{J}(\Lambda)$ *is a morphism of complete semilattices* $(\mathbb{L}(\Gamma), \vee) \longrightarrow (\mathbb{L}(\mathfrak{G}), +)$, *satisfying* $\mathscr{J}(\sigma\Lambda\sigma^{-1}) = {}^{\sigma}\mathscr{J}(\Lambda)$ *for all* $\sigma \in \Gamma$, $\Lambda \in \mathbb{L}(\Gamma)$.

(3) *The continuous (not necessarily coherent) map* $\mathbf{a} \mapsto \mathscr{S}(\mathbf{a})$ *is a morphism of complete semilattices* $(\mathbb{L}(\mathfrak{G}), \cap) \longrightarrow (\mathbb{L}(\Gamma), \cap)$, *satisfying* $\mathscr{S}({}^{\sigma}\mathbf{a}) = \sigma\mathscr{S}(\mathbf{a})\sigma^{-1}$ *for all* $\sigma \in \Gamma, \mathbf{a} \in \mathbb{L}(\mathfrak{G})$.

(4) *The order-preserving maps* $\mathscr{J}$ *and* $\mathscr{S}$ *induce by restriction a coGalois connection between the posets* $\mathbb{L}_{\mathrm{n}}(\Gamma)$ *and* $\mathbb{L}_{\Gamma}(\mathfrak{G})$.

4.4 A Nondegenerate Pairing and the Induced Galois Connection

By Propositions 4.3, 4.6, the lattice $\mathbb{L}(\Gamma)$ is related to the lattices $\mathbb{L}(Z^1(\Gamma, A))$ and $\mathbb{L}(\mathfrak{G})$ through canonical maps defining a Galois connection and a coGalois connection respectively. It is also natural to consider the relation between the lattices $\mathbb{L}(Z^1(\Gamma, A))$ and $\mathbb{L}(\mathfrak{G})$.

First, with data and notation from Sect. 4.3, we obtain

Proposition 4.7. *There is a canonical nondegenerate pairing*

$$\langle\, ,\, \rangle : \mathfrak{G} \times Z^1(\Gamma, A) \longrightarrow A,\ (\varphi, \alpha) \mapsto \langle \varphi, \alpha \rangle := \varphi(\alpha)$$

satisfying the identities

$$\langle \sigma\varphi, \alpha \rangle = \sigma \langle \varphi, \alpha \rangle = \langle {}^{\sigma}\varphi, \sigma\alpha \rangle$$

for all $\sigma \in \Gamma, \varphi \in \mathfrak{G}, \alpha \in Z^1(\Gamma, A)$.

The pairing above induces a canonical isomorphism of discrete torsion Γ*-modules* $\lambda : Z^1(\Gamma, A) \longrightarrow \mathrm{Hom}_{\Gamma}(\mathfrak{G}, A)$, *defined by* $\lambda(\alpha)(\varphi) := \varphi(\alpha)$, *with its inverse* $\mu : \mathrm{Hom}_{\Gamma}(\mathfrak{G}, A) \longrightarrow Z^1(\Gamma, A)$ *defined by* $\mu(\psi) := \psi \circ \eta$; *here* $\mathrm{Hom}_{\Gamma}(\mathfrak{G}, A)$ *denotes the discrete torsion abelian group of all continuous homomorphisms* $\psi : \mathfrak{G} \longrightarrow A$ *satisfying* $\psi(\sigma\varphi) = \sigma\psi(\varphi)$ *for* $\sigma \in \Gamma, \varphi \in \mathfrak{G}$, *together with the continuous action of* Γ *defined by* $({}^{\sigma}\psi)(\varphi) := \sigma \cdot \psi({}^{\sigma^{-1}}\varphi) = \psi(\sigma \cdot ({}^{\sigma^{-1}}\varphi))$.

Proof. The statement follows easily from the definitions given in Sect. 4.3 of $\mathfrak{G}$ and of the actions $\Gamma \times \mathfrak{G} \longrightarrow \mathfrak{G}, (\sigma, \varphi) \mapsto \sigma\varphi, {}^{\sigma}\varphi$. To prove that λ and μ are isomorphisms inverse to one another we have to show that $\mu \circ \lambda = 1_{Z^1(\Gamma,A)}$ and μ is injective. The first condition is satisfied since $(\mu \circ \lambda)(\alpha) = \lambda(\alpha) \circ \eta$ for all $\alpha \in Z^1(\Gamma, A)$, and $(\lambda(\alpha) \circ \eta)(\gamma) = \lambda(\alpha)(\eta_\gamma) = \eta_\gamma(\alpha) = \alpha(\gamma)$ for all $\gamma \in \Gamma$. To check the injectivity of μ, let $\psi \in \mathrm{Hom}_\Gamma(\mathfrak{G}, A)$ be such that $\mu(\psi) = \psi \circ \eta = 0$. As $\mathfrak{G}$ is topologically generated by the set $\eta(\Gamma)$, it follows that $\psi = 0$ as desired.

Next we note that the posets $\mathbb{L}(\mathfrak{G})$ and $\mathbb{L}(Z^1(\Gamma, A))$ are related through the canonical order-reversing maps

$$\mathbb{L}(\mathfrak{G}) \longrightarrow \mathbb{L}(Z^1(\Gamma, A)),\ \mathbf{a} \mapsto \mathbf{a}_\perp := \bigcap_{\varphi \in \mathbf{a}} \varphi_\perp,$$

where $\varphi_\perp := \mathrm{Ker}\,(\varphi)$ for $\varphi \in \mathfrak{G} \subseteq \mathrm{Hom}(Z^1(\Gamma, A), A)$, and

$$\mathbb{L}(Z^1(\Gamma, A)) \longrightarrow \mathbb{L}(\mathfrak{G}),\ G \mapsto G_\perp := \bigcap_{\alpha \in G} \alpha_\perp,$$

where $\alpha_\perp := \mathrm{Ker}\,(\lambda(\alpha)) = \{\varphi \in \mathfrak{G} \mid \varphi(\alpha) = 0\}$.

The following result is an analogue of Proposition 4.3.

Proposition 4.8. *The following assertions hold.*

(1) *The pair of order-reversing maps*

$$\mathbb{L}(\mathfrak{G}) \longrightarrow \mathbb{L}(Z^1(\Gamma, A)),\ \mathbf{a} \mapsto \mathbf{a}_\perp,\ \mathbb{L}(Z^1(\Gamma, A)) \longrightarrow \mathbb{L}(\mathfrak{G}),\ G \mapsto G_\perp,$$

establishes a Galois connection, i.e., $X \subseteq X_{\perp\perp}$ *for any element* X *of* $\mathbb{L}(\mathfrak{G})$ *or* $\mathbb{L}(Z^1(\Gamma, A))$.

(2) *The coherent map* $\mathbf{a} \mapsto \mathbf{a}_\perp$ *is a morphism of complete semilattices* $(\mathbb{L}(\mathfrak{G}), +) \longrightarrow (\mathbb{L}(Z^1(\Gamma, A)), \cap)$, *satisfying* $({}^{\sigma}\mathbf{a})_\perp = \sigma \cdot (\mathbf{a}_\perp)$ *for all* $\sigma \in \Gamma$, $\mathrm{a} \in \mathbb{L}(\mathfrak{G})$.

(3) *The continuous (not necessarily coherent) map* $G \mapsto G_\perp$ *is a morphism of complete semilattices* $(\mathbb{L}(Z^1(\Gamma, A)), +) \longrightarrow (\mathbb{L}(\mathfrak{G}), \cap)$, *satisfying* $(\sigma G)_\perp = {}^{\sigma}(G_\perp)$ *for all* $\sigma \in \Gamma$, $G \in \mathbb{L}(Z^1(\Gamma, A))$.

(4) *The maps above induce by restriction a Galois connection between the posets* $\mathbb{L}_\Gamma(\mathfrak{G})$ *and* $\mathbb{L}_\Gamma(Z^1(\Gamma, A))$.

Corollary 4.9. *The Galois connection between* $\mathbb{L}(\Gamma)$ *and* $\mathbb{L}(Z^1(\Gamma, A))$ (cf. Proposition 4.3) *is obtained by composing the coGalois connection between* $\mathbb{L}(\Gamma)$ *and* $\mathbb{L}(\mathfrak{G})$ (cf. Proposition 4.6) *with the Galois connection between* $\mathbb{L}(\mathfrak{G})$ *and* $\mathbb{L}(Z^1(\Gamma, A))$ (cf. Proposition 4.8): $\Lambda^\perp = \mathscr{J}(\Lambda)_\perp, G^\perp = \mathscr{S}(G_\perp)$ *for* $\Lambda \in \mathbb{L}(\Gamma)$, $G \in \mathbb{L}(Z^1(\Gamma, A))$.

4.5 Examples

Example 4.10 (The cyclotomic abstract coGalois theory [3, 11, 12]). Let Γ be a profinite group, A a discrete quasi-cyclic group identified with a subgroup of $\mathbb{Q}/\mathbb{Z}$, $A^\vee := \text{Hom}(A, \mathbb{Q}/\mathbb{Z})$, its Pontryagin dual, and $\Gamma \times A \longrightarrow A$, $(\sigma, a) \mapsto \sigma a := \chi(\sigma)a$, a continuous action given by the continuous *cyclotomic character* $\chi : \Gamma \longrightarrow (A^\vee)^\times$. With the notation from Sect. 4.3, we obtain $\mathfrak{G} = \mathfrak{H} = \text{Hom}(Z^1(\Gamma, A), A) = \text{Hom}(Z^1(\Gamma, A), \mathbb{Q}/\mathbb{Z}) = Z^1(\Gamma, A)^\vee$, the Pontryagin dual of the discrete torsion abelian group $Z^1(\Gamma, A)$. $\mathfrak{G}$ is a profinite Γ-module with respect to the continuous action $\Gamma \times \mathfrak{G} \longrightarrow \mathfrak{G}$, $(\sigma, \varphi) \mapsto \chi(\sigma)\varphi$, and the canonical map $\eta : \Gamma \longrightarrow \mathfrak{G}$, $\gamma \mapsto (\eta_\gamma : Z^1(\Gamma, A) \longrightarrow A)$, defined by $\eta_\gamma(\alpha) := \alpha(\gamma)$, is a continuous g-cocycle. Assuming without loss that η is injective, i.e., the closed normal subgroup $Z^1(\Gamma, A)^\perp$ is trivial, it follows that Γ is metabelian as an extension of the abelian profinite quotient group $\Gamma/\text{Fix}_\Gamma(A) \cong \chi(\Gamma) \subseteq (A^\vee)^\times$ by the abelian profinite group $B^1(\Gamma, A)^\perp = \text{Fix}_\Gamma(A)$. It follows also that any closed subgroup of $\mathfrak{G}$ is invariant under the action of Γ, any continuous homomorphism $\psi : \mathfrak{G} \longrightarrow A$ is Γ-equivariant, and, by the Pontryagin duality, the Galois connection between $\mathbb{L}(Z^1(\Gamma, A))$ and $\mathbb{L}(\mathfrak{G})$ is perfect, i.e., the maps $\mathbb{L}(Z^1(\Gamma, A)) \longrightarrow \mathbb{L}(\mathfrak{G})$, $G \mapsto G_\perp$, and $\mathbb{L}(\mathfrak{G}) \longrightarrow \mathbb{L}(Z^1(\Gamma, A))$, $\mathbf{a} \mapsto \mathbf{a}_\perp$, as defined in Sect. 4.4, are lattice anti-isomorphisms inverse to one another. The last-mentioned fact and Corollary 4.9 are key ingredients used in [3, 11, 12] in the investigation of the Galois connection between the lattices $\mathbb{L}(\Gamma)$ and $\mathbb{L}(Z^1(\Gamma, A))$.

In particular, if the action of Γ on A is trivial then $Z^1(\Gamma, A) = \text{Hom}(\Gamma, A)$ and $\eta : \Gamma \longrightarrow \mathfrak{G} = \text{Hom}(\Gamma, A)^\vee$ is an isomorphism, therefore the Galois connection between the lattices $\mathbb{L}(\Gamma)$ and $\mathbb{L}(\text{Hom}(\Gamma, A))$ is perfect. However, according to [12, Proposition 5.6, Remarks 5.5, 5.8] there exist nontrivial actions $\Gamma \times A \longrightarrow A$ called *strongly coGalois* such that the Galois connection between the lattices $\mathbb{L}(\Gamma)$ and $\mathbb{L}(Z^1(\Gamma, A))$ is perfect and $\Gamma \cong \mathfrak{G}$ (non-canonically) provided Γ is abelian.

Example 4.11 (The coGalois theory of separable radical extensions). Let L/K be a Galois extension with $\Gamma := \text{Gal}(L/K)$. In addition we assume that the extension L/K is *radical*, i.e., $L = K(\text{Rad}(L/K))$, where

$$\text{Rad}(L/K) = \{x \in L^\times \mid x^n \in K \text{ for some } n \in \mathbb{N} \setminus \{0\}\},$$

so $\text{Rad}(L/K)^\Gamma = (L^\times)^\Gamma = K^\times$; as L/K is separable, the exponents n above may be assumed to be prime with the characteristic exponent of K. Note that any element $x \in \text{Rad}(L/K)$ is an n-th radical $\sqrt[n]{a}$ of an element $a \in K^\times$ for some $n \geq 1$; thus, $\text{Rad}(L/K)$ is precisely the set of all radicals belonging to L of elements of $K^\times$.

With notation from Sects. 4.1 and Example 4.10, we consider the discrete Γ-module $E := L^\times$, the multiplicative group of the field L with the Galois action, and let $A := t(L^\times) = t(E) = \mu_L$, with $A^\Gamma = t(K^\times) = \mu_K$, be the quasi-cyclic multiplicative group of the roots of unity contained in L, identified through a non-canonical monomorphism $\mu_L \longrightarrow \mathbb{Q}/\mathbb{Z}$ with a subgroup of $\mathbb{Q}/\mathbb{Z}$. Thus $\text{coG}(E) = \text{coG}(L/K) := t(L^\times/K^\times) = \text{Rad}(L/K)/K^\times$ is a discrete torsion abelian group with the induced Galois action. Since $H^1(\Gamma, L^\times) = 0$ by Hilbert's Theorem 90, it follows by Lemma 4.1 that the canonical map $\theta : \mathrm{Rad}(L/K) \longrightarrow Z^1(\Gamma, A)$ defined by $\theta(x)(\gamma) = \frac{\gamma x}{x}$, for $x \in \mathrm{Rad}(L/K)$, $\gamma \in \Gamma$, induces an isomorphism of the Γ-modules $\mathrm{coG}(E) = \mathrm{coG}(L/K)$ and $Z^1(\Gamma, A)$.

Composing the perfect standard Galois connection between the lattice $\mathbb{L}(L/K)$ of all intermediate fields $K \subseteq F \subseteq L$ and the lattice $\mathbb{L}(\Gamma)$ with the Galois connection between the lattices $\mathbb{L}(\Gamma)$ and $\mathbb{L}(Z^1(\Gamma, A)) \cong \mathbb{L}(\mathrm{coG}(L/K))$ described in Sect. 4.2, Example 4.10, we obtain the coGalois connection between the lattices $\mathbb{L}(L/K)$ and $\mathbb{L}(\mathrm{coG}(L/K))$, the main object of investigation of the *coGalois theory of separable radical extensions*; consequently, all the results of this theory, in particular, the *Kummer theory*, could be obtained easily by transferring the corresponding results from its abstract version (Example 4.10), [3, 11, 12]. See [12, 6. Sect. 6: Two examples] for a detailed analysis of the abstract version of the coGalois theory of the Galois extensions L/K, where K is a finite field, $L = K^s$, respectively K is a local field and L its maximal tamely ramified extension.

Example 4.12 (The additive analogue of Example 4.11*).* Using *Witt calculus* and *higher Artin-Schreier theory* [6, 24, 40], we present an additive analogue of the multiplicative framework discussed in Example 4.11.

Let K be an arbitrary field of characteristic $p > 0$. Letting V be the *shift operator* on the *Witt ring* $W(K)$, we consider the quotient rings $W_n(K) := W(K)/V^n W(K)$ consisting of *Witt vectors of length* $n \geq 1$ over K; in particular, $W_1(K) = K$. Since the shift operator $V : W(K) \longrightarrow W(K)$ is additive, the family of abelian groups $W_n(K)^+ := (W_n(K), +)$, indexed by the totally ordered set $\mathbb{N} \setminus \{0\}$, forms a direct system with the canonical connecting monomorphisms $W_n(K)^+ \longrightarrow W_{n+1}(K)^+$ induced by V. We denote $W_\infty(K) := \bigcup_{n\geq 1} W_n(K)^+$, the direct limit, identifying $W_n(K)^+$, $n \geq 1$, with an increasing sequence of subgroups of $W_\infty(K)$. On the other hand, the *Frobenius operator* F is an endomorphism of the ring $W(K)$, and $FV = VF$, whence F induces an endomorphism of the ring $W_n(K)$ for all $n \geq 1$ and so gives rise to an endomorphism, denoted also by F, of the abelian group $W_\infty(K)$ such that $F W_n(K)^+ \subseteq W_n(K)^+$ for all $n \geq 1$. Setting $e_n := (1, 0, \ldots, 0)$, the unit element of the ring $W_n(K)$, and using the identity $px = FVx$ in the ring $W(K)$, we deduce that e_n has order p^n. Consequently, $W_n(K)^+$ is a torsion abelian group of exponent p^n containing $W_n(\mathbb{F}_p)^+ \cong p^{-n}\mathbb{Z}/\mathbb{Z}$, so $W_\infty(K)$ is a discrete $\mathbb{Z}_p$-module containing $W_\infty(\mathbb{F}_p) \cong (\mathbb{Q}/\mathbb{Z})(p) = \mathbb{Q}_p/\mathbb{Z}_p$.

Now consider the family $(\mathfrak{P}_m)_{m\geq 1}$ of endomorphisms of the discrete $\mathbb{Z}_p$-module $W_\infty(K)$, defined by $\mathfrak{P}_m(x) = F^m x - x$, and note that $\mathfrak{P}_m W_n(K)^+ \subseteq W_n(K)^+$ for $n, m \geq 1$. For $n = m = 1$ we obtain $\mathfrak{P}_1(x) = \mathfrak{P}(x) = x^p - x$ in $K^+ = W_1(K)^+$, the well-known operator from the *Artin-Schreier theory* of abelian extensions of exponent p. It follows that $\mathrm{Ker}\,(\mathfrak{P}_m) = W_\infty(K \cap \mathbb{F}_{p^m})$, in particular, $\mathrm{Ker}\,(\mathfrak{P}_1) = W_\infty(\mathbb{F}_p)$.

Letting L/K be a Galois extension with $\Gamma := \mathrm{Gal}(L/K)$, we denote by $k = K \cap \tilde{\mathbb{F}}_p$, $l := L \cap \tilde{\mathbb{F}}_p$ the relative algebraic closure of the prime field $\mathbb{F}_p$ in K and L respectively. Let $\overline{\Gamma}$ be the procyclic group $\mathrm{Gal}(l/k)$, and $\Gamma' := \mathrm{Gal}(L/(K \cdot l))$, the kernel of the epimorphism $\Gamma \longrightarrow \overline{\Gamma}$, $\gamma \mapsto \overline{\gamma} := \gamma\,|_l$. The afore defined additive operators $\mathfrak{P}_m$ are seen as endomorphisms of $W_\infty(L)$ inducing by restriction endomorphisms of $W_\infty(E)$ for any subfield E of L. The Galois group Γ acts continuously on the discrete $\mathbb{Z}_p$-module $W_\infty(L)$, $\gamma \cdot W_n(L) = W_n(L)$ for all $\gamma \in \Gamma, n \geq 1$, $W_\infty(L)^\Gamma = W_\infty(K)$, $W_\infty(L)^{\Gamma'} = W_\infty(K \cdot l)$, and $\gamma \mathfrak{P}_m = \mathfrak{P}_m \gamma$ for all $\gamma \in \Gamma, m \geq 1$. For any $a = (a_0, \cdots, a_{n-1}) \in W_n(L)$, we set $K(a) := K(a_0, \cdots, a_{n-1})$, a finite subextension of L/K; more generally, for any subset S of $W_\infty(L)$, we denote by $K(S)$ the composite of all $K(a)$ with $a \in S$.

The additive version of the multiplicative group of radicals $\mathrm{Rad}(L/K)$ defined in Example 4.11 is the $\mathbb{Z}_p[[\Gamma]]$-submodule of $W_\infty(L)$ lying over $W_\infty(K) + W_\infty(l)$

$$\mathrm{Rad}_+(L/K) := \{x \in W_\infty(L) \mid \mathfrak{P}_m(x) \in W_\infty(K) \text{ for some } m \geq 1\},$$

the direct limit of the family of submodules $(\mathfrak{P}_m^{-1}(W_\infty(K)))_{m\geq 1}$ of $W_\infty(L)$, with $\mathfrak{P}_n^{-1}(W_\infty(K)) \subseteq \mathfrak{P}_m^{-1}(W_\infty(K))$ for $n \mid m$. Note that the $\mathbb{Z}_p[[\Gamma]]$-submodule $W_\infty(l) = \bigcup_{m\geq 1} \mathrm{Ker}\,(\mathfrak{P}_m)$ plays the role of the multiplicative group $\mu_L = t(L^\times)$ of the roots of unity contained in L.

The subextension $K(\mathrm{Rad}_+(L/K))$ of the Galois extension L/K is Galois; it is the maximal subextension N/K satisfying $N = K(\mathrm{Rad}_+(N/K))$. By analogy with the radical extensions considered in Example 4.11, we say that L/K is an *additive-radical extension* (for short *a-radical extension*) if $L = K(\mathrm{Rad}_+(L/K))$. In particular, the usually called *Artin-Schreier extensions*, i.e., cyclic extensions of degree p over a field of characteristic p, are minimal a-radical extensions.

Note that for any $n \geq 1$, $M_n := K(\mathfrak{P}_1^{-1}(W_n(K)))$ is the maximal abelian subextension of L/K whose exponent is a divisor of p^n, and hence $M_\infty := \bigcup_{n\geq 1} M_n = K(\mathfrak{P}_1^{-1}(W_\infty(K)))$ is the maximal abelian p-extension of K contained in L; in particular, $M_\infty \cap (K \cdot l) = K \cdot l(p)$, where $l(p)$ is the maximal (procyclic) p-subextension of l/k. Using [16, Propositions 1.1, 1.2] as the first step of induction on n, we deduce that $\mathrm{Rad}_+(M_n/K) = \bigcup_{\mathbb{F}_{p^m} \subseteq k} \mathfrak{P}_m^{-1}(W_n(K))$ for all $n \geq 1$, whence $\mathrm{Rad}_+(M_\infty/K) = \bigcup_{\mathbb{F}_{p^m} \subseteq k} \mathfrak{P}_m^{-1}(W_\infty(K))$.

As an additive analogue of the coGalois group of a radical extension defined in Example 4.11, we take the quotient $\mathbb{Z}_p[[\Gamma]]$-module $\mathrm{coG}_+(L/K) := \mathrm{Rad}_+(L/K)/W_\infty(K)$, the "torsion" of $W_\infty(L)/W_\infty(K)$ with respect to the endomorphisms $\mathfrak{P}_m$, $m \geq 1$. Considering the canonical action of Γ on $Z^1(\Gamma, W_\infty(l))$ as defined in Remark 4.2, the map $\theta : \mathrm{Rad}_+(L/K) \longrightarrow Z^1(\Gamma, W_\infty(l))$, defined by $\theta(x)(\gamma) := \gamma x - x$, induces an injective morphism of discrete $\mathbb{Z}_p[[\Gamma]]$-modules $\tilde{\theta} : \mathrm{coG}_+(L/K) \hookrightarrow Z^1(\Gamma, W_\infty(l))$,

Moreover, using the exact sequence of cohomology groups in low dimensions associated with the exact sequence of discrete Γ-modules

$$0 \longrightarrow W_\infty(l) \longrightarrow W_\infty(L) \longrightarrow W_\infty(L)/W_\infty(l) \longrightarrow 0,$$

we deduce that $\tilde{\theta}$ is an isomorphism since $H^1(\Gamma, W_\infty(L)) = 0$ follows by induction from the additive Hilbert's Theorem 90. Thus, as in Example 4.11, we have arrived to the abstract framework described in Sects. 4.2–4.4: given a continuous action of a profinite group Γ on a discrete torsion $\mathbb{Z}_p$-module of the form $A := W_\infty(l) = Q(W(l))/W(l) = \bigcup_{n\geq 1} p^{-n}W(l)/W(l)$, where l is an algebraic extension of $\mathbb{F}_p$ and the fraction field $Q(W(l))$ of $W(l)$ is the unique, up to isomorphism, complete unramified discrete valued field of characteristic 0 having l as residue field, investigate the Galois connection between the lattices $\mathbb{L}(\Gamma)$ and $\mathbb{L}(Z^1(\Gamma, A)$. Note that we can replace the Witt ring $W(l)$ by the integral closure of $\mathbb{Z}_p$ in $W(l)$, the discrete valuation ring of the unramified field extension of $\mathbb{Q}_p$ having l as residue field, provided the residue field extension $l/\mathbb{F}_p$ is infinite.

Example 4.13 (The coGalois theory of Galois algebras over fields). Let K be a field, and $\mathfrak{L}$ be a discrete commutative K-algebra together with a continuous action of a profinite group Γ. Thus $\mathfrak{L}$ is the union of its subalgebras $\mathfrak{L}^\Lambda = \{f \in \mathfrak{L} \mid \gamma f = f \text{ for all } \gamma \in \Lambda\}$, where Λ ranges over the open normal subgroups of Γ. In addition we assume that $\mathfrak{L}$ is a *Galois Γ-algebra*, i.e., for every open normal subgroup Λ of Γ, the subalgebra $\mathfrak{L}^\Lambda$ is *semisimple* (a finite product of field extensions of K), and also a free $K[\Gamma/\Lambda]$-module of rank 1 (there exists a *normal basis* for $\mathfrak{L}^\Lambda \mid K$). With slight adaptations of the definitions and the arguments from [25, Sect. 5: Appendix: On Galois algebras], we obtain

Lemma 4.14. *$\mathfrak{L} \mid K$ is a Galois Γ-algebra if and only if the following three conditions are satisfied.*

(1) *The commutative ring $\mathfrak{L}$ is* regular *(in the sense of von Neumann), so $\mathfrak{L}$ is canonically represented as a subdirect product of the family of the residue field extensions $(\mathfrak{L}/\mathbf{m})/K$, where $\mathbf{m}$ ranges over all prime (= maximal) ideals of $\mathfrak{L}$,*
(2) *$\mathfrak{L}^\Gamma = K$, and*
(3) *the continuous action of Γ on the profinite space* $\mathrm{Spec}(\mathfrak{L}) = \mathrm{Max}(\mathfrak{L})$, *canonically identified with the* Stone dual *of the boolean algebra $B(\mathfrak{L})$ of all idempotents of $\mathfrak{L}$, is transitive, and for some (for all) maximal ideal(s) $\mathbf{m}$, the stabilizer $\Gamma_{\mathbf{m}}$ acts faithfully on the residue field $\mathfrak{L}/\mathbf{m}$.*

For a maximal ideal $\mathbf{m}$ of $\mathfrak{L}$ (uniquely determined by a primitive idempotent provided Γ is finite), the residue field $L = \mathfrak{L}/\mathbf{m}$ is a Galois extension of K, whose Galois group $\mathrm{Gal}(L/K)$ is isomorphic with the stabilizer $\Gamma_{\mathbf{m}}$, the *decomposition group of $\mathfrak{L} \mid K$ associated with the maximal ideal* $\mathbf{m}$. Thus the Galois Γ-algebra $\mathfrak{L} \mid K$ determines a conjugation class of closed subgroups of Γ as its decomposition groups. As stressed in [25, A.7], the theory of Galois Γ-algebras over a field K is essentially the same as the theory of Galois field extensions L/K, together with an embedding $\mathrm{Gal}(L/K) \hookrightarrow \Gamma$. Identifying $\mathrm{Gal}(L/K)$ with a closed subgroup Γ' of Γ, the corresponding Galois Γ-algebra $\mathfrak{L}$ is up to isomorphism the *induced algebra* $M_\Gamma^{\Gamma'}(L)$ of all continuous (locally constant) functions $f : \Gamma \longrightarrow L$ satisfying $f(\sigma\gamma) = \sigma f(\gamma)$ for all $\sigma \in \Gamma', \gamma \in \Gamma$, with the ring operations pointwise induced from the field L, the base field K identified with the subfield of all constant functions, and the faithful action of Γ defined by $(\sigma f)(\gamma) = f(\gamma\sigma)$ for $f \in \mathfrak{L}$ and $\sigma, \gamma \in \Gamma$.

For every $\gamma \in \Gamma$, let $\varphi_\gamma \in \mathrm{Hom}_K(\mathfrak{L}, L)$ be the epimorphism defined by $\varphi_\gamma(f) = f(\gamma)$ for $f \in \mathfrak{L}$. The map $\varphi : \Gamma \longrightarrow \mathrm{Hom}_K(\mathfrak{L}, L)$, $\gamma \mapsto \varphi_\gamma$, is a homeomorphism, considering on $\mathrm{Hom}_K(\mathfrak{L}, L)$ the Zariski (= Stone) topology whose closed sets have the form $Z(I) := \{\psi \in \mathrm{Hom}_K(\mathfrak{L}, L) \mid \psi|_I = 0\}$, where I is an ideal of $\mathfrak{L}$; note that $Z(I) = Z(I \cap B(\mathfrak{L}))$. The canonical action from the right of Γ on itself induced by multiplication determines via the homeomorphism φ a continuous simple transitive action from the right of Γ on the profinite space $\mathrm{Hom}_K(\mathfrak{L}, L)$, defined by $(\psi\gamma)(f) = \psi(\gamma f)$ for $\gamma \in \Gamma, \psi \in \mathrm{Hom}_K(\mathfrak{L}, L), f \in \mathfrak{L}$. The canonical continuous projection $\Gamma \longrightarrow \mathrm{Max}(\mathfrak{L}),\ \gamma \mapsto \mathrm{Ker}\,(\varphi_{\gamma^{-1}}) = \gamma\,\mathbf{m}$, with $\mathbf{m} = \mathrm{Ker}\,(\varphi_1)$, induces an isomorphism of profinite Γ-spaces $\Gamma/\Gamma_{\mathbf{m}} \longrightarrow \mathrm{Max}(\mathfrak{L})$, with $\Gamma_{\mathbf{m}} = \Gamma' = \mathrm{Gal}(L/K)$.

By analogy with Example 4.11, call *coGalois group of the Galois Γ-algebra* $\mathfrak{L}\,|\,K$ the coGalois group $\mathrm{coG}(\mathfrak{L}^\times) := t(\mathfrak{L}^\times/K^\times)$ of the discrete Γ-module $\mathfrak{L}^\times = M_\Gamma^{\Gamma'}(L^\times)$ of units of the K-algebra $\mathfrak{L}$. Since $H^1(\Gamma, \mathfrak{L}^\times) = H^1(\Gamma, M_\Gamma^{\Gamma'}(L^\times)) \cong H^1(\Gamma', L^\times) = 0$ by Shapiro's lemma [28, Proposition 1.6.3] and Hilbert's Theorem 90, it follows by Lemma 4.1 that the discrete torsion Γ-module $\mathrm{coG}(\mathfrak{L}\,|\,K) := \mathrm{coG}(\mathfrak{L}^\times)$ is canonically isomorphic with $Z^1(\Gamma, \mu_{\mathfrak{L}})$, where $\mu_{\mathfrak{L}} := t(\mathfrak{L}^\times) = M_\Gamma^{\Gamma'}(\mu_L)$.

In abstract terms, we have arrived to a framework extending Example 4.10 as follows. Let Γ be a profinite group, Γ' a closed subgroup of Γ, A a discrete quasi-cyclic group identified with a subgroup of $\mathbb{Q}/\mathbb{Z}$, $A^\vee := \mathrm{Hom}(A, \mathbb{Q}/\mathbb{Z})$ its Pontryagin dual, and $\Gamma' \times A \longrightarrow A,\ (\sigma, a) \mapsto \sigma a := \chi(\sigma)a$, a continuous action given by a continuous homomorphism $\chi : \Gamma' \longrightarrow (A^\vee)^\times$. Let $\mathfrak{A} := M_\Gamma^{\Gamma'}(A)$ be the discrete torsion Γ-module of continuous functions $f : \Gamma \longrightarrow A$ satisfying $f(\sigma\gamma) = \chi(\sigma)\cdot f(\gamma)$ for $\sigma \in \Gamma', \gamma \in \Gamma$, with the action $\Gamma \times \mathfrak{A} \longrightarrow \mathfrak{A}$ defined by $(\sigma f)(\gamma) := f(\gamma\sigma)$ for $f \in \mathfrak{A}, \sigma, \gamma \in \Gamma$; in particular, the map $\mathfrak{A} \longrightarrow A,\ f \mapsto f(1)$ is an isomorphism provided $\Gamma = \Gamma'$. The object of investigation is the Galois connection between the lattices $\mathbb{L}(\Gamma)$ and $\mathbb{L}(Z^1(\Gamma, \mathfrak{A}))$ using the technic described in Sects. 4.2–4.4.

5 Kneser and Minimal Non-Kneser Triples

In this section we study a remarkable class of g-cocycles and associated structures, extending to the more general framework introduced in Sect. 3 the main results on *Kneser groups of cocycles* from [3].

5.1 Surjective Cocycles

Let Γ be a profinite group together with a continuous action on a profinite group $\mathfrak{G}$ and a continuous 1-cocycle $\eta : \Gamma \longrightarrow \mathfrak{G}$. Setting $\Delta := \mathrm{Ker}\,(\eta)$, $\mathfrak{E} := \mathfrak{G} \rtimes \Gamma$, $p : \mathfrak{E} \longrightarrow \Gamma$ the natural projection, $s_1 : \Gamma \longrightarrow \mathfrak{E}$ the canonical homomorphic section, $s_2 : \Gamma \longrightarrow \mathfrak{E}$ the homomorphic section induced by the cocycle η, and $\Gamma_i := s_i(\Gamma) \cong \Gamma,\ i = 1, 2$, we obtain

Lemma 5.1. *The following assertions are equivalent.*

(1) *The cocycle* $\eta : \Gamma \longrightarrow \mathfrak{G}$ *is surjective.*
(2) *For any open ideal* $\mathbf{a}$ *of* $\mathfrak{G}$*, the induced cocycle* $\eta_{\mathbf{a}} : \Gamma \longrightarrow \mathfrak{G}/\mathbf{a}$ *is surjective.*
(3) *The map* $\Gamma/\Delta \longrightarrow \mathfrak{G}$ *induced by* η *is a homeomorphism.*
(4) *The map* $\Gamma_1/(\Gamma_1 \cap \Gamma_2) \longrightarrow \mathfrak{E}/\Gamma_2$ *induced by the inclusion* $\Gamma_1 \subseteq \mathfrak{E}$ *is a homeomorphism.*
(5) *The map* $\Gamma_2/(\Gamma_1 \cap \Gamma_2) \longrightarrow \mathfrak{E}/\Gamma_1$ *induced by the inclusion* $\Gamma_2 \subseteq \mathfrak{E}$ *is a homeomorphism.*
(6) $\mathfrak{E} = \Gamma_1 \cdot \Gamma_2 = \{\gamma_1\gamma_2 \mid \gamma_i \in \Gamma_i, i = 1, 2\}$.
(7) $\mathfrak{E} = \Gamma_2 \cdot \Gamma_1$.
(8) *The map* $\Gamma \times \mathfrak{G} \longrightarrow \mathfrak{G}$, $(\gamma, g) \mapsto \eta(\gamma) \cdot \gamma g$, *is a transitive continuous action of* Γ *on the underlying profinite space of* $\mathfrak{G}$*, and the closed subgroup* $\Delta \subseteq \Gamma$ *is the stabilizer of the neutral element of* $\mathfrak{G}$.

Proof. It suffices to note that the isomorphisms $\Gamma \longrightarrow \Gamma_i$, $i = 1, 2$ induce the homeomorphisms of profinite spaces $\Gamma/\Delta \longrightarrow \Gamma_i/(\Gamma_1 \cap \Gamma_2)$, $i = 1, 2$, while the inclusion $\mathfrak{G} \subseteq \mathfrak{E}$ induces the homeomorphisms $\mathfrak{G} \longrightarrow \mathfrak{E}/\Gamma_i$, $i = 1, 2$. On the other hand, the continuous maps from (3) to (5) are obviously injective.

Corollary 5.2. *With notation above, assume that the* Γ*-group* $\mathfrak{G}$ *is finite, so* Δ *is open. The following assertions are equivalent.*

(1) *The cocycle* $\eta : \Gamma \longrightarrow \mathfrak{G}$ *is surjective.*
(2) $(\Gamma : \Delta) = (\Gamma_1 : \Gamma_1 \cap \Gamma_2) = (\Gamma_2 : \Gamma_1 \cap \Gamma_2) \geq |\mathfrak{G}| = (\mathfrak{E} : \Gamma_1) = (\mathfrak{E} : \Gamma_2)$.
(3) $(\Gamma : \Delta) = (\Gamma_1 : \Gamma_1 \cap \Gamma_2) = (\Gamma_2 : \Gamma_1 \cap \Gamma_2) = |\mathfrak{G}| = (\mathfrak{E} : \Gamma_1) = (\mathfrak{E} : \Gamma_2)$.

Definition 5.3. With notation above, we call $(\Gamma, \mathfrak{G}, \eta)$ a *Kneser triple* if the continuous cocycle $\eta : \Gamma \longrightarrow \mathfrak{G}$ is surjective.

Denote by $\mathscr{K}\mathscr{Z}^1$ the full subcategory of the category $\mathscr{Z}^1$ defined in Sect. 3.1, having the Kneser triples as objects.

Remark 5.4. According to Lemma 5.1,(8), the category $\mathscr{K}\mathscr{Z}^1$ is equivalent with the category of the systems $(\Gamma, \Delta, \bullet)$, termed *Kneser structures*, consisting of a profinite group Γ, a closed subgroup Δ, and a continuous group operation $\bullet$ on the profinite space $X := \Gamma/\Delta = \{\hat{\gamma} := \gamma\Delta \mid \gamma \in \Gamma\}$, with $\hat{1}$ as neutral element, such that the canonical transitive action of Γ on X and the group operation $\bullet$ are related through the following condition

$$\gamma \cdot (x \bullet y) = (\gamma \cdot x) \bullet I(\hat{\gamma}) \bullet (\gamma \cdot y) \text{ for } \gamma \in \Gamma, x, y \in X,$$

where $I(x)$ denotes the inverse of $x \in X$ with respect to the group operation $\bullet$; in particular, $\gamma \cdot I(x) = \hat{\gamma} \bullet I(\gamma \cdot x) \bullet \hat{\gamma}$ for $\gamma \in \Gamma, x \in X$, and the restricted action of Δ on X is compatible with the group operation $\bullet$. A morphism $(\Gamma, \Delta, \bullet) \longrightarrow (\Gamma', \Delta', \bullet)$ is a continuous homomorphism $\varphi : \Gamma \longrightarrow \Gamma'$ inducing a homomorphism of profinite groups $\tilde{\varphi} : (\Gamma/\Delta, \bullet) \longrightarrow (\Gamma'/\Delta', \bullet')$.

5.1.1 Surjectivity Criteria for Cocycles

The next result, extending [3, Proposition 1.14, Corollary 1.16, Corollary 1.17], provides a criterion for cocycles taking values in pronilpotent groups to be surjective.

Proposition 5.5. *Let $\eta \in Z^1(\Gamma, \mathfrak{G})$, where $\mathfrak{G} \cong \prod_p \mathfrak{G}_p$ is a pronilpotent Γ-group, and $\mathfrak{G}_p$ denotes the maximal pro-p-quotient of $\mathfrak{G}$ for any prime p, with the induced action of Γ. Then the cocycle $\eta : \Gamma \longrightarrow \mathfrak{G}$ is surjective if and only if the induced cocycle $\eta_p : \Gamma \longrightarrow \mathfrak{G}_p$ is surjective for all prime numbers p.*

The criterion above is a consequence of Lemma 5.1 (2) and the next lemma.

Lemma 5.6. *Let $\eta \in Z^1(\Gamma, \mathfrak{G})$, where $\mathfrak{G} = \prod_{i=1}^{n} \mathfrak{G}_i$ is a direct product of finite Γ-groups such that $\gcd(|\mathfrak{G}_i|, |\mathfrak{G}_j|) = 1$ for $i \neq j$. Then $\eta : \Gamma \longrightarrow \mathfrak{G}$ is surjective if and only if the induced cocycle $\eta_i : \Gamma \longrightarrow \mathfrak{G}_i$ is surjective for $i = 1, \ldots, n$.*

Proof. An implication is obvious. Conversely, assuming that $\eta_i : \Gamma \longrightarrow \mathfrak{G}_i$ is surjective for $i = 1, \ldots, n$, we obtain $|\mathfrak{G}| = \prod_{i=1}^{n} |\mathfrak{G}_i| = \prod_{i=1}^{n} (\Gamma : \mathrm{Ker}\,(\eta_i))$ by Corollary 5.2,(3). As $\mathrm{Ker}\,(\eta) \subseteq \mathrm{Ker}\,(\eta_i)$, we get $(\Gamma : \mathrm{Ker}\,(\eta_i)) \mid (\Gamma : \mathrm{Ker}\,(\eta))$, for $i = 1, \ldots, n$. Since $(\Gamma : \mathrm{Ker}\,(\eta_i)) = |\mathfrak{G}_i|$, $i = 1, \ldots, n$, are mutually relatively prime by hypothesis, it follows that $\prod_{i=1}^{n} (\Gamma : \mathrm{Ker}\,(\eta_i)) \mid (\Gamma : \mathrm{Ker}\,(\eta))$, whence $|\mathfrak{G}| \mid (\Gamma : \mathrm{Ker}\,(\eta))$. Consequently, the cocycle η is surjective by Corollary 5.2,(2).

Another surjectivity criterion for g-cocycles is given by the next lemma.

Lemma 5.7. *Given a g-cocycle $\eta : \Gamma \longrightarrow \mathfrak{G}$, the following assertions are equivalent.*

(1) *$\eta : \Gamma \longrightarrow \mathfrak{G}$ is surjective.*
(2) *There exists a closed subgroup $\Lambda \subseteq \Gamma$ such that $\mathbf{a} := \eta(\Lambda)$ is an ideal of $\mathfrak{G}$, and the induced cocycle $\eta_{\mathbf{a}} : \Gamma \longrightarrow \mathfrak{G}/\mathbf{a}$ is surjective.*

Proof. The implication (1)$\Longrightarrow$(2) is obvious. Conversely, assume $\Lambda \in \mathbb{L}(\Gamma)$ satisfies (2), with $\mathbf{a} = \eta(\Lambda) \in \mathbb{L}(\mathfrak{G})$, and let $g \in \mathfrak{G}$. As $\eta_{\mathbf{a}} : \Gamma \longrightarrow \mathfrak{G}/\mathbf{a}$ is surjective by assumption, there exists $\sigma \in \Gamma$ such that $\eta(\sigma)^{-1} g \in \mathbf{a}$, and hence $\eta(\sigma^{-1}) \cdot (\sigma^{-1} g) = \sigma^{-1}(\eta(\sigma)^{-1} g) \in \mathbf{a}$ since $\mathbf{a}$ is Γ-invariant. On the other hand, as $\mathbf{a} = \eta(\Lambda)$, we obtain $\eta(\sigma^{-1}) \cdot (\sigma^{-1} g) = \eta(\tau)$ for some $\tau \in \Lambda$. Consequently, acting with σ, it follows that $g = \eta(\sigma) \cdot (\sigma\eta(\tau)) = \eta(\sigma\tau)$, so $\eta(\Gamma) = \mathfrak{G}$ as desired.

Given a g-cocycle $\eta : \Gamma \longrightarrow \mathfrak{G}$, let $\overline{\Delta} := \Delta' \cap \Delta''$, where $\Delta' := \mathrm{Fix}_{\Gamma}(\mathfrak{G})$, $\Delta'' := \{\sigma \in \Gamma \mid \forall \gamma \in \Gamma, \eta(\gamma\sigma\gamma^{-1}) = \gamma\eta(\sigma)\}$. According to Lemma 3.8, $\overline{\Delta}$ is a closed normal subgroup of Γ, and $\eta(\overline{\Delta})$ is a central ideal of $\mathfrak{G}$. Applying Lemma 5.7 to $\mathbf{a} := \eta(\overline{\Delta})$, we obtain

Corollary 5.8. *With data above, the g-cocycle $\eta : \Gamma \longrightarrow \mathfrak{G}$ is surjective provided the induced cocycle $\eta_{\mathbf{a}} : \Gamma \longrightarrow \mathfrak{G}/\mathbf{a}$ is surjective.*

Remark 5.9. Though not stated explicitly, a particular form of Corollary 5.8 plays implicitly a key role in the proof of [3, Lemma 1.18, Theorem 1.20].

5.1.2 Bijective Cocycles Induced by Self-Actions

We describe a procedure for obtaining Kneser triples $(\Gamma, \mathfrak{G}, \eta)$ where the cocycle η is bijective. Let Γ be a profinite group, and $\omega : \Gamma \longrightarrow \mathrm{Aut}(\Gamma)$ a continuous action of Γ on itself. Setting $\tilde{\Gamma} := \Gamma \rtimes \omega(\Gamma)$, $\tilde{\Gamma}$ becomes a profinite Γ-group via the canonical continuous action

$$\Gamma \times \tilde{\Gamma} \longrightarrow \tilde{\Gamma},\ (\gamma, (\delta, \theta)) \mapsto (\omega(\gamma)(\delta),\ \omega(\gamma) \circ \theta \circ \omega(\gamma)^{-1}),$$

and the continuous map $\eta_\omega : \Gamma \longrightarrow \tilde{\Gamma},\ \gamma \mapsto (\gamma, \omega(\gamma)^{-1})$ is an injective cocycle.

Definition 5.10. The continuous self-action $\omega : \Gamma \longrightarrow \mathrm{Aut}(\Gamma)$ is said to be *adequate* if the image $\eta_\omega(\Gamma)$ is a (closed) subgroup of $\tilde{\Gamma}$, i.e., $\omega(\theta(\gamma)) = \theta \circ \omega(\gamma) \circ \theta^{-1}$ for all $\gamma \in \Gamma, \theta \in \omega(\Gamma)$; we say that the profinite Γ-group $\Gamma_\omega := \eta_\omega(\Gamma)$ is a *deformation of* Γ via the adequate self-action ω.

In other words, the self-action $\omega : \Gamma \longrightarrow \mathrm{Aut}(\Gamma)$ is adequate if and only if the continuous binary operation on Γ, defined by $\gamma \bullet \delta = \gamma \cdot \omega(\gamma)^{-1}(\delta)$, is associative; in this case, $(\Gamma, \bullet)$ is a profinite Γ-group isomorphic to Γ_ω, and the identity $(\Gamma, \cdot) \longrightarrow (\Gamma, \bullet)$ is a cocycle, i.e., $\gamma \cdot \delta = \gamma \bullet \omega(\gamma)(\delta)$ for $\gamma, \delta \in \Gamma$.

For any adequate self-action $\omega : \Gamma \longrightarrow \mathrm{Aut}(\Gamma)$, $(\Gamma, \Gamma_\omega, \eta_\omega : \Gamma \longrightarrow \Gamma_\omega)$ is a Kneser triple with a bijective cocycle η_ω. Note that $\mathrm{Ker}\,(\omega)$ is identified with a common closed normal subgroup of Γ and Γ_ω on which the cocycle η_ω is the identity. Moreover Γ becomes a profinite Γ_ω-group via the continuous action

$$\Gamma_\omega \times \Gamma \longrightarrow \Gamma,\ ((\gamma, \omega(\gamma)^{-1}), \delta) \mapsto \omega(\gamma)^{-1}(\delta),$$

whose kernel is identified with $\mathrm{Ker}\,(\omega)$, and $(\Gamma_\omega, \Gamma, \eta_\omega^{-1} : \Gamma_\omega \longrightarrow \Gamma)$ is also a Kneser triple with a bijective cocycle η_ω^{-1}. We obtain in this way a symmetric relation on profinite groups, weaker than the isomorphism-relation and finer than the cardinality equivalence, defined as follows.

Definition 5.11. For profinite groups Γ and Λ, (Γ, Λ) is a *deformation pair* if the following equivalent conditions are satisfied.

(1) There exists an adequate self-action $\omega : \Gamma \longrightarrow \mathrm{Aut}(\Gamma)$ such that $\Lambda \cong \Gamma_\omega$.
(2) There exist continuous actions $\Gamma \times \Lambda \longrightarrow \Lambda$, $\Lambda \times \Gamma \longrightarrow \Gamma$, and a continuous bijective cocycle $\eta \in Z^1(\Gamma, \Lambda)$ such that $\eta^{-1} \in Z^1(\Lambda, \Gamma)$.

Problem 5.12. Is the deformation relation transitive, i.e., an equivalence relation?

A profinite group Γ is termed *rigid*, if $\Gamma \cong \Lambda$ whenever (Γ, Λ) is a deformation pair. Among the rigid profinite groups, we mention the pro-p-cyclic groups, where p is an odd prime number, and the finite simple groups. By contrast, the pro-2-cyclic groups of order ≥ 4 are not rigid; indeed, for $\Gamma = \mathbb{Z}/2^n\mathbb{Z}$, with $n \geq 2$, $(\Gamma, \mathbb{D}_{2^n} = \mathbb{Z}/2^{n-1}\mathbb{Z} \rtimes \mathbb{Z}/2\mathbb{Z})$ is a deformation pair, while $\mathbb{D}_{2\infty} = \mathbb{Z}_2 \rtimes \mathbb{Z}/2\mathbb{Z}$ is, up to isomorphism, the unique profinite group $\Lambda \ncong \mathbb{Z}_2$ such that $(\mathbb{Z}_2, \Lambda)$ is a deformation pair (see also Example 5.15).

Note that not all bijective cocycles are induced by adequate self-actions.

5.1.3 Examples

Example 5.13. Let K be a local field with discrete valuation v, valuation ring $\mathcal{O}$, maximal ideal $\mathfrak{p}$, finite residue field $k := \mathcal{O}/\mathfrak{p}$, and $U^{(n)} := 1 + \mathfrak{p}^n$, the group of n-units of K for some natural number $n \geq 1$. Fix a local uniformizer π, and consider the faithful continuous action by multiplication of $U^{(n)}$ on $\mathcal{O}^+$. The map $Z^1(U^{(n)}, \mathcal{O}^+) \longrightarrow \mathcal{O}^+$, $\eta \mapsto \eta(1 + \pi^n)$ is an isomorphism whose inverse sends an element $a \in \mathcal{O}$ to the cocycle η_a defined by $\eta_a(x) := \pi^{-n}(x-1)a$. The isomorphism above maps the subgroup $B^1(U^{(n)}, \mathcal{O}^+)$ onto $\mathfrak{p}^n$ inducing an isomorphism $H^1(U^{(n)}, \mathcal{O}^+) \cong \mathcal{O}/\mathfrak{p}^n$. For $a \neq 0$, η_a is injective, while η_a is a g-cocycle if and only if it is bijective, i.e., $a \in \mathcal{O}^\times$; one checks easily that for all $a \in O^\times$, the bijective cocycle η_a is not induced by an adequate self-action. Note that two cocycles η_a and η_b are cohomologous if and only if $a \equiv b \bmod \mathfrak{p}^n$, while they are isomorphic as objects in the category $\mathcal{Z}^1$ if and only if $a\mathcal{O} = b\mathcal{O}$, i.e., $v(a) = v(b)$. In particular, $\eta_a \cong \eta_1$ for all $a \in \mathcal{O}^\times$, with the associated Kneser structure $(U^{(n)}, 1, \bullet)$, with $x \bullet y = x + y - 1$, neutral element 1, and inverse $I(x) = 2 - x$. For any natural number $m \geq 1$, the bijective cocycle $\eta := \eta_1$ induces a bijective cocycle $\eta^{(m)} \in Z^1(U^{(n)}/U^{(n+m)}, \mathcal{O}/\mathfrak{p}^m)$ and $\eta = \varprojlim \eta^{(m)}$.

Example 5.14. This example plays a key role in the proof of Proposition 7.8. With the data from Example 5.13, let $m > n \geq 1$, and consider the continuous action of $U^{(n)}$ on the quotient abelian group $\mathcal{O}_m := (\mathcal{O}/\mathfrak{p}^m)^+$ with $\mathrm{Fix}_{U^{(n)}}(\mathcal{O}_m) = U^{(m)}$ and $H^0(U^{(n)}, \mathcal{O}_m) = \mathfrak{p}^{m-n}\mathcal{O}_m$. For any $a \in \mathcal{O}_m$, put $v_m(a) := \max\{i \in \{0, 1, \dots, m\} \mid a \in \mathfrak{p}^i\mathcal{O}_m\}$, so $\mathfrak{p}^i\mathcal{O}_m = \pi^i\mathcal{O}_m = \{a \in \mathcal{O}_m \mid v_m(a) \geq i\}$ for $i = 0, 1, \dots, m$. Denote by V the closed subgroup of $U^{(n)}$ generated by $1 + \pi^n$. We obtain an isomorphism

$$Z^1(U^{(n)}, \mathcal{O}_m) \longrightarrow \mathcal{O}_m \times \mathrm{Hom}(U^{(n)}/V, \mathfrak{p}^{m-n}\mathcal{O}_m)$$

sending a cocycle η to the pair $(\eta(1 + \pi^n), \tilde{\eta})$, where the continuous homomorphism $\tilde{\eta}$ is defined by $\tilde{\eta}(x) = \eta(x) - \pi^{-n}(x-1)\eta(1 + \pi^n)$ for all $x \in U^{(n)}$. The isomorphism above maps $B^1(U^{(n)}, \mathcal{O}_m)$ onto $\mathfrak{p}^n\mathcal{O}_m$ inducing an isomorphism

$$\begin{aligned} H^1(U^{(n)}, \mathcal{O}_m) &\cong \mathcal{O}_n \times \mathrm{Hom}(U^{(n)}/V, \mathfrak{p}^{m-n}\mathcal{O}_m) \\ &= H^1(U^{(n)}, \mathcal{O}_n) \oplus \mathrm{Hom}(U^{(n)}/V, H^0(U^{(n)}, \mathcal{O}_m)). \end{aligned}$$

For $a \in \mathcal{O}_m, \alpha \in \mathrm{Hom}(U^{(n)}/V, \mathfrak{p}^{m-n}\mathcal{O}_m)$, let $\eta_{a,\alpha}$ denote the unique cocycle η satisfying $\eta(1+\pi^n) = a, \tilde{\eta} = \alpha$, i.e., $\eta_{a,\alpha}(x) = \alpha(x) + \pi^{-n}(x-1)a$ for all $x \in U^{(n)}$. It follows that

$$\mathrm{Ker}\,(\eta_{a,\alpha}) = \mathrm{Eq}(\alpha, \eta_{-a,0}) = \{x \in U^{(n)} \mid \alpha(x) = \pi^{-n}(1-x)a\} \subseteq U^{(r(a))},$$

where $r(a) := \max(n, m - v_m(a))$. Note that $U^{(r(a))} = \eta_{a,\alpha}^{-1}(\mathfrak{p}^{m-n}\mathcal{O}_m)$ is the maximal open subgroup H of $U^{(n)}$ for which the restriction map $\eta_{a,\alpha}|_H$ is a homomorphism, so a cocycle η is a homomorphism if and only if $\eta(1+\pi^n) \in \mathfrak{p}^{m-n}\mathcal{O}_m$, and $\mathrm{Ker}\,(\eta_{a,\alpha}) = \mathrm{Ker}\,(\eta_{a,\alpha}|_{U^{(r(a))}})$.

Example 5.15. Let $\Gamma = (\mathbb{Z}/n\mathbb{Z}, +)$ be a finite cyclic group of order $n \geq 2$. As $\mathrm{Aut}(\Gamma) \cong (\mathbb{Z}/n\mathbb{Z})^\times$, the self-actions of Γ on itself are identified through the map $u \mapsto \omega_u : \Gamma \longrightarrow \mathrm{Aut}(\Gamma)$, where $\omega_u(x)(y) = u^x \cdot y$, with the multiplicative subgroup $\mathcal{U}_n$ of $(\mathbb{Z}/n\mathbb{Z})^\times$ consisting of those units $u \in (\mathbb{Z}/n\mathbb{Z})^\times$ satisfying $u^n \equiv 1 \bmod n$. The adequate self-actions ω_u are in 1-1 correspondence to those $u \in \mathcal{U}_n$ satisfying the supplementary condition $u^{u-1} \equiv 1 \bmod n$. Thus the adequate self-actions of $\mathbb{Z}/n\mathbb{Z}$ are identified with the set $\mathcal{U}_n^{\mathrm{ad}} = \{u \in (\mathbb{Z}/n\mathbb{Z})^\times \mid u^{r_u} \equiv 1 \bmod n\}$, where $r_u := (n, u-1)$ for $u \in (\mathbb{Z}/n\mathbb{Z})^\times$. We obtain a more explicit description of the set $\mathcal{U}_n^{\mathrm{ad}}$ as follows.

Lemma 5.16. *Let $\mathcal{P}_n$ be the set of odd prime divisors of n, to which we add the prime number 2 provided $4 \mid n$. For $u \in (\mathbb{Z}/n\mathbb{Z})^\times$ and $p \in \mathcal{P}_n$, put $u_p := u \bmod p^{v_p(n)}$, and denote by $o(u_p)$ the order of u_p. Then $u \in \mathcal{U}_n^{\mathrm{ad}}$ if, and only if, the following conditions are satisfied.*

(i) $r_u \neq 1$, *i.e.*, $u - 1 \notin (\mathbb{Z}/n\mathbb{Z})^\times$.
(ii) $p \in \mathcal{P}_{r_u} \Longrightarrow v_p(n) \leq 2v_p(r_u)$.
(iii) $2 \neq p \in \mathcal{P}_n \setminus \mathcal{P}_{r_u} \Longrightarrow 2 \leq o(u_p) \mid (r_u, p-1)$.
(iv) $2 \in \mathcal{P}_n \setminus \mathcal{P}_{r_u} \Longrightarrow u \equiv -1 \bmod 2^{v_2(n)-1}$, *in particular,* $o(u_2) = 2$.

Lemma 5.16 remains valid for arbitrary procyclic groups of order n -a supernatural number. In particular, $\mathcal{U}_{p^\infty}^{\mathrm{ad}} = \mathcal{U}^{\mathrm{ad}}((\mathbb{Z}_p)^\times) = \{1\}$ for $p \neq 2$, while $\mathcal{U}_{2^\infty}^{\mathrm{ad}} = \mathcal{U}^{\mathrm{ad}}((\mathbb{Z}_2)^\times) = \{1, -1\}$; in the latter case, for $\Gamma = \mathbb{Z}_2, u = -1$, we obtain $\Gamma_{-1} \cong \mathbb{Z}_2 \rtimes \mathbb{Z}/2\mathbb{Z} \cong \mathbb{Z}/2\mathbb{Z} \,\hat{*}_2\, \mathbb{Z}/2\mathbb{Z}$ (the free product of two copies of the cyclic group of order 2 in the category of pro-2-groups),[2] the coGalois triple[3] $(\Gamma, \Gamma_{-1}, \eta_{-1})$, and the Kneser, but not coGalois, triple $(\Gamma_{-1}, \Gamma, \eta_{-1}^{-1})$.

---

[2]The procedure described in Sect. 5.1.2 works for arbitrary topological groups, in particular, for discrete ones. For $\Gamma = (\mathbb{Z}, +)$, we obtain $\mathcal{U}^{\mathrm{ad}} = \mathbb{Z}^\times = \{\pm 1\}$, and $\Gamma_{-1} = \mathbb{D}_\infty = \mathbb{Z} \rtimes \mathbb{Z}/2\mathbb{Z} \cong (\mathbb{Z}/2\mathbb{Z}) * (\mathbb{Z}/2\mathbb{Z})$.

[3]see Definition 6.1.

For $n = 4, \Gamma = \mathbb{Z}/4\mathbb{Z}$, $\mathcal{U}_4^{\mathrm{ad}} = (\mathbb{Z}/4\mathbb{Z})^\times \cong \mathbb{Z}/2\mathbb{Z}$. Setting $\hat{i} := i \bmod 4$, we obtain $\Gamma_{\hat{1}} \cong \Gamma$, $\Gamma_{\hat{3}} \cong \mathbb{Z}/2\mathbb{Z} \times \mathbb{Z}/2\mathbb{Z}$, and $(\Gamma, \Gamma_{\hat{3}}, \eta_{\hat{3}})$ is a coGalois^3 triple, while $(\Gamma_{\hat{3}}, \Gamma, \eta_{\hat{3}}^{-1})$ is a Kneser minimal non-coGalois triple.[4]

For $n = 8, \Gamma = \mathbb{Z}/8\mathbb{Z}$, $\mathcal{U}_8^{\mathrm{ad}} = (\mathbb{Z}/8\mathbb{Z})^\times \cong \mathbb{Z}/2\mathbb{Z} \times \mathbb{Z}/2\mathbb{Z}$. Setting $\hat{i} := i \bmod 8$, it follows that $\Gamma = \Gamma_{\hat{1}} \cong \Gamma_{\hat{5}}$, while $\Gamma_{\hat{7}} \cong \mathbb{D}_8 \cong \mathbb{Z}/4\mathbb{Z} \rtimes \mathbb{Z}/2\mathbb{Z}$, and $\Gamma_{\hat{3}} \cong Q$ (the quaternion group). The triples $(\Gamma, \Gamma_{\hat{7}}, \eta_{\hat{7}})$ and $(\Gamma, \Gamma_{\hat{3}}, \eta_{\hat{3}})$ are coGalois^3, but the Kneser triples $(\Gamma_{\hat{7}}, \Gamma, \eta_{\hat{7}}^{-1})$ and $(\Gamma_{\hat{3}}, \Gamma, \eta_{\hat{3}}^{-1})$ are not. Note also that $(\mathbb{D}_8, Q)$ is a deformation pair [see Remarks 6.9,(3)].

5.2 Kneser Ideals

Let Γ be a profinite group, $\mathfrak{G}$ a profinite Γ-group, and $\eta : \Gamma \longrightarrow \mathfrak{G}$ a continuous g-cocycle, with $\Delta := \mathrm{Ker}(\eta)$, $\Delta' := \mathrm{Fix}_\Gamma(\mathfrak{G})$. We may assume without loss that the triple $(\Gamma, \mathfrak{G}, \eta)$ is normalized, i.e., $\Delta \cap \Delta' = \{1\}$. Let $\mathbb{L}(\Gamma \mid \Delta)$, $\mathbb{L}(\mathfrak{G})$ be the lattices connected through the operators $\mathcal{J} : \mathbb{L}(\Gamma \mid \Delta) \longrightarrow \mathbb{L}(\mathfrak{G})$, $\mathcal{S} : \mathbb{L}(\mathfrak{G}) \longrightarrow \mathbb{L}(\Gamma \mid \Delta)$ defined in Sect. 3.2.

Definition 5.17. $\mathbf{a} \in \mathbb{L}(\mathfrak{G})$ is called a *Kneser ideal* (with respect to η) if the cocycle $\eta_{\mathbf{a}} : \Gamma \longrightarrow \mathfrak{G}/\mathbf{a}$ induced by η is surjective, while $\mathfrak{G}/\mathbf{a}$ is called a *Kneser quotient* (with respect to η) of the profinite Γ-group $\mathfrak{G}$.

We denote by $\mathbb{K}(\mathfrak{G})$ the set of all Kneser ideals of $\mathfrak{G}$, partially ordered by inclusion, with $\mathfrak{G}$ as the last element, canonically anti-isomorphic to the poset of all Kneser quotients of $\mathfrak{G}$. Note that $\mathbb{K}(\mathfrak{G})$ is an upper subset of the lattice $\mathbb{L}(\mathfrak{G})$, i.e., $\forall \mathbf{a} \in \mathbb{K}(\mathfrak{G})$, $\forall \mathbf{b} \in \mathbb{L}(\mathfrak{G})$, $\mathbf{a} \subseteq \mathbf{b} \Longrightarrow \mathbf{b} \in \mathbb{K}(\mathfrak{G})$.

Lemma 5.18. *If $(\Gamma, \mathfrak{G}, \eta)$ is a Kneser triple then $\mathbb{K}(\mathfrak{G}) = \mathbb{L}(\mathfrak{G})$, and the continuous map $\mathcal{S} : \mathbb{L}(\mathfrak{G}) \longrightarrow \mathbb{L}(\Gamma \mid \Delta)$ is a section of the coherent map $\mathcal{J} : \mathbb{L}(\Gamma \mid \Delta) \longrightarrow \mathbb{L}(\mathfrak{G})$.*

Proof. By assumption $\eta : \Gamma \longrightarrow \mathfrak{G}$ is surjective, so $\{1\} \in \mathbb{K}(\mathfrak{G})$, therefore $\mathbb{K}(\mathfrak{G}) = \mathbb{L}(\mathfrak{G})$ since $\mathbb{K}(\mathfrak{G})$ is an upper subset of $\mathbb{L}(\mathfrak{G})$. Let $\mathbf{a} \in \mathbb{L}(\mathfrak{G})$ and put $\Lambda := \mathcal{S}(\mathbf{a}) \in \mathbb{L}(\Gamma \mid \Delta)$, $\mathbf{b} := \mathcal{J}(\Lambda) \subseteq \mathbf{a}$. As $\mathcal{S} \circ \mathcal{J} \circ \mathcal{S} = \mathcal{S}$, we obtain $\mathcal{S}(\mathbf{b}) = \Lambda$. On the other hand, the canonical maps $\eta'_{\mathbf{a}} : \Gamma/\Lambda \longrightarrow \mathfrak{G}/\mathbf{a}$ and $\eta'_{\mathbf{b}} : \Gamma/\Lambda \longrightarrow \mathfrak{G}/\mathbf{b}$ induced by η are bijective. Since $\eta'_{\mathbf{a}}$ is the composition of $\eta'_{\mathbf{b}}$ with the natural projection $\mathfrak{G}/\mathbf{b} \longrightarrow \mathfrak{G}/\mathbf{a}$, it follows that $\mathbf{b} = \mathbf{a}$, hence $\mathcal{J} \circ \mathcal{S} = 1_{\mathbb{L}(\mathfrak{G})}$ as desired.

[4] see Definition 6.7 and Example 5.15 (i).

Proposition 5.19. *Given a triple* $(\Gamma, \mathfrak{G}, \eta)$, *where* $\eta : \Gamma \longrightarrow \mathfrak{G}$ *is a g-cocycle, the following assertions hold.*

(1) *Let* $\mathbf{a} \in \mathbb{L}(\mathfrak{G})$. *Then* $\mathbf{a} \in \mathbb{K}(\mathfrak{G})$ *if and only if* $(\Gamma : \mathscr{S}(\mathbf{b})) = (\mathfrak{G} : \mathbf{b})$ *for all* open *ideals* $\mathbf{b} \in \mathbb{L}(\mathfrak{G} \mid \mathbf{a})$.
(2) $\mathbb{K}(\mathfrak{G})$ *is a closed subspace of the spectral space* $\mathbb{L}(\mathfrak{G})$.
(3) *For every Kneser ideal* $\mathbf{a}$, *the set of minimal Kneser ideals contained in* $\mathbf{a}$ *is nonempty.*
(4) *The subspace* $\mathbb{K}(\mathfrak{G})_{\min}$ *of the spectral space* $\mathbb{K}(\mathfrak{G})$ *consisting of all minimal Kneser ideals of* $\mathfrak{G}$ *is Hausdorff.*

Proof. (1) follows by Lemma 5.1,(2) and Corollary 5.2 applied to the surjective cocycle $\eta_{\mathbf{a}} : \Gamma \longrightarrow \mathfrak{G}/\mathbf{a}$ for any $\mathbf{a} \in \mathbb{K}(\mathfrak{G})$.
(2) Let $\mathbf{a} \in \mathbb{L}(\mathfrak{G}) \setminus \mathbb{K}(\mathfrak{G})$. By (1) there exists an open ideal $\mathbf{b} \in \mathbb{L}(\mathfrak{G} \mid \mathbf{a})$ such that $\mathbf{b} \notin \mathbb{K}(\mathfrak{G})$. It follows that $\mathbb{L}(\mathbf{b})$ is an open neighborhood of $\mathbf{a}$ and $\mathbb{L}(\mathbf{b}) \cap \mathbb{K}(\mathfrak{G}) = \varnothing$, so $\mathbb{K}(\mathfrak{G})$ is closed in the spectral space $\mathbb{L}(\mathfrak{G})$; in particular, $\mathbb{K}(\mathfrak{G})$ is a spectral space with respect to the induced topology.
(3) follows by (1) and Zorn's lemma.
(4) Let $\mathbf{a}_i \in \mathbb{K}(\mathfrak{G})_{\min}$, $i = 1, 2$, be such that $\mathbf{a}_1 \neq \mathbf{a}_2$. Consequently, there exist open ideals $\mathbf{b}_i \in \mathbb{L}(\mathfrak{G} \mid \mathbf{a}_i)$, $i = 1, 2$, such that $\mathbf{b}_1 \cap \mathbf{b}_2 \notin \mathbb{K}(\mathfrak{G})$, since otherwise it would follow by compactness that each open ideal lying over $\mathbf{a}_1 \cap \mathbf{a}_2$ is Kneser, whence $\mathbf{a}_1 \cap \mathbf{a}_2 \in \mathscr{K}(G)$ by (1), contrary to the minimality condition satisfied by the distinct Kneser ideals $\mathbf{a}_1$ and $\mathbf{a}_2$. For such a pair $(\mathbf{b}_1, \mathbf{b}_2)$, $\mathbb{L}(\mathbf{b}_i)$ is an open neighborhood of $\mathbf{a}_i$ for $i = 1, 2$, and $\mathbb{L}(\mathbf{b}_1) \cap \mathbb{L}(\mathbf{b}_2) \cap \mathbb{K}(\mathfrak{G}) = \varnothing$, so $\mathbb{K}(\mathfrak{G})_{\min}$ is a Hausdorff space with respect to the topology induced from the spectral space $\mathbb{K}(\mathfrak{G})$.

As shown in Remark 3.7, the continuous map $\mathscr{S} : \mathbb{L}(\mathfrak{G}) \longrightarrow \mathbb{L}(\Gamma \mid \Delta)$ is not necessarily a coherent map. Related to this fact, the following question arises naturally.

Problem 5.20. Is the continuous restriction map

$$\mathscr{S}\,|_{\mathbb{K}(\mathfrak{G})} : \mathbb{K}(\mathfrak{G}) \longrightarrow \mathbb{L}(\Gamma \mid \Delta), \mathbf{a} \mapsto \eta^{-1}(\mathbf{a}),$$

necessarily coherent?

A positive answer to the question above is given in [11, Theorem 2.13] in the particular framework of cyclotomic abstract coGalois theory presented in Example 4.10.

Remarks 5.21. (1) Let $\mathbb{K}(\Gamma)$ denote the image of the restriction map $\mathscr{S}\,|_{\mathbb{K}(\mathfrak{G})} : \mathbb{K}(\mathfrak{G}) \longrightarrow \mathbb{L}(\Gamma \mid \Delta)$, and call its members *Kneser subgroups* of the profinite group Γ. The map $\mathscr{S}\,|_{\mathbb{K}(\mathfrak{G})}$ is not necessarily injective; moreover $\mathscr{S}(\mathbf{a}) = \mathscr{S}(\mathbf{b})$, with $\mathbf{a}, \mathbf{b} \in \mathbb{K}(\mathfrak{G})$, does not imply $\mathbf{a} \cong \mathbf{b}$ as profinite groups (see [11, Remarks 2.6 (1)] for a simple example). Note also that, in general, $\mathbb{K}(\Gamma)$ is not an upper subset of $\mathbb{L}(\Gamma \mid \Delta)$, and hence not necessarily a closed subset of the spectral space $\mathbb{L}(\Gamma \mid \Delta)$ (for an example see [11, Remarks 2.6 (2)]);

however $\mathbb{K}(\Gamma)$ is closed with respect to the profinite topology as image through a continuous map of $\mathbb{K}(\mathfrak{G})$ with its profinite topology.

(2) Let $\mathbb{HK}(\Gamma)$ denote the subset of $\mathbb{K}(\Gamma)$ consisting of those $\Lambda \in \mathbb{L}(\Gamma \mid \Delta)$ for which $\mathbb{L}(\Gamma \mid \Lambda) \subseteq \mathbb{K}(\Gamma)$, termed *hereditarily Kneser subgroups* of Γ. In the particular framework of cyclotomic abstract coGalois theory, $\mathbb{HK}(\Gamma)$ is a closed subspace of the spectral space $\mathbb{L}(\Gamma \mid \Delta)$ [11, Corollary 2.15], and an explicit *hereditarily-Kneser criterion* for closed subgroups of Γ is provided by [11, Lemma 3.1, Theorem 3.2]; field theoretic interpretations are provided by [11, Remark 2.16].

5.3 A General Kneser Criterion

We end this section with a general omitting-type criterion for an ideal to be Kneser and with an open problem concerning the classification of certain finite algebraic structures arising from this criterion.

Given a g-cocycle $\eta \in Z^1(\Gamma, \mathfrak{G})$, we denote by $\mathbb{NK}(\mathfrak{G})$ the complementary set of $\mathbb{K}(\mathfrak{G})$ in the set $\mathbb{L}(\mathfrak{G})$ of all ideals of the profinite Γ-group $\mathfrak{G}$. By Proposition 5.19,(2), $\mathbb{NK}(\mathfrak{G})$ is an open and hence a lower subset of the spectral space $\mathbb{L}(\mathfrak{G})$. Note that $\mathbb{NK}(\mathfrak{G}) \neq \varnothing \iff \eta(\Gamma) \neq \mathfrak{G}$. We denote by $\mathbb{NK}(\mathfrak{G})_{\max}$ the set of all maximal members of the poset $\mathbb{NK}(\mathfrak{G})$. By Proposition 5.19,(1), $\mathbf{m}$ is an open ideal of $\mathfrak{G}$ provided $\mathbf{m} \in \mathbb{NK}(\mathfrak{G})_{\max}$, and $\mathbb{NK}(\mathfrak{G})$ is the union of the basic open compact sets $\mathbb{L}(\mathbf{m})$ for $\mathbf{m}$ ranging over $\mathbb{NK}(\mathfrak{G})_{\max}$. Consequently, we obtain

Proposition 5.22 (Abstract Kneser Criterion). *Let $\eta : \Gamma \longrightarrow \mathfrak{G}$ be a g-cocycle. Then the following assertions are equivalent for any ideal $\mathbf{a}$ of the profinite Γ-group $\mathfrak{G}$.*

(1) $\mathbf{a} \in \mathbb{K}(\mathfrak{G})$*, i.e., the induced cocycle $\eta_{\mathbf{a}} : \Gamma \longrightarrow \mathfrak{G}/\mathbf{a}$ is surjective.*
(2) $\mathbf{a} \not\subseteq \mathbf{m}$ *for all* $\mathbf{m} \in \mathbb{NK}(\mathfrak{G})_{\max}$*, i.e.,* $\mathbb{L}(\mathfrak{G} \mid \mathbf{a}) \cap \mathbb{NK}(\mathfrak{G})_{\max} = \varnothing$.

The following class of finite algebraic structures arises naturally from the abstract Kneser criterion above, considering the quotient Γ-groups $\mathfrak{G}/\mathbf{m}$ with $\mathbf{m} \in \mathbb{NK}(\mathfrak{G})_{\max}$.

Definition 5.23. A triple $(\Gamma, \mathfrak{G}, \eta)$ consisting of a finite group Γ, a finite Γ-group $\mathfrak{G}$ and a g-cocycle $\eta : \Gamma \longrightarrow \mathfrak{G}$, with $\Delta := \operatorname{Ker}(\eta)$, is called a *minimal non-Kneser triple* (for short *mnK triple*) if the following conditions are satisfied.

(1) $\eta(\Gamma) \neq \mathfrak{G}$, i.e., the cocycle η is not surjective,
(2) for every ideal $\mathbf{a} \neq \{1\}$ of the Γ-group $\mathfrak{G}$, $(\Gamma : \eta^{-1}(\mathbf{a})) = (\mathfrak{G} : \mathbf{a})$, i.e., the induced cocycle $\eta_{\mathbf{a}} : \Gamma \longrightarrow \mathfrak{G}/\mathbf{a}$ is surjective, and
(3) $\bigcap_{\gamma \in \Gamma} \gamma \Delta \gamma^{-1} = \operatorname{Fix}_\Gamma(G) \cap \Delta = \{1\}$, i.e., the triple $(\Gamma, \mathfrak{G}, \eta)$ is normalized.

Note that the conjunction of conditions (1) and (2) above is equivalent with the sentence $\mathbb{NK}(\mathfrak{G}) = \mathbb{NK}(\mathfrak{G})_{\mathbf{max}} = \{1\}$.

In its full generality, the following problem is far from being trivial.

Problem 5.24. Classify (up to isomorphism in $\mathscr{Z}^1$) the minimal non-Kneser triples.

Partial answers are given in Sect. 7.

6 CoGalois and Minimal Non-coGalois Triples

In this section we introduce two remarkable classes of surjective cocycles extending to a more general framework some notions and results from [3] on *coGalois groups of cocycles.*

Definition 6.1. A triple $(\Gamma, \mathfrak{G}, \eta)$ is called *coGalois* if $\eta : \Gamma \longrightarrow \mathfrak{G}$ is surjective and the maps $\mathscr{J} : \mathbb{L}(\Gamma \mid \mathrm{Ker}\,(\eta)) \longrightarrow \mathbb{L}(\mathfrak{G})$, $\mathscr{S} : \mathbb{L}(G) \longrightarrow \mathbb{L}(\Gamma \mid \mathrm{Ker}\,(\eta))$ are lattice isomorphisms inverse to one another, i.e., the coGalois connection between the lattices $\mathbb{L}(\Gamma \mid \mathrm{Ker}\,(\eta))$ and $\mathbb{L}(\mathfrak{G})$ is perfect.

Characterizations of coGalois triples are given by the next lemma whose proof is straightforward.

Lemma 6.2. *Let $\eta \in Z^1(\Gamma, \mathfrak{G})$ be a g-cocycle with $\Delta := \mathrm{Ker}\,(\eta)$. The following assertions are equivalent.*

(1) *$(\Gamma, \mathfrak{G}, \eta)$ is a coGalois triple.*
(2) *η is surjective and $\mathscr{S} \circ \mathscr{J} = 1_{\mathbb{L}(\Gamma \mid \Delta)}$.*
(3) *η is surjective and $\mathscr{J} : \mathbb{L}(\Gamma \mid \Delta) \longrightarrow \mathbb{L}(\mathfrak{G})$ is injective.*
(4) *η and $\mathscr{S} : \mathbb{L}(\mathfrak{G}) \longrightarrow \mathbb{L}(\Gamma \mid \Delta)$ are surjective.*
(5) *$\eta(\Lambda) = \mathscr{J}(\Lambda)$ for all $\Lambda \in \mathbb{L}(\Gamma \mid \Delta)$.*
(6) *$\eta(\Lambda) \in \mathbb{L}(\mathfrak{G})$ for all $\Lambda \in \mathbb{L}(\Gamma \mid \Delta)$.*

Definition 6.3. Let $\eta \in Z^1(\Gamma, \mathfrak{G})$ be a g-cocycle. An ideal $\mathbf{a} \in \mathbb{L}(\mathfrak{G})$ is called *coGalois* if the induced triple $(\Gamma, \mathfrak{G}/\mathbf{a}, \eta_{\mathbf{a}} : \Gamma \longrightarrow \mathfrak{G}/\mathbf{a})$ is coGalois.

We denote by $\mathbb{CG}(\mathfrak{G})$ the poset with the last element $\mathfrak{G}$ of all coGalois ideals of $\mathfrak{G}$. Some properties of coGalois ideals are collected together in the next result.

Proposition 6.4. *Let $\eta \in Z^1(\Gamma, \mathfrak{G})$ be a g-cocycle. The following assertions hold.*

(1) *$\mathbb{CG}(\mathfrak{G})$ is an upper subset of $\mathbb{L}(\mathfrak{G})$, contained in $\mathbb{K}(\mathfrak{G})$.*
(2) *For any ideal $\mathbf{a} \in \mathbb{L}(\mathfrak{G})$, $\mathbf{a} \in \mathbb{CG}(\mathfrak{G})$ if and only if $\mathbf{b} \in \mathbb{CG}(\mathfrak{G})$ for all open ideals $\mathbf{b}$ containing $\mathbf{a}$.*
(3) *$\mathbb{CG}(\mathfrak{G})$ is a closed subspace of the spectral space $\mathbb{K}(\mathfrak{G})$.*
(4) *For every $\mathbf{a} \in \mathbb{CG}(\mathfrak{G})$ there exists at least one minimal coGalois ideal $\mathbf{b} \subseteq \mathbf{a}$.*
(5) *The space $\mathbb{CG}(\mathfrak{G})_{\min}$ of all minimal members of $\mathbb{CG}(\mathfrak{G})$ is Hausdorff with respect to the topology induced from the spectral space $\mathbb{CG}(\mathfrak{G})$.*

Proof. The proof is similar with the proof of Proposition 5.19.

Remark 6.5. The restriction map $\mathscr{S}|_{\mathbb{CG}(\mathfrak{G})} : \mathbb{CG}(\mathfrak{G}) \longrightarrow \mathbb{L}(\Gamma \mid \Delta)$ is injective and coherent, inducing a homeomorphism of the spectral space $\mathbb{CG}(\mathfrak{G})$ onto the closed subspace $\mathbb{CG}(\Gamma) := \mathscr{S}(\mathbb{CG}(\mathfrak{G}))$ of the spectral space $\mathbb{L}(\Gamma \mid \Delta)$, contained in the subspace $\mathbb{HK}(\Gamma)$ defined in Remarks 5.21; call its members *coGalois subgroups* of the profinite group Γ. An explicit *coGalois criterion for hereditarily Kneser subgroups* of Γ is provided by [12, Theorem 3.2] in the particular framework of cyclotomic abstract coGalois theory.

Given a surjective cocycle $\eta \in Z^1(\Gamma, \mathfrak{G})$, we denote by $\mathbb{NCG}(\mathfrak{G})$ the complementary set of $\mathbb{CG}(\mathfrak{G})$ in the set $\mathbb{L}(\mathfrak{G}) = \mathbb{K}(\mathfrak{G})$ of all ideals of the profinite Γ-group $\mathfrak{G}$. By Proposition 6.4,(3), $\mathbb{NCG}(\mathfrak{G})$ is an open and hence a lower subset of the spectral space $\mathbb{L}(\mathfrak{G})$. By Lemma 6.2,(6), $\mathbb{NCG}(\mathfrak{G}) \neq \varnothing \Longleftrightarrow \eta(\Lambda) \notin \mathbb{L}(\mathfrak{G})$ for some $\Lambda \in \mathbb{L}(\Gamma \mid \mathrm{Ker}\,(\eta)) \setminus \{\mathrm{Ker}\,(\eta), \Gamma\}$. We denote by $\mathbb{NCG}(\mathfrak{G})_{\max}$ the set of all maximal members of the poset $\mathbb{NCG}(\mathfrak{G})$. By Proposition 6.4,(4), $\mathbf{m}$ is an open ideal of $\mathfrak{G}$ provided $\mathbf{m} \in \mathbb{NCG}(\mathfrak{G})_{\max}$, and $\mathbb{NCG}(\mathfrak{G})$ is the union of the basic open compact sets $\mathbb{L}(\mathbf{m})$ for $\mathbf{m}$ ranging over $\mathbb{NCG}(\mathfrak{G})_{\max}$. Consequently, we obtain the following analogue of Proposition 5.22.

Proposition 6.6 (Abstract coGalois Criterion). *Let $\eta \in Z^1(\Gamma, \mathfrak{G})$ be a surjective cocycle, and let $\mathbf{a} \in \mathbb{L}(\mathfrak{G}) = \mathbb{K}(\mathfrak{G})$. The following assertions are equivalent.*

(1) $\mathbf{a} \in \mathbb{CG}(\mathfrak{G})$, *i.e., the induced triple* $(\Gamma, \mathfrak{G}/\mathbf{a}, \eta_{\mathbf{a}} : \Gamma \longrightarrow \mathfrak{G}/\mathbf{a})$ *is coGalois.*
(2) $\mathbf{a} \nsubseteq \mathbf{m}$ *for all* $\mathbf{m} \in \mathbb{NCG}(\mathfrak{G})_{\max}$, *i.e.,* $\mathbb{L}(\mathfrak{G} \mid \mathbf{a}) \cap \mathbb{NCG}(\mathfrak{G})_{\max} = \varnothing$.

Considering the quotient Γ-groups $\mathfrak{G}/\mathbf{m}$ for $\mathbf{m} \in \mathbb{NCG}(\mathfrak{G})_{\max}$, we obtain the following class of finite algebraic structures.

Definition 6.7. A triple $(\Gamma, \mathfrak{G}, \eta)$ consisting of a finite group Γ, a finite Γ-group $\mathfrak{G}$ and a surjective cocycle $\eta : \Gamma \longrightarrow \mathfrak{G}$, with $\Delta := \mathrm{Ker}\,(\eta)$, is called a (Kneser) *minimal non-coGalois triple* (for short *mncG triple*) if the following conditions are satisfied.

(1) $\eta(\Lambda)$ is not an ideal of $\mathfrak{G}$ for some $\Lambda \in \mathbb{L}(\Gamma \mid \Delta) \setminus \{\Delta, \Gamma\}$, i.e., $(\Gamma, \mathfrak{G}, \eta)$ is not a coGalois triple,
(2) for every ideal $\mathbf{a} \neq \{1\}$ of $\mathfrak{G}$, the triple $(\Gamma, \mathfrak{G}/\mathbf{a}, \eta_{\mathbf{a}} : \Gamma \longrightarrow \mathfrak{G}/\mathbf{a})$ is coGalois, and
(3) $\bigcap_{\gamma \in \Gamma} \gamma \Delta \gamma^{-1} = \mathrm{Fix}_\Gamma(G) \cap \Delta = \{1\}$, i.e., the triple $(\Gamma, \mathfrak{G}, \eta)$ is normalized.

Note that the conjunction of conditions (1) and (2) above is equivalent with the sentence $\mathbb{NCG}(\mathfrak{G}) = \mathbb{NCG}(\mathfrak{G})_{\max} = \{1\}$.

Problem 6.8. Classify (up to isomorphism in $\mathscr{Z}^1$) the (Kneser) minimal non-coGalois triples.

Remarks 6.9. (1) Problem 6.8 is solved in [3] in the particular framework described in Example 4.10. Assume $\mathfrak{G}$ is a finite Γ-module of exponent k, and the action of Γ is given by a character $\chi : \Gamma \longrightarrow (\mathbb{Z}/k\mathbb{Z})^\times$. Let $\eta \in Z^1(\Gamma, \mathfrak{G})$. According to [3, Lemma 2.17, Corollary 2.18], $(\Gamma, \mathfrak{G}, \eta)$ is a

Kneser mncG triple if and only if the triple $(\Gamma, \mathfrak{G}, \eta)$ is, up to isomorphism, of one of the following three types.

(i) $k = 4$, $\Gamma = \langle \sigma, \tau \,|\, \sigma^2 = \tau^2 = (\sigma\tau)^2 = 1 \rangle \cong \mathbb{Z}/2\mathbb{Z} \times \mathbb{Z}/2\mathbb{Z}$, $\mathfrak{G} = \mathbb{Z}/4\mathbb{Z}$, $\chi(\sigma) = -1 \bmod 4$, $\chi(\tau) = 1 \bmod 4$, $\eta(\sigma) = 1 \bmod 4$, $\eta(\tau) = 2 \bmod 4$.

(ii) $k = 4$, $\Gamma = \mathbb{D}_8 = \langle \sigma, \tau \,|\, \sigma^2 = \tau^4 = (\sigma\tau)^2 = 1 \rangle$, $\mathfrak{G} = (\mathbb{Z}/2\mathbb{Z})e_1 \oplus (\mathbb{Z}/4\mathbb{Z})e_2$, $\chi(\sigma) = -1 \bmod 4$, $\chi(\tau) = 1 \bmod 4$, $\eta(\sigma) = e_2$, $\eta(\tau) = e_1 + e_2$.

(iii) $k = pr$, $p \neq 2$ prime, $1 \neq r \,|\, (p-1)$, $\Gamma = \langle \sigma, \tau \,|\, \sigma^r = \tau^p = \sigma\tau\sigma^{-1}\tau^{-u} = 1 \rangle \cong \mathbb{Z}/p\mathbb{Z} \rtimes_u \mathbb{Z}/r\mathbb{Z}$, where $u \in (\mathbb{Z}/pr\mathbb{Z})^\times$ such that the order of $u \bmod p \in (\mathbb{Z}/p\mathbb{Z})^\times$ is r and $l \,|\, (u-1)$ for all $l \,|\, r$ with $l \neq 2$ prime or $l = 4$, $\mathfrak{G} = \mathbb{Z}/pr\mathbb{Z}$, $\chi(\sigma) = u$, $\chi(\tau) = 1 \bmod pr$, $\eta(\sigma) = p \bmod pr$, $\eta(\tau) = r \bmod pr$.

Based on the classification of mncG triples above, [3, Theorem 2.19] provides an explicit form of Proposition 6.6 in the framework of cyclotomic abstract coGalois theory.

(2) The bijective cocycle $\mathbb{Z}/2\mathbb{Z} \times \mathbb{Z}/2\mathbb{Z} \longrightarrow \mathbb{Z}/4\mathbb{Z}$ from (1)(i) is induced by an action of $\mathbb{Z}/2\mathbb{Z} \times \mathbb{Z}/2\mathbb{Z}$ on itself, while the bijective cocycle $\mathbb{D}_8 \longrightarrow \mathbb{Z}/2\mathbb{Z} \oplus \mathbb{Z}/4\mathbb{Z}$ from (1)(ii) is not since $(\mathbb{D}_8, \mathbb{Z}/2\mathbb{Z} \oplus \mathbb{Z}/4\mathbb{Z})$ is not a deformation pair.

(3) The dihedral group $\mathbb{D}_8 = \mathbb{Z}/4\mathbb{Z} \rtimes \mathbb{Z}/2\mathbb{Z} \cong \langle \sigma, \tau \,|\, \sigma^2 = \tau^4 = (\sigma\tau)^2 = 1 \rangle$ and the quaternion group $Q \cong \langle \rho, \theta \,|\, \rho^4 = 1, \rho^2 = \theta^2, \rho\theta\rho^{-1} = \theta^{-1} \rangle$ form a deformation pair: consider the actions $\mathbb{D}_8 \times Q \longrightarrow Q$, $Q \times \mathbb{D}_8 \longrightarrow D_8$, defined by

$$^{\sigma}\rho = \rho^{-1},\ ^{\sigma}\theta = \theta,\ ^{\tau}\rho = \rho,\ ^{\tau}\theta = \theta;\ ^{\rho}\sigma = \tau^2\sigma,\ ^{\rho}\tau = \tau,\ ^{\theta}\sigma = \sigma,\ ^{\theta}\tau = \tau.$$

Setting $\eta(\sigma) = \rho$, $\eta(\tau) = \theta$, we obtain a bijective cocycle $\eta \in Z^1(\mathbb{D}_8, Q)$ such that $\eta^{-1} \in Z^1(Q, \mathbb{D}_8)$ as desired. Since $\mathbb{D}_8/C(\mathbb{D}_8) \cong Q/C(Q) \cong \mathbb{Z}/2\mathbb{Z} \times \mathbb{Z}/2\mathbb{Z}$, it follows easily that the Kneser triples $(\mathbb{D}_8, Q, \eta)$ and $(Q, \mathbb{D}_8, \eta^{-1})$ are both mncG, providing examples of Kneser mncG triples in a purely noncommutative framework.

Similarly, for $m \geq 3$, the dihedral group $\mathbb{D}_{2^m} \cong \langle \sigma, \tau \,|\, \sigma^2 = \tau^{2^{m-1}} = (\sigma\tau)^2 = 1 \rangle$ and the generalized quaternion group $Q_{2^m} \cong \langle \rho, \theta \,|\, \rho^4 = 1, \rho^2 = \theta^{2^{m-2}}, \rho\theta\rho^{-1} = \theta^{-1} \rangle$ form a deformation pair inducing Kneser non-coGalois triples, but the minimality condition is satisfied only for $m = 3$.

(4) Other remarkable Kneser mncG triples in a purely noncommutative framework are obtained as follows. Let $K = \mathbb{F}_q$, $q = p^f$, $f \geq 1$, $p \neq 2$, $L = \mathbb{F}_{q^2}$, $\mathrm{Gal}(L/K) = \langle \alpha \rangle \cong \mathbb{Z}/2\mathbb{Z}$, with $\alpha(x) = x^q$, $\chi : L^\times/(L^\times)^2 \longrightarrow \mathbb{Z}/2\mathbb{Z}$ the unique isomorphism. Setting $\Gamma := PGL(2, q^2) = PGL(2, L)$, extend α to an automorphism of Γ, and consider the self-action $\omega : \Gamma \longrightarrow \mathrm{Aut}(\Gamma)$, defined by $\omega(A \bmod L^\times) := \alpha^{\chi(\det(A)\cdot(L^\times)^2)}$ for $A \in GL(2, L)$. We get $\omega(\Gamma) = \langle \alpha \rangle \cong \mathbb{Z}/2\mathbb{Z}$, and $\mathrm{Ker}\,(\omega) = PSL(2, L) = PSL(2, q^2)$. The self-action ω is adequate, and the deformation of Γ via ω is the *Zassenhaus group* $\Gamma_\omega = M(q^2)$. The non-isomorphic finite groups Γ and Γ_ω of the same

order $(q^2-1)q^2(q^2+1)$, having the *simple* group Ker $(\omega) = PSL(2, q^2)$ as a common (normal) subgroup of index 2, act both *faithfully* and *sharply three-transitive* on the projective line $\mathbb{P}^1(\mathbb{F}_{q^2})$. The induced Kneser triples $(\Gamma, \Gamma_\omega, \eta_\omega)$ and $(\Gamma_\omega, \Gamma, \eta_\omega^{-1})$ are both mncG.

7 Partial Answers to Classification Problem 5.24

This last section, devoted to partial answers to Problem 5.24, provides some classes of mnK triples including the very simple ones which occur in [3, Lemma 1.18, Theorem 1.20] in the framework of cyclotomic abstract coGalois theory described in Example 4.10.

First some useful lemmas. A proper subclass of mnK triples is provided by the next obvious lemma.

Lemma 7.1. *Let $\eta \in Z^1(\Gamma, \mathfrak{G})$ be a normalized g-cocycle with $\eta(\Gamma) \neq \mathfrak{G}$. Then $(\Gamma, \mathfrak{G}, \eta)$ is a mnK triple whenever the Γ-group $\mathfrak{G}$ is* simple, *i.e., $\{1\}$ and $\mathfrak{G}$ are its only ideals.*

Remark 7.2. Not all mnK triples are of the type above. The simplest example of a mnK triple $(\Gamma, \mathfrak{G}, \eta)$, where the Γ-group $\mathfrak{G}$ is not simple, is obtained by taking $\Gamma := \mathbb{Z}/2\mathbb{Z}$, $\mathfrak{G} := \mathbb{Z}/4\mathbb{Z}$ with the nontrivial action of Γ on $\mathfrak{G}$. Then $Z^1(\Gamma, \mathfrak{G}) \cong \mathbb{Z}/4\mathbb{Z}$ is generated by the injective g-cocycle η defined by $\eta(1 \bmod 2) = 1 \bmod 4$. Note that $\eta(\Gamma) = \{0 \bmod 4,\ 1 \bmod 4\} \neq \mathfrak{G}$. As 0, $\mathfrak{G}$ and $\mathbf{a} := \{0 \bmod 4,\ 2 \bmod 4\} \cong \mathbb{Z}/2\mathbb{Z}$ are the only submodules of $\mathfrak{G}$, and the induced cocycle $\eta_{\mathbf{a}} : \Gamma \longrightarrow \mathfrak{G}/\mathbf{a} \cong \mathbb{Z}/2\mathbb{Z}$ is an isomorphism, it follows that $(\Gamma, \mathfrak{G}, \eta)$ is a mnK triple but the Γ-module $\mathfrak{G}$ is not simple.

Corollary 7.3. *Assume $\Gamma = \mathfrak{G}$ is a non-abelian finite simple group acting on itself by inner automorphisms $(\gamma, g) \mapsto \gamma g \gamma^{-1}$. The following assertions hold.*

(1) *Every nontrivial cocycle $\eta \in Z^1(\Gamma, \mathfrak{G})$ is a normalized g-cocycle.*
(2) *$(\Gamma, \mathfrak{G}, \eta_g)$, with $1 \neq g \in \mathfrak{G}$, is a mnK triple, where $\eta_g : \Gamma \longrightarrow \mathfrak{G}$ is the coboundary $\gamma \mapsto [g, \gamma] := g\gamma g^{-1}\gamma^{-1}$, with* Ker $(\eta_g) = C_\Gamma(g)$, *the centralizer of g.*
(3) *For $g, h \in \mathfrak{G} \setminus \{1\}$, the mnK triples $(\Gamma, \mathfrak{G}, \eta_g)$ and $(\Gamma, \mathfrak{G}, \eta_h)$ are isomorphic if and only if $\varphi(g) = h$ for some automorphism φ of $\mathfrak{G} = \Gamma$.*

Proof. (1) For a nontrivial cocycle $\eta \in Z^1(\Gamma, G)$, let $\mathfrak{H}$ denote the subgroup of $\mathfrak{G}$ generated by $\eta(\Gamma)$. As the cocycle η is nontrivial, $\mathfrak{H}$ is a proper subgroup of $\mathfrak{G}$. Since $\sigma\eta(\tau)\sigma^{-1} = \eta(\sigma)^{-1}\eta(\sigma\tau) \in \mathfrak{H}$ for all $\sigma, \tau \in \Gamma$, it follows that $\mathfrak{H}$ is a normal subgroup of $\mathfrak{G}$, and hence $\mathfrak{H} = \mathfrak{G} = \Gamma$. Consequently, η is a g-cocycle. It is also normalized since Ker $(\eta) \neq \Gamma$ implies $\bigcap_{\gamma \in \Gamma} \gamma \mathrm{Ker}\,(\eta)\gamma^{-1} = \{1\}$ as the group Γ is simple.

(2) By (1), η_g is a normalized g-cocycle. As $1 \neq g \in C_\Gamma(g) = \mathrm{Ker}(\eta_g)$, it follows that $\eta_g(\Gamma) \neq \mathfrak{G}$ since $\Gamma = \mathfrak{G}$ is finite. The conclusion is immediate by Lemma 7.1.
(3) follows from the definition of isomorphic cocycles.

Lemma 7.4. *Let* $(\Gamma, \mathfrak{G}, \eta)$ *be a mnK triple,* $\Delta := \mathrm{Ker}(\eta)$. *Then the following assertions hold.*

(1) $\mathrm{Fix}_\Gamma(\mathfrak{G}) \cap \eta^{-1}(C(\mathfrak{G})) = \{1\}$.
(2) *If* $\mathfrak{G}$ *is abelian then* $\mathrm{Fix}_\Gamma(\mathfrak{G}) = \{1\}$, *i.e.,* Γ *acts faithfully on* $\mathfrak{G}$.
(3) *If* $\mathfrak{G}$ *is nilpotent then it is a* p*-group for some prime number* p. *If* $p \not| (\Gamma : \Delta)$, *then* $\mathfrak{G}$ *is a simple* $\mathbb{F}_p[\Gamma]$*-module, the action of* Γ *is faithful, and* η *is a coboundary.*

Proof. (1) Let $\overline{\Delta} := \mathrm{Fix}_\Gamma(\mathfrak{G}) \cap \eta^{-1}(C(\mathfrak{G}))$. It suffices to show that $\overline{\Delta} \subseteq \Delta$ since $\Delta \cap \mathrm{Fix}_\Gamma(\mathfrak{G}) = \{1\}$ by assumption. By Lemma 3.8,(3, 4), $\mathbf{a} := \eta(\overline{\Delta})$ is an ideal of the Γ-group $\mathfrak{G}$. On the other hand, since η is not surjective by assumption, it follows by Corollary 5.8 that the cocycle $\eta_\mathbf{a} : \Gamma \longrightarrow \mathfrak{G}/\mathbf{a}$ is not surjective, so $\mathbf{a} = \{1\}$ by the minimality property of η, and hence $\overline{\Delta} \subseteq \Delta$ as desired.
(2) is an immediate consequence of (1).
(3) Since the cocycle $\eta : \Gamma \longrightarrow \mathfrak{G}$ is not surjective and $\mathfrak{G}$ is nilpotent by assumption, it follows by Proposition 5.5 that there exists a prime number p such that the induced cocycle $\eta_p : \Gamma \longrightarrow \mathfrak{G}_p$ is not surjective, therefore the kernel of the natural projection $\mathfrak{G} \longrightarrow \mathfrak{G}_p$ is trivial by the minimality property of η, and hence $\mathfrak{G} \cong \mathfrak{G}_p$ as required.
Assuming $p \not| (\Gamma : \Delta)$, it follows that the Γ-group $\mathfrak{G}$ is simple. In particular, since the center $C(\mathfrak{G})$ of the p-group $\mathfrak{G}$ is a nontrivial ideal of the Γ-group $\mathfrak{G}$, we obtain $\mathfrak{G} = C(\mathfrak{G})$, so $\mathfrak{G}$ is an abelian p-group. On the other hand, $p\mathfrak{G} = 0$ since $p\mathfrak{G} \neq \mathfrak{G}$. Thus $\mathfrak{G}$ is a simple $\mathbb{F}_p[\Gamma]$-module as desired. The faithfulness of the action of Γ follows by (2), while $\eta(\gamma) = (\gamma - 1)g$ for $\gamma \in \Gamma$, with $g = -\frac{1}{(\Gamma:\Delta)} \sum_{\sigma \in \Gamma/\Delta} \eta(\sigma)$.

The proof of the next lemma is straightforward.

Lemma 7.5. *Let* p *be a prime number,* Γ *a nontrivial finite group, and* $\mathfrak{G}$ *a simple* $\mathbb{F}_p[\Gamma]$*-module such that the action of* Γ *on* $\mathfrak{G}$ *is faithful. Let* $K \cong \mathbb{F}_q$, $q = p^f$, $f \geq 1$, *denote the field of the endomorphisms of the simple* $\mathbb{F}_p[\Gamma]$*-module* $\mathfrak{G}$. *The following assertions hold.*

(1) $\mathfrak{G}$ *is a simple* $K[\Gamma]$*-module,* $\mathrm{End}_{K[\Gamma]}(\mathfrak{G}) = K$, *the group* Γ, *identified with a subgroup of the linear group* $GL_K(\mathfrak{G})$, *is not a* p*-group, and* $C(\Gamma) = \Gamma \cap K^\times \cong \mathbb{Z}/r\mathbb{Z}$, $r \mid (q-1)$.
(2) *Every nontrivial cocycle* $\eta : \Gamma \longrightarrow \mathfrak{G}$ *is a normalized g-cocycle.*
(3) $(\Gamma, \mathfrak{G}, \eta)$ *is a mnK triple provided* η *is a nontrivial coboundary.*
(4) *Every cocycle* $\eta : \Gamma \longrightarrow \mathfrak{G}$ *for which* $p \not| (\Gamma : \mathrm{Ker}(\eta))$ *is a coboundary.*

7.1 The Abelian Case of Problem 5.24

Now we approach the classification Problem 5.24 in the special case when Γ and $\mathfrak{G}$ are abelian.

Lemma 7.6. *Let $(\Gamma, \mathfrak{G}, \eta)$ be a mnK triple, with abelian Γ and $\mathfrak{G}$. Then the following assertions hold.*

(1) *$\mathfrak{G}$ is an abelian p-group for some prime number p, Γ acts faithfully on $\mathfrak{G}$, and the g-cocycle η is injective. Let $p^n, n \geq 1$, be the exponent of $\mathfrak{G}$.*
(2) *Γ is a p-group $\Longleftrightarrow \mathfrak{G}^\Gamma \neq 0 \Longleftrightarrow \mathfrak{G}^\Gamma \cong \mathbb{Z}/p\mathbb{Z} \Longleftrightarrow |\mathfrak{G}| = p|\Gamma|$.*
(3) *There exists a unique minimal nonzero Γ-submodule of $\mathfrak{G}$.*
(4) *The image R of the canonical ring morphism $(\mathbb{Z}/p^n\mathbb{Z})[\Gamma] \longrightarrow \mathrm{End}(\mathfrak{G})$ is a finite commutative local ring of characteristic p^n, and Γ, identified with a subgroup of $R^\times$, generates R as $(\mathbb{Z}/p^n\mathbb{Z})$-module, in particular as ring.*
(5) *If Γ is a p-group then the residue field $k := R/\mathfrak{m} \cong \mathbb{F}_p$, $\Gamma = 1 + \mathfrak{m}$, and the R-module $\mathfrak{G}$ is (non-canonically) isomorphic to the Pontryagin dual $R^\vee = \mathrm{Hom}(R^+, 1/p^n\,\mathbb{Z}/\mathbb{Z})$ with the induced structure of R-module. In particular, if the local ring R is principal then the R-module $\mathfrak{G}$ is free of rank* 1.

Proof. The assertion (1) is immediate from Lemma 7.4.

(2) We have to show that $|\mathfrak{G}| = p|\Gamma|$ and $\mathfrak{G}^\Gamma \cong \mathbb{Z}/p\mathbb{Z}$ provided $\mathfrak{G}^\Gamma \neq 0$. Put $\Lambda := \eta^{-1}(\mathfrak{G}^\Gamma)$. As $(\Gamma, \mathfrak{G}, \eta)$ is a mnK triple, it follows $|\mathfrak{G}| = (\Gamma : \Lambda)|\mathfrak{G}^\Gamma|$. Assuming $\Lambda \neq 1$, the injective cocycle η induces by restriction a nontrivial monomorphism $\eta|_\Lambda : \Lambda \longrightarrow \mathfrak{G}^\Gamma$ whose image $\mathfrak{M} := \eta(\Lambda) \cong \Lambda$ is a nonzero Γ-submodule of $\mathfrak{G}$, therefore the induced cocycle $\eta_{\mathfrak{M}} \in Z^1(\Gamma, \mathfrak{G}/\mathfrak{M})$ is surjective since $(\Gamma, \mathfrak{G}, \eta)$ is a mnK triple. By Lemma 5.7 we deduce that the cocycle η is surjective, whence a contradiction. Thus $\Lambda = 1$, so it remains to show that $\mathfrak{G}^\Gamma \cong \mathbb{Z}/p\mathbb{Z}$.

Choose an element $h \in \mathfrak{G}^\Gamma$ of order p, and let $\mathfrak{H} \cong \mathbb{Z}/p\mathbb{Z}$ be the Γ-submodule generated by h. As $\mathfrak{H} \subseteq \mathfrak{G}^\Gamma$, we obtain $\eta^{-1}(\mathfrak{H}) \subseteq \eta^{-1}(\mathfrak{G}^\Gamma) = 1$, and hence

$$(\mathfrak{G} : \mathfrak{G}^\Gamma) = (\Gamma : \eta^{-1}(\mathfrak{G}^\Gamma)) = |\Gamma| = (\Gamma : \eta^{-1}(\mathfrak{H})) = (\mathfrak{G} : \mathfrak{H}),$$

so $\mathfrak{G}^\Gamma = \mathfrak{H} \cong \mathbb{Z}/p\mathbb{Z}$ as desired.

(3) If Γ is a p-group then $\mathfrak{G}^\Gamma \cong \mathbb{Z}/p\mathbb{Z}$ (by (2)) is the unique minimal nonzero Γ-submodule of $\mathfrak{G}$. Assume Γ is not a p-group and $\mathfrak{H}_i, i = 1, 2$, are distinct minimal nonzero Γ-submodules of $\mathfrak{G}$, so $\mathfrak{H}_1 \cap \mathfrak{H}_2 = 0$. Put $\Lambda_i := \eta^{-1}(\mathfrak{H}_i), i = 1, 2$. Since $(\Gamma, \mathfrak{G}, \eta)$ is mnK, it follows that $(\Gamma : \Lambda_i) = (\mathfrak{G} : \mathfrak{H}_i)$ is a p-th power and hence $\Lambda_i \neq 1, i = 1, 2$, as Γ is not a p-group. On the other hand, $\Lambda_1 \cap \Lambda_2 = \eta^{-1}(\mathfrak{H}_1 \cap \mathfrak{H}_2) = \eta^{-1}(0) = \mathrm{Ker}\,(\eta) = 1$, therefore the inclusion $\Lambda_1 \hookrightarrow \Gamma$ induces a monomorphism $\Lambda_1 \longrightarrow \Gamma/\Lambda_2$, so Λ_1 is a p-group and hence $|\Gamma| = |\Lambda_1|(\Gamma : \Lambda_1)$ is a p-th power, which is a contradiction.

(4) We have only to show that the finite commutative ring R is local, i.e., the only idempotents of R are the elements 0 and 1. Let e_i, $i = 1, \dots, s$, denote the minimal nonzero idempotents of R, the atoms of the boolean algebra $B(R) := \{e \in R \mid e^2 = e\}$ of idempotents of R with respect to the partial order $e \leq f \iff ef = e$. Put $R_i := Re_i$, $\mathfrak{G}_i := R_i\mathfrak{G} = e_i\mathfrak{G}, i = 1, \dots, s$. It follows that $\sum_{i=1}^{s} e_i = 1, e_ie_j = 0$ for $i \neq j$, the R_i's are local rings, $B(R_i) = \{0, 1_{R_i} := e_i\}$, $R \cong \prod_{i=1}^{s} R_i$ is a semi-local ring, and $R = \bigoplus_{1\leq i\leq s} R_i, \mathfrak{G} = \bigoplus_{1\leq i\leq s} \mathfrak{G}_i$ are R-module direct sums. Note that $\mathfrak{G}_i \neq 0$ for $i = 1, \dots, s$, since $0 \neq e_i \in R \subseteq \mathrm{End}(\mathfrak{G})$. By (3), we conclude that $s = 1$, i.e., R is a local ring as desired.

(5) Let h be a generator of $\mathfrak{H} := \mathfrak{G}^{\Gamma} \cong \mathbb{Z}/p\mathbb{Z}$, the unique minimal nonzero R-submodule of $\mathfrak{G}$. The kernel of the surjective homomorphism of R-modules $R \longrightarrow \mathfrak{H}$, $\lambda \mapsto \lambda h$, is the maximal ideal $\mathfrak{m}$ of R, so $k := R/\mathfrak{m} \cong (\mathbb{Z}/p^n\mathbb{Z})/p(\mathbb{Z}/p^n\mathbb{Z}) \cong \mathbb{F}_p$ and $\Gamma \subseteq 1 + \mathfrak{m}$.

As $|\mathfrak{G}| = |\Gamma| \cdot p \,||1 + \mathfrak{m}| \cdot p = |\mathfrak{m}| \cdot |R/\mathfrak{m}| = |R|$, we deduce that $|\mathfrak{G}| \leq |R|$. Consider the Pontryagin dual $\mathfrak{G}^{\vee} = \mathrm{Hom}(\mathfrak{G}, 1/p^n\,\mathbb{Z}/\mathbb{Z})$ of the finite abelian group $\mathfrak{G}$ of exponent p^n with its natural structure of R-module given by $(\lambda\psi)(g) = \psi(\lambda g)$ for $\lambda \in R, \psi \in \mathfrak{G}^{\vee}, g \in \mathfrak{G}$. Choose $\varphi \in \mathfrak{G}^{\vee}$ such that $\varphi\mid_{\mathfrak{H}} \neq 0$. The homomorphism of R-modules $R \longrightarrow \mathfrak{G}^{\vee}, \lambda \mapsto \lambda\varphi$, is injective: Indeed, assuming the contrary, let $0 \neq \lambda \in R$ be such that $\lambda\varphi = 0$, i.e., $\varphi\mid_{\lambda\mathfrak{G}} = 0$. As $R \subseteq \mathrm{End}(\mathfrak{G})$, it follows that $\lambda\mathfrak{G}$ is a nonzero R-submodule of $\mathfrak{G}$ and hence $\mathfrak{H} \subseteq \lambda\mathfrak{G}$, so $\varphi\mid_{\mathfrak{H}} = 0$, which is a contradiction. As we already know that $|\mathfrak{G}^{\vee}| = |\mathfrak{G}| \leq |R|$, we deduce that $\Gamma = 1 + \mathfrak{m}$ and the injective map above is an isomorphism, inducing by duality an isomorphism of R-modules (depending on the choice of φ)

$$\mathfrak{G} \longrightarrow R^{\vee} = \mathrm{Hom}(R^{+}, 1/p^n\,\mathbb{Z}/\mathbb{Z}),\ g \mapsto \psi_g \text{ with } \psi_g(\lambda) = \varphi(\lambda g).$$

The isomorphism $\mathfrak{G} \longrightarrow R^{\vee}$ maps $\mathfrak{H} = \mathfrak{G}^{\Gamma}$ onto $(R^{\vee})^{\Gamma} = (R/\mathfrak{m})^{\vee} \cong k^{+}$, inducing an isomorphism $\mathfrak{G}/\mathfrak{H} \cong \mathfrak{m}^{\vee} = \{\psi\mid_{\mathfrak{m}} : \psi \in R^{\vee}\}$. Composing successively the bijective map $\mathfrak{m} \longrightarrow \Gamma, \lambda \mapsto 1 + \lambda$, (the inverse of the canonical cocycle $\gamma \mapsto \gamma - 1$), the cocycle $\eta : \Gamma \longrightarrow \mathfrak{G}$, the isomorphism above $\mathfrak{G} \longrightarrow R^{\vee}$ and the natural projection $R^{\vee} \longrightarrow \mathfrak{m}^{\vee}$, we get an isomorphism of R-modules $\mathfrak{m} \longrightarrow \mathfrak{m}^{\vee}, \lambda \mapsto \lambda^{\vee}$, with $\lambda^{\vee}(\mu) = \varphi(\mu\eta(1+\lambda)) = \varphi(\lambda\eta(1+\mu))$, inducing a nondegenerate pairing $\mathfrak{m} \times \mathfrak{m} \longrightarrow \mathbb{Q}/\mathbb{Z}$, $(\lambda, \mu) \mapsto \lambda^{\vee}(\mu) = \mu^{\vee}(\lambda)$, which is compatible with the canonical action of R on $\mathfrak{m}$.

Assuming that R is principal, $\mathfrak{m} = R\theta$ for any $\theta \in \mathfrak{m}\setminus\mathfrak{m}^2$, and every ideal of R is of the form $\mathfrak{m}^i = R\theta^i$, whence for each $a \in R$, there exists a unique nonnegative integer $i \leq m$ such that $a = u\theta^i$, where $u \in R^{\times}$ and m is the nilpotency index of $\mathfrak{m}$. Putting $S := \mathbb{Z}/p^n\mathbb{Z}$, $R = S \oplus S\theta \oplus \cdots \oplus S\theta^{e-1}$ is an S-module direct sum where e is the greatest integer $i \leq m$ such that $p \in \mathfrak{m}^i$.

It follows that θ satisfies an *Eisenstein polynomial*

$$f(x) := x^e - p(a_{e-1}x^{e-1} + \cdots + a_0), \text{ with } a_i \in S, a_0 \in S^\times.$$

Thus, $R \cong S[x]/(f(x), p^{n-1}x^t)$, where $1 \le t := m - (n-1)e \le e$. As S-module, R is free of rank e if and only if $t = e$, i.e., $m = ne$.

As we already know that $\mathfrak{G}$ and $R^\vee$ are isomorphic R-modules, we deduce that $\mathfrak{G}$ is free of rank 1 since the finite commutative ring R is principal by assumption.

The next two results add new information to that given by Lemma 7.6.

Proposition 7.7. *Let p be a prime number, and $\eta \in Z^1(\Gamma, \mathfrak{G})$ be such that $\mathfrak{G}$ is a finite abelian p-group, while the finite abelian group Γ is not a p-group. With notation from* Lemma 7.6, *the following assertions are equivalent.*

(1) $(\Gamma, \mathfrak{G}, \eta)$ *is a mnK triple.*
(2) $R \cong \mathbb{F}_q$ *is a finite field with $q = p^f$, $f \ge 1$ for $p \ne 2$, $f \ge 2$ for $p = 2$, $\mathfrak{G}$ is a one-dimensional R-vector space identified with R^+, the group Γ, identified with a subgroup of the multiplicative group $R^\times$, is cyclic of order $1 \ne r \mid (q-1)$ such that f is the order of $p \bmod r \in (\mathbb{Z}/r\mathbb{Z})^\times$, and $\eta \in Z^1(\Gamma, R^+) = B^1(\Gamma, R^+) \cong (\mathbb{Z}/p\mathbb{Z})^f$ is, up to multiplication by elements in $R^\times$, the coboundary $u \in \Gamma \mapsto u - 1 \in R^+$.*

Proof. (1) $\Longrightarrow$ (2). By Lemma 7.6,(4), R is a finite local ring; $k := R/\mathfrak{m}$ its residue field. By Lemma 7.6,(3) again, there exists a unique nonzero minimal R-submodule $\mathfrak{H}$ of $\mathfrak{G}$, so $\mathfrak{H} = Rh$ for any $0 \ne h \in \mathfrak{H}$, and the surjective morphism of R-modules $R \longrightarrow \mathfrak{H}, \lambda \mapsto \lambda h$ induces an isomorphism $\mathfrak{H} \cong k^+$. Put $\Lambda := \eta^{-1}(\mathfrak{H})$. As $(\Gamma, \mathfrak{G}, \eta)$ is a mnK triple, it follows that $(\Gamma : \Lambda) = (\mathfrak{G} : \mathfrak{H})$ is a p-th power, and hence Λ is not a p-group. In particular, $\Lambda \ne 1$, and $\mathfrak{H} = R\eta(\sigma)$ for any $1 \ne \sigma \in \Lambda$. As $\Gamma \subseteq R^\times$ is not a p-group, its p-primary component $\Gamma(p) := \Gamma \cap (1 + \mathfrak{m}) = \mathrm{Fix}_\Gamma(\mathfrak{H})$ is a proper subgroup of Γ, therefore its complement $\Gamma(p') \cong \Gamma/\Gamma(p)$ is identified with a nontrivial subgroup of the multiplicative group $k^\times$; in particular, $|k| := q = p^f > 2$, so $f \ge 2$ for $p = 2$. Thus $1 \ne r := |\Gamma(p')| \mid (q-1)$, and $\Gamma(p') \cong \mathbb{Z}/r\mathbb{Z}$. Moreover, since Γ generates the local ring R, it follows that $k = \mathbb{F}_p(\Gamma(p'))$, and hence f is the order of $p \bmod r \in (\mathbb{Z}/r\mathbb{Z})^\times$.
We show that $\Gamma(p) \cap \Lambda = 1$, so $\Lambda = \Gamma(p') \cong \mathbb{Z}/r\mathbb{Z}$, $|\Gamma(p)| = (\mathfrak{G} : \mathfrak{H})$. Assuming the contrary, let $h := \eta(\sigma)$ for some $1 \ne \sigma \in \Gamma(p) \cap \Lambda$, so $\mathfrak{H} = Rh$. For any $\tau \in \Lambda$, we obtain

$$\tau h = \eta(\tau\sigma) - \eta(\tau) = \eta(\sigma\tau) - \eta(\tau) = \eta(\sigma) + (\sigma - 1)\eta(\tau) = \eta(\sigma) = h,$$

i.e., Λ acts trivially on $\mathfrak{H}$. Consequently, the restriction map $\eta|_\Lambda : \Lambda \longrightarrow \mathfrak{H}$ is a monomorphism, therefore $\Lambda \cong \eta(\Lambda)$ is a p-group, which is a contradiction.
It remains to show that $\Gamma(p) = 1$, so $\Gamma = \Lambda$, $R \cong k$, $\mathfrak{G} = \mathfrak{H} \cong k^+$ as required. Assuming the contrary, let $\mathfrak{G}'$ be the subgroup of $\mathfrak{G}$ generated by

$\eta(\Gamma(p)) \neq 0$. $\mathfrak{G}'$ is obviously stable under the action of $\Gamma(p)$ and also fixed by $\Gamma(p') = \Lambda$ since, acting on generators, we obtain

$$\tau\eta(\sigma) = \eta(\sigma) + (\sigma - 1)\eta(\tau) = \eta(\sigma)$$

for $\sigma \in \Gamma(p) = \mathrm{Fix}_\Gamma(\mathfrak{H}), \tau \in \Lambda = \eta^{-1}(\mathfrak{H})$.
Thus $\mathfrak{G}'$ is a nonzero Γ-submodule of $\mathfrak{G}$, and hence $\mathfrak{H} \subseteq \mathfrak{G}'$. On the other hand, the cocycle $\tilde{\eta} : \Gamma(p) \longrightarrow \mathfrak{G}/\mathfrak{H}$ induced by η is bijective since $\mathrm{Ker}\,(\tilde{\eta}) = \Gamma(p) \cap \eta^{-1}(\mathfrak{H}) = 1$ and $|\Gamma(p)| = (\mathfrak{G} : \mathfrak{H})$, therefore $\mathfrak{G}' = \mathfrak{G}$ and $1 \neq \Lambda \subseteq \mathrm{Fix}_\Gamma(\mathfrak{G}') = \mathrm{Fix}_\Gamma(\mathfrak{G}) = 1$, which is a contradiction.

(2) $\Longrightarrow$ (1). For an arbitrary prime number p and an arbitrary integer $r \geq 2$ such that $(p, r) = 1$ and $f \geq 2$ for $p = 2$, where f is the order of $p \bmod r \in (\mathbb{Z}/r\mathbb{Z})^\times$, it follows easily that (Γ, k^+, η) is a mnK triple, where $k = \mathbb{F}_q$, $q := p^f$, $\Gamma \cong \mathbb{Z}/r\mathbb{Z}$ is the unique subgroup of order r of the multiplicative group $k^\times$, acting canonically on k^+, and $0 \neq \eta \in Z^1(\Gamma, k^+) = B^1(\Gamma, k^+) \cong (\mathbb{Z}/p\mathbb{Z})^f$. Consequently, the pairs (p, r) as above classify, up to isomorphism, the mnK triples (Γ, G, η) with abelian Γ and $\mathfrak{G}$ such that $(|\Gamma|, |\mathfrak{G}|) = 1$.

Proposition 7.8. *Let $\eta \in Z^1(\Gamma, \mathfrak{G})$, where Γ and $\mathfrak{G}$ are finite abelian p-groups for some prime number p. With the notation from* Lemma 7.6, *the following assertions are equivalent.*

(1) *$(\Gamma, \mathfrak{G}, \eta)$ is a mnK triple, and the finite local ring R of characteristic p^n is principal.*
(2) *One of the following conditions is satisfied.*

   (i) *$n = 1$, i.e., $\mathrm{char}\, R = p$, and $R \cong \mathbb{F}_p[x]/(x^m)$ with $m \geq 2$ and $(m, p) = 1$.*
   (ii) *$p = n = 2$, i.e., $\mathrm{char}\, R = 4$, $R \cong (\mathbb{Z}/4\mathbb{Z})[x]/(f(x), 2x^e)$, where $f \in (\mathbb{Z}/4\mathbb{Z})[x]$ is an Eisenstein polynomial of degree $e \geq 1$, the nilpotency index $m = 2e$ is even, and R is free of rank e as $\mathbb{Z}/4\mathbb{Z}$-module.*
   (iii) *$p = n = 2$, i.e., $\mathrm{char} R = 4$, $R \cong (\mathbb{Z}/4\mathbb{Z})[x]/(f(x), 2x^t)$, where $f \in (\mathbb{Z}/4\mathbb{Z})[x]$ is an Eisenstein polynomial of degree $e \geq 2, 0 < t < e$, and the nilpotency index $m = e + t$ is odd.*

In all three cases above, the R-module $\mathfrak{G}$ is identified with R^+, $R/\mathfrak{m} \cong \mathbb{F}_p$, $\Gamma = 1 + \mathfrak{m}$, and the injective cocycle $\eta \in Z^1(\Gamma, R^+)$ is unique up to multiplication with elements from $R^\times$ and summation with homomorphisms defined on Γ with values in $(R^+)^\Gamma \cong \mathbb{Z}/p\mathbb{Z}$.

Proof. (1) $\Longrightarrow$ (2). By Lemma 7.6,(5), we can identify the R-module $\mathfrak{G}$ with R^+, $\Gamma = 1 + \mathfrak{m}$, $\eta \in Z^1(\Gamma, R^+)$ is injective, and $R = (\mathbb{Z}/p^n\mathbb{Z})[\theta] \cong \mathbb{Z}/p^n\mathbb{Z}[x]/(f(x), p^{n-1}x^t)$, where $\theta R = \mathfrak{m}$ and θ is a root of the Eisenstein polynomial of degree e

$$f(x) = x^e - p(a_{e-1}x^{e-1} + \cdots + a_0),\ a_i \in \mathbb{Z}/p^n\mathbb{Z}, a_0 \in (\mathbb{Z}/p^n\mathbb{Z})^\times,$$

and $1 \le t = m-(n-1)e \le e$, where m is the nilpotency index of the maximal ideal $\mathfrak{m}$. Choose $\widetilde{a_i} \in \mathbb{Z}_p$ such that $\widetilde{a_i} \equiv a_i \bmod p^n, i = 0,\dots,e-1$. The Eisenstein polynomial

$$\tilde{f}(x) = x^e - p(\widetilde{a_{e-1}}x^{e-1} + \cdots + \widetilde{a_0}) \in \mathbb{Z}_p[x]$$

is irreducible over $\mathbb{Q}_p$. Let $\pi \in \widetilde{\mathbb{Q}_p}$ be a root of $\tilde{f}$. Then $K = \mathbb{Q}_p(\pi)$ is a totally ramified extension of $\mathbb{Q}_p$ of degree e, with valuation ring $\mathcal{O} = \mathbb{Z}_p[\pi]$, maximal ideal $\mathfrak{p} = \pi\mathcal{O}$ and residue field $k = \mathcal{O}/\mathfrak{p} \cong \mathbb{Z}_p/p\mathbb{Z}_p \cong \mathbb{F}_p$. Sending π to θ, we obtain an epimorphism $\mathcal{O} \longrightarrow R$, therefore $R \cong \mathcal{O}_m := \mathcal{O}/\mathfrak{p}^m$, $\mathfrak{m} = \mathfrak{p}\mathcal{O}_m, \Gamma \cong U^{(1)}/U^{(m)}$. The cocycle $\eta : \Gamma \longrightarrow \mathcal{O}_m$ extends to a cocycle $\tilde{\eta} : U^{(1)} \longrightarrow \mathcal{O}_m$ with $\mathrm{Ker}\,(\tilde{\eta}) = U^{(m)}$. According to Example 5.14,

$$\tilde{\eta}(x) = \tilde{\eta}_{u,\alpha}(x) \equiv \pi^{m-1}\alpha(x) + \frac{x-1}{\pi}u \bmod \mathfrak{p}^m,$$

for some $u \in \mathcal{O}_m^{\times}, \alpha \in T_{m+1}^{\vee} = \mathrm{Hom}(T_{m+1}, \mathbb{F}_p^{+})$, with $T_{m+1} := \frac{U^{(1)}}{U^{(m+1)}V(U^{(1)})^p}$, where V is the closed subgroup of $U^{(1)}$ generated by $1+\pi$. As $U^{(m+1)} \subseteq \mathrm{Ker}\,(\tilde{\eta}) \subseteq U^{(m)}$ and $m \ge 2$, the necessary and sufficient condition for the required equality $\mathrm{Ker}\,(\tilde{\eta}) = U^{(m)}$ is that $U^{(m)} \cap (U^{(1)})^p \subseteq U^{(m+1)}$ and the restriction $\alpha \,|\, _{U^{(m)}}$ is the nontrivial character $\chi_{\overline{u}} : U^{(m)}/U^{(m+1)} \longrightarrow \mathbb{F}_p^{+}$, defined by

$$\chi_{\overline{u}}(x) = \frac{1-x}{\pi^m}\overline{u} \text{ for } x \in U^{(m)}, \text{ with } \overline{u} := u \bmod \mathfrak{m}.$$

Assuming $m > e$, whence $n \ge 2$, put $a := (1+\pi^{m-e})^p \in (U^{(1)})^p$. Since

$$v(a-1) \ge \min\,(v(p\pi^{m-e}), v(\pi^{(m-e)p})) = \min\,(m, (m-e)p),$$

the condition above implies the inequality $(m-e)p \le m$, therefore $(n-1)e < m \le \frac{pe}{p-1}$. It follows that $2 \le n \le \frac{p}{p-1} = 1 + \frac{1}{p-1}$, so $n = p = 2$. Consequently, if $p \ne 2$ then $n = 1$, i.e., char $R = p$, and $m = e = t \ge 2$, $R \cong \mathbb{F}_p[x]/(x^m)$, so we may take $\tilde{f}(x) = x^m - p$, $K = \mathbb{Q}_p(p^{\frac{1}{m}})$.
Next assume $p \,|\, m$, and put $b := (1+\pi^{\frac{m}{p}})^p \in U^{(1)}$. As

$$v(b-1) \ge \min\,(v(p\pi^{\frac{m}{p}}), v(\pi^m)) = \min\,(e + \frac{m}{p}, m),$$

we deduce that $e + \frac{m}{p} \le m$, i.e., $m \ge \frac{pe}{p-1}$, since otherwise we get the contradiction $b \in (U^{(m)} \cap (U^{(1)})^p) \setminus U^{(m+1)}$. Consequently, if $p \ne 2$ then $(m,p) = 1$ as stated by (i), i.e., the totally ramified extension $K/\mathbb{Q}_p$ is tame, since otherwise we obtain $e = m \ge \frac{pe}{p-1}$, a contradiction. If $p = 2 \,|\, m$ then $m \ge 2e$, in particular, $m > e$, and hence $n = 2$ and $m \le 2e$, so $m = 2e$, as stated by (ii). If $p = 2 \nmid m$, in particular, $m \geq 3$, then either $n = 1$ [case (i)] or $n = 2, e < m < 2e$ [case (iii)].

According to [15, Theorem 2], the principal local ring $R \cong \mathcal{O}_m$ is determined up to isomorphism by its invariants p, n, m only in the case (i), the cases (ii) and (iii) with e odd, and the case (iii) with $t = 1$, i.e., e even and $m = e + 1$. As shown above, the cocycles $\eta : \Gamma = 1 + \mathfrak{m} \longrightarrow \mathfrak{G} = R^+$ for which $(\Gamma, \mathfrak{G}, \eta)$ is a mnK triple, i.e., η is injective, are in 1-1 correspondence with the pairs $(u, \alpha) \in R^\times \times T_{m+1}^\vee$ satisfying $\alpha|_{U^{(m)}} = \chi_{\overline{u}}$. The condition $U^{(m)} \cap (U^{(1)})^p \subseteq U^{(m+1)}$ is equivalent with the fact that $U^{(m)}/U^{(m+1)} \cong \mathbb{Z}/p\mathbb{Z}$ is the kernel of the canonical projection $T_{m+1} \longrightarrow T_m$, so $T_{m+1}^\vee \cong T_m^\vee \oplus (U^{(m)}/U^{(m+1)})^\vee$. Thus the cocycles above are parametrized by the elements of the direct product $R^\times \times T_m^\vee = R^\times \times (\Gamma/\langle 1+\theta\rangle\Gamma^p)^\vee$, and hence they are, up to isomorphism in $\mathscr{L}^1$, in 1-1 correspondence to the elements of the elementary p-group $(\Gamma/\langle 1+\theta\rangle\Gamma^p)^\vee$. Consequently, the cocycle η is unique up to multiplication by elements of $R^\times$ and summation with homomorphisms from Γ to $\mathfrak{G}^\Gamma \cong \mathbb{Z}/p\mathbb{Z}$ as desired.

(2) $\Longrightarrow$ (1). Assume that the finite principal local ring R satisfies one of the conditions (i)–(iii). As in the first part of the proof, we choose a suitable finite extension K of $\mathbb{Q}_p$ with valuation ring $\mathcal{O}$, maximal ideal $\mathfrak{p} = \pi\mathcal{O}$, and residue field $k \cong \mathbb{F}_p$, such that $R \cong \mathcal{O}_m = \mathcal{O}/\mathfrak{p}^m$, where m is the nilpotency index of the maximal ideal $\mathfrak{m}$ of R. Thanks to the arguments from the first part of the proof, it suffices to check the condition $U^{(m)} \cap (U^{(1)})^p \subseteq U^{(m+1)}$. Assuming the contrary, let $x \in \mathfrak{p}, c := (1+x)^p - 1$ be such that $v(c) = m$. We consider separately the cases (i)–(iii).

Case (i). It follows that

$$m = v(c) \geq \min(v(px), v(x^p)) = \min(m+v(x), pv(x)) \geq \min(m+1, pv(x)),$$

therefore $m = pv(x)$, contrary to the hypothesis $(m, p) = 1$.

Case (ii). We obtain $c = (1+x)^2 - 1 = x(x+2)$, so

$$2e = m = v(c) = v(x) + v(x+2) \geq v(x) + \min(v(x), e),$$

hence $v(x) \leq e$. Assuming $v(x) < v(2) = e$, it follows that $2e = v(c) = 2v(x)$, which is a contradiction. Thus $v(x) = e$, therefore $v(1 + \frac{2}{x}) = v(c) - 2v(x) = 0$, contrary to the fact that $1 + \frac{2}{x} \equiv 2 \equiv 0 \bmod \mathfrak{p}$ since $\mathcal{O}/\mathfrak{p} \cong \mathbb{F}_2$.

Case (iii). It follows that $m = v(c) \geq v(x) + \min(e, v(x))$, therefore

$$e > t = m - e \geq v(x) + \min(0, v(x) - e),$$

so $v(x) < e$. Consequently, $m = v(c) = 2v(x)$, again a contradiction, as m is odd by hypothesis.

Remarks 7.9. (1) For large enough prime numbers p, an effective description of the corresponding mnK triples is easy. Indeed, let us consider the case (i) above with $p > m$. Then $\Gamma = 1 + \mathfrak{m} \cong (\mathbb{Z}/p\mathbb{Z})^{m-1}$ with a base consisting of the elements $\gamma_i := 1 + \theta^i, i = 1, \ldots, m-1$, and $R^\times \cong \Gamma \times \mathbb{F}_p^\times$. The canonical injective cocycle $\eta : \Gamma \longrightarrow R^+$ is completely determined by its values $\eta(\gamma_i) = \theta^{i-1}, i = 1, \ldots, m-1$, and every injective cocycle $\Gamma \longrightarrow R^+$ has the form $u(\eta + \theta^{m-1}\beta)$ with $u \in R^\times$, $\beta \in \mathrm{Hom}\,(\Gamma/\langle\gamma_1\rangle, \mathbb{F}_p^+) \cong (\mathbb{Z}/p\mathbb{Z})^{m-2}$.

(2) The classification of the mnK triples for which the local ring R is not principal seems to be a more difficult task. We give only a simple example of such triples. Let W be a vector space over the prime field $\mathbb{F}_p, p \neq 2$, with the base $\{\theta_i \mid i = 1, \ldots, s\}, s \geq 2$, and let $R = \mathbb{F}_p \oplus W$ with the multiplication given by

$$(x \oplus y) \cdot (x' \oplus y') = xx' \oplus (xy' + x'y).$$

R is a local ring with the maximal ideal $\mathfrak{m} = W$ whose nilpotency index is 2. The canonical map $\mathfrak{m} \longrightarrow \Gamma = 1 + \mathfrak{m}, x \mapsto 1 + x$ is an isomorphism, the elements $\gamma_i := 1 + \theta_i, i = 1, \ldots, s$, form a base of the elementary p-group Γ, and $R^\times \cong \mathbb{F}_p^\times \times \Gamma \cong \mathbb{Z}/(p-1)\mathbb{Z} \times (\mathbb{Z}/p\mathbb{Z})^s$. Setting $\theta_0 = 1$, the Pontryagin dual $\mathfrak{G} := R^\vee = \mathrm{Hom}\,(R^+, \mathbb{F}_p)$ is a vector space over $\mathbb{F}_p$ with the dual base $\{\theta_i^\vee \mid i = 0, \ldots, s\}$. The canonical R-module structure on $\mathfrak{G}$ is defined by the relations $\theta_0\theta_j^\vee = \theta_j^\vee, \theta_i\theta_j^\vee = \delta_{i\,j}\theta_0^\vee, i = 1, \ldots, s, j = 0, \ldots, s$, where $\delta_{i\,j}$ is the Kronecker symbol. It follows that $\mathfrak{H} := \mathfrak{G}^\Gamma = \mathbb{F}_p\theta_0^\vee \cong \mathbb{Z}/p\mathbb{Z}$, and the induced action of Γ on the quotient $\mathfrak{G}/\mathfrak{H} \cong (\mathbb{Z}/p\mathbb{Z})^s$ is trivial. Every cocycle $\eta \in Z^1(\Gamma, \mathfrak{G})$ is completely determined by its values $\eta(\gamma_i), i = 1, \ldots, s$, which must satisfy the condition $\theta_i\eta(\gamma_j) = \theta_j\eta(\gamma_i)$ for $1 \leq i, j \leq s$. Writing $\eta(\gamma_i) = \sum_{j=0}^{s} \lambda_{i\,j}\theta_j^\vee$, the cocycles η are in 1-1 correspondence to the pairs (λ_0, Λ) consisting of a homomorphism $\lambda_0 : \Gamma \longrightarrow \mathbb{F}_p, \gamma_i \mapsto \lambda_{i\,0}$, and a symmetric matrix $\Lambda = (\lambda_{i\,j})_{1 \leq i,j \leq s}$ with entries in $\mathbb{F}_p$.

$(\Gamma, \mathfrak{G}, \eta)$ is a mnK triple $\Longleftrightarrow$ η is injective $\Longleftrightarrow$ the matrix Λ is invertible $\Longleftrightarrow$ $\det(\Lambda) \neq 0$ $\Longleftrightarrow$ the quadratic form $Q(x) = \sum_{1 \leq i,j \leq s} \lambda_{i\,j}x_ix_j$ in $x_1, \ldots, x_s$ is nondegenerate. Consider the group $\mathcal{G} := \mathrm{Aut}\,(\Gamma, \mathfrak{G})$ consisting of the pairs (φ, ψ), where $\varphi \in \mathrm{Aut}(\Gamma)$ and $\psi \in \mathrm{Aut}(\mathfrak{G})$ satisfy the condition $\psi(\varphi(\gamma)g) = \gamma\psi(g)$ for $\gamma \in \Gamma, g \in \mathfrak{G}$, with the composition law $(\varphi, \psi) \circ (\varphi', \psi') = (\varphi' \circ \varphi, \psi \circ \psi')$. The elements of $\mathcal{G}$ are in 1-1 correspondence with the triples

$$(A = (a_{i\,j})_{1 \leq i,j \leq s} \in GL_s(\mathbb{F}_p), b_0 \in \mathbb{F}_p^\times, \mathbf{b} = (b_1, \ldots, b_s) \in \mathbb{F}_p^s),$$

assigning to such a triple $(A, b_0, \mathbf{b})$ the pair (φ, ψ) defined by

$$\varphi(\gamma_i) = \prod_{j=1}^{s} \gamma_j^{a_{ij}}, \psi(\theta_0^\vee) = b_0\theta_0^\vee, \psi(\theta_i^\vee) = b_0(b_i\theta_0^\vee + \sum_{j=1}^{s} a_{ji}\theta_j^\vee), i = 1, \ldots, s.$$

The group $\mathscr{G}$ acts on the injective cocycles, sending a cocycle η to the cocycle $\psi \circ \eta \circ \varphi$. The action is transitive whenever s is odd and hence, in this case, the mnK triple $(\Gamma, \mathfrak{G}, \eta)$ is unique up to isomorphism. Indeed let η_0 be the canonical injective cocycle defined by $\eta_0(\gamma_i) = \theta_i^\vee, i = 1, \ldots, s$, and η be an arbitrary injective cocycle defined by the pair (λ_0, Λ). Put

$$b_0 = \begin{cases} 1 & \text{if } \det(\Lambda) \in \mathbb{F}_p^2 \\ u \in \mathbb{F}_p \setminus \mathbb{F}_p^2 & \text{if } \det(\Lambda) \notin \mathbb{F}_p^2 \end{cases}$$

By [36, 1.7, Proposition 5], there exists $A \in GL_s(\mathbb{F}_p)$ such that $\Lambda = A \cdot (b_0 I_s) \cdot {}^t A$, where ${}^t A$ denotes the transpose of A. Let $\mathbf{b} = (b_1, \ldots, b_s) \in \mathbb{F}_p^s$ be the unique solution of the linear system in $x_1, \ldots, x_s$

$$\sum_{j=1}^{s} a_{ij}x_j = b_0^{-1}\lambda_{i0} + \sum_{j=1}^{s} \frac{a_{ij}(1 - a_{ij})}{2}, i = 1, \ldots, s.$$

One checks easily that $\eta = \psi \circ \eta_0 \circ \varphi$, where the pair $(\varphi, \psi) \in \mathscr{G}$ is determined by the triple $(A, b_0, \mathbf{b})$ as defined above. Similarly, it follows for s even that $(\Gamma, \mathfrak{G}, \eta_i), i = 0, 1$, are up to isomorphism the only two mnK triples associated with the pair $(\Gamma, \mathfrak{G})$, where the injective cocycle η_1 is defined by $\eta_1(\gamma_i) = \theta_i^\vee, i = 1, \ldots, s-1, \eta_1(\gamma_s) = u\theta_s^\vee$ with $u \in \mathbb{F}_p \setminus \mathbb{F}_p^2$.

Corollary 7.10. *Let $(\Gamma, \mathfrak{G}, \eta)$ be a mnK triple, and assume $\mathfrak{G}$ is abelian of exponent k and the action of Γ is induced by a character $\chi : \Gamma \longrightarrow (\mathbb{Z}/k\mathbb{Z})^\times$, i.e., $\gamma g = \chi(\gamma)g$ for $\gamma \in \Gamma, g \in \mathfrak{G}$. Then either $k = p$ is an odd prime number or $k = 4$, and $\mathfrak{G} \cong \mathbb{Z}/k\mathbb{Z}$.*

If $k = p \neq 2$ then $(\Gamma, \mathfrak{G}, \eta) \cong (U, \mathbb{F}_p^+, u \mapsto u - 1)$, where $U \cong \Gamma \cong \mathbb{Z}/r\mathbb{Z}$, $2 \leq r \mid (p-1)$, is the unique subgroup of order $r = |\Gamma|$ of the multiplicative group $\mathbb{F}_p^\times$ acting by multiplication on $\mathbb{F}_p^+$.

If $k = 4$ then $(\Gamma, \mathfrak{G}, \eta) \cong (\mathbb{Z}/2\mathbb{Z}, \mathbb{Z}/4\mathbb{Z}, 1 \bmod 2 \mapsto 1 \bmod 4)$ with the nontrivial action of $\mathbb{Z}/2\mathbb{Z}$ on $\mathbb{Z}/4\mathbb{Z}$.

Proof. Since $(\Gamma, \mathfrak{G}, \eta)$ is a mnK triple and $\mathfrak{G}$ is abelian, we get $\mathrm{Fix}_\Gamma(\mathfrak{G}) = 1$ by Lemma 7.4,(2), and hence $\Gamma \cong \chi(\Gamma) \subseteq (\mathbb{Z}/k\mathbb{Z})^\times$ is abelian. It remains to apply Lemma 7.6, Proposition 7.7, and Proposition 7.8.

As a consequence of Corollary 7.10 and Proposition 5.22, we find again [3, Lemma 1.18, Theorem 1.20], the abstract version in the framework described in Example 4.10 of the classical Kneser criterion for separable radical extensions [21], [2, Theorem 11.1.5].

References

1. Albu, T.: Infinite field extensions with Cogalois correspondence. Commun. Algebra **30**, 2335–2353 (2002)
2. Albu, T.: Cogalois Theory. A Series of Monographs and Textbooks, vol. 252. Marcel Dekker, New York (2003)
3. Albu, T., Basarab, Ş.A.: An abstract Cogalois Theory for profinite groups. J. Pure Appl. Algebra **200**, 227–250 (2005)
4. Albu, T., Nicolae, F.: Kneser field extensions with Cogalois correspondence. J. Number Theory **52**, 299–318 (1995)
5. Albu, T., Nicolae, F.: Finite radical field extensions and crossed homomorphisms. J. Number Theory **60**, 291–309 (1996)
6. Artin, E., Schreier, O.: Eine Kennzeichnung der reel abgeschlossenen Körper. Hamb. Abh. **5**, 225–231 (1927)
7. Barbilian, D.: Solution exhaustive du problème du Steinitz (Romanian). Stud. Cerc. Mat. **2**, 1–55 (1951)
8. Barrera-Mora, F., Lam-Estrada, P.: Radical extensions and crossed homomorphisms. Bull. Aust. Math. Soc. **64**, 107–119 (2001)
9. Barrera-Mora, F., Rzedowski-Calderón, M., Villa-Salvador, G.: On Cogalois extensions. J. Pure Appl. Algebra **76**, 1–11 (1991)
10. Basarab, Ş.A.: The dual of the category of generalized trees. An. Ştiinţ. Univ. Ovidius Constanţa Ser. Mat. **9**(1), 1–20 (2001)
11. Basarab, Ş.A.: Kneser and hereditarily Kneser subgroups of a profinite group. Serdica Math. J. **30**, 325–348 (2004)
12. Basarab, Ş.A.: CoGalois and strongly coGalois actions. J. Pure Appl. Algebra **212**, 1674–1694 (2008)
13. Basarab, Ş.A.: A new setting for co-Galois theory, I: Universal cocycles and Kneser structures. Preprint, 2006, 77 pp
14. Besicovitch, A.S.: On the linear independence of fractional powers of integers. J. Lond. Math. Soc. **15**, 3–6 (1940)
15. Clark, W.E., Liang, J.J.: Enumeration of finite commutative chain rings. J. Algebra **27**, 445–453 (1973)
16. Garcia, A., Stichtenoth, H.: Elementary abelian p-extensions of algebraic function fields. Manuscr. Math. **72**, 67–79 (1991)
17. Greither, C., and Harrison, D.K.: A Galois correspondence for radical extensions of fields. J. Pure Appl. Algebra **43**, 257–270 (1986)
18. Hasse, H.: Bericht über neuere Untersuchungen und Probleme aus der Theorie der algebraischen Zahlkörper, Teil II: Reziprozitätsgesetz, Physica-Verlag, Würzburg Wien (1965)
19. Hochster, M.: Prime ideal structure in commutative rings. Trans. Am. Math. Soc. **142**, 43–60 (1969)
20. Johnstone, P.T.: Stone Spaces. Cambridge University Press, Cambridge (1982)
21. Kneser, M.: Lineare Abhängigkeit von Wurzeln. Acta Arith. **26**, 307–308 (1975)
22. Krull, W.: Galoissche Theorie der unendlichen algebraischen Erweiterungen. Math. Ann. **100**, 687–698 (1928)
23. Krull, W.: Über eine Verallgemeinerung des Normalkörperbegriffs. J. Reine Angew. Math. **191**, 54–63 (1953)
24. Lorenz, F.: Algebra. vol. 2. Springer, New York (2008)
25. Lorenz, F., Roquette, P.: The theorem of Grunwald-Wang in the setting of valuation theory. In: Kuhlmann, F-V., Kuhlmann, S., Marshall, M. (eds.) Valuation Theory and Its Applications, vol. 2, pp.175–212. Fields Institute Communications, vol. 33. American Mathematical Society, Providence (2003)
26. Mordell, L.J.: On the linear independence of algebraic numbers. Pac. J. Math. **3**, 625–630 (1953)

27. Neukirch, J.: Algebraische Zahlentheorie. Springer, Berlin (1992)
28. Neukirch, J., Schmidt, A., Wingberg, K.: Cohomology of Number Fields. Springer, Berlin (2000)
29. Ore, O.: Galois connexions. Trans. Am. Math. Soc. **55**, 493–513 (1944)
30. Orozco, M.A., Velez, W.Y.: The torsion group of a field defined by radicals. J. Number Theory **19**, 283–294 (1984)
31. Pop, F.: Classicaly projective groups and pseudo classicaly closed groups. In: Kuhlmann, F-V., Kuhlmann, S., Marshall, M. (eds.) Valuation Theory and Its Applications, Vol. II, pp.251–283. Fields Institute Communications, Volume **33**, American Mathematical Society, Providence (2003)
32. Ribes, L., Zalesskii, P.: Profinite Groups. Springer, Berlin (2000)
33. Rovinsky, M.: On certain isomorphisms between absolute Galois groups. Compos. Math. **136**(1), 61–67 (2003)
34. Schinzel, A.: On linear dependence of roots. Acta Arith. **28**, 161–175 (1975)
35. Serre, J-P.: Cohomologie Galoisienne. Lecture Notes in Mathematics, vol. 5. Springer, Berlin (1964)
36. Serre, J-P.: A Course in Arithmetic. Springer, New York (1973)
37. Shimura, G.: Introduction to the Arithmetic Theory of Automorphic Functions. Iwanami Shoten Publishers, Tokyo and Princeton University Press, Princeton (1971)
38. Siegel, C.L.: Algebraische Abhängigkeit von Wurzeln. Acta Arith. **21**, 59–64 (1972)
39. Steinitz, E.: Algebraische Theorie der Körper. J. Reine Angew. Math. **137**, 167–300 (1910)
40. Witt, E.: Zyklische Körper und Algebren der Charakteristik p vom Grad p^n. Struktur diskret bewerteter perfekter Körper mit vollkommenen Restklassenkörper der Charakteristik p. J. Reine Angew. Math. **167**, 126–140 (1937)

Connectivity and a Problem of Formal Geometry

Lucian Bădescu

To Alexandru Dimca and Ştefan Papadima on the occasion of their sixtieth anniversaries

Abstract Let $P = \mathbb{P}^m(e) \times \mathbb{P}^n(h)$ be a product of weighted projective spaces, and let Δ_P be the diagonal of $P \times P$. We prove an algebraization result for formal-rational functions on certain closed subvarieties X of $P \times P$ along the intersection $X \cap \Delta_P$.

Keywords Rational functions • Formal functions • Meromorphic functions Connectivity • Algebraization of formal entities

1 Introduction

Let P be a projective irreducible variety and let $f: X \to P \times P$ be a morphism from a complete irreducible variety X over an algebraically closed field k. Denote by Δ_P the diagonal of $P \times P$. Then one may ask under which conditions the inverse image $f^{-1}(\Delta_P)$ is connected (resp. nonempty). Here by a connected scheme we shall mean a nonempty scheme whose underlying topological space is connected. The first result in this direction is the famous theorem of Fulton and Hansen [13] which states that the answer to this question is affirmative if $P = \mathbb{P}^n$ and if $\dim f(X) > n$ (resp. if $\dim f(X) \geq n$). That result has a lot of interesting geometric applications (see [14]).

The connectivity result of Fulton and Hansen has been generalized in various directions by Hansen [15], Faltings [11,12], Debarre [7–9], Bădescu [1,3], Bădescu–Repetto [4], and others.

On the other hand, in [1] and [3] the connectivity results of Fulton–Hansen [14] and Debarre [9] have been improved to get stronger conclusions involving the $G3$ condition of Hironaka–Matsumura [17] on the extension of formal-rational

L. Bădescu (✉)
Dipartimento di Matematica, Università degli Studi di Genova,
Via Dodecaneso 35, 16146 Genova, Italy
e-mail: badescu@dima.unige.it

D. Ibadula, W. Veys (eds.), *Bridging Algebra, Geometry, and Topology*,
Springer Proceedings in Mathematics & Statistics 96,
DOI 10.1007/978-3-319-09186-0_5

functions on X along $f^{-1}(\Delta_P)$ (see Definition 1 below). The aim of the present paper is to improve the connectivity result of Bădescu and Repetto [4] in the same spirit.

To state our main result, let P denote the product $\mathbb{P}^m(e) \times \mathbb{P}^n(h)$ of weighted projective spaces $\mathbb{P}^m(e)$ and $\mathbb{P}^n(h)$ of weights $e = (e_0, \dots, e_m)$ and $h = (h_0, \dots, h_n)$ respectively, with $e_i, h_j \geq 1$, $i = 0, \dots, m$ and $j = 0, \dots, n$. Let $f: X \to P \times P$ be a morphism from a complete irreducible variety X. Denote by $X_{13} \subseteq \mathbb{P}^m(e) \times \mathbb{P}^m(e)$ (resp. by $X_{24} \subseteq \mathbb{P}^n(h) \times \mathbb{P}^n(h)$) the image of $f(X)$ under the projection p_{13} of $P \times P = \mathbb{P}^m(e) \times \mathbb{P}^n(h) \times \mathbb{P}^m(e) \times \mathbb{P}^n(h)$ onto $\mathbb{P}^m(e) \times \mathbb{P}^m(e)$ (resp. under the projection p_{24} onto $\mathbb{P}^n(h) \times \mathbb{P}^n(h)$). For the basic properties of weighted projective spaces, see [10] or [6].

Precisely, our aim is to prove the following strengthening of the connectivity result of Bădescu and Repetto [4] (see Theorem 1 below), and, under a slightly stronger hypothesis, also a generalization of the main result of Bădescu and Repetto [3]:

Theorem (=Theorem 6 below). *Under the above notation, let $f: X \to P \times P$ be a morphism from a complete irreducible variety X, with $P := \mathbb{P}^m(e) \times \mathbb{P}^n(h)$, the product of the weighted projective spaces $\mathbb{P}^m(e)$ and $\mathbb{P}^n(h)$ over an algebraically closed field k. Let Δ_P be the diagonal of $P \times P$ and set $a := \max\{m + \dim X_{24}, n + \dim X_{13}\}$. If $\dim f(X) > a$ then $f^{-1}(\Delta_P)$ is $G3$ in X, i.e., the canonical injective map $\alpha: K(X) \to K(X_{/f^{-1}(\Delta_P)})$, from the field $K(X)$ of rational functions of X to the ring $K(X_{/f^{-1}(\Delta_P)})$ of formal-rational functions of X along $f^{-1}(\Delta_P)$, is an isomorphism (see the Definition 1 below).*

In other words, this theorem is an extension result of the formal rational functions on X along $X \cap \Delta_P$ to rational functions on X. Let me explain why this theorem is an improvement of the connectivity result proved in [4]. By Theorem 3 below the connectivity result of [4] (see Theorem 1 below) is equivalent to saying that the ring $K(f(X)_{/f(X)\cap\Delta_P})$ is a field and the subfield $K(f(X))$ is algebraically closed in $K(f(X)_{/f(X)\cap\Delta_P})$, while the above theorem is equivalent to saying that the natural map $K(f(X)) \to K(f(X)_{/f(X)\cap\Delta_P})$ is an isomorphism.

To prove this result we use an extension theorem for formal-rational functions for the case $P = \mathbb{P}^m \times \mathbb{P}^n$ proved in [3] (see Theorem 4 below) and the connectivity result proved in [4] (Theorem 1 below), via some basic known results on formal-rational functions.

Here are two consequences of the above Theorem:

Corollary 1. *Let $f: X \to P \times P$ be a morphism from a complete irreducible variety X, with $P = \mathbb{P}^m(e) \times \mathbb{P}^n(h)$ over an algebraically closed field of arbitrary characteristic such that $m \geq n \geq 1$ and $\operatorname{codim}_{P\times P} f(X) < n$. Then $f^{-1}(\Delta)$ is $G3$ in X.*

In the special case when $P = \mathbb{P}^m \times \mathbb{P}^n$ is a product of two ordinary projective spaces over an algebraically closed field of characteristic zero and f is a closed embedding, Corollary 1 also follows from an old general result of Faltings (see [12],

Satz 8, p. 161) proved in the case when P is a complex projective rational homogeneous space. In general $P = \mathbb{P}^m(e) \times \mathbb{P}^n(h)$ is singular, so that Corollary 1 (to our best knowledge) is new.

Corollary 2. *Let X and Y be two closed irreducible subvarieties of $P = \mathbb{P}^m(e) \times \mathbb{P}^n(h)$ such that $m \geq n \geq 1$ and $\dim X + \dim Y > 2m + n$. Then $X \cap Y$ is $G3$ in X and in Y.*

Corollary 2 extends to the case $P = \mathbb{P}^m(e) \times \mathbb{P}^n(h)$ an old result of Faltings [11] proved (by local methods) if $P = \mathbb{P}^n$.

The paper is organized as follows. In the first section we recall some known results that will be needed in Sect. 2. In the second section we prove the theorem and the two corollaries stated above.

Terminology and notation. Unless otherwise specified, we shall use the standard terminology and notation in algebraic geometry. We shall work over an algebraically closed ground field k of arbitrary characteristic.

2 Background Material

In this section we gather together the known results which are going to be used in Sect. 2.

Theorem 1 (Bădescu–Repetto [4]). *Under the notation of the introduction, let $f: X \to P \times P$ be a morphism from a complete irreducible variety X, with $P = \mathbb{P}^m(e) \times \mathbb{P}^n(h)$ the product of the weighted projective spaces $\mathbb{P}^m(e)$ and $\mathbb{P}^n(h)$ over an algebraically closed field k. Let Δ_P be the diagonal of $P \times P$ and set $a := \max\{m + \dim X_{24}, n + \dim X_{13}\}$. Then the following statements hold true:*

i) *If $\dim f(X) \geq a$ then $f^{-1}(\Delta_P)$ is nonempty, and*
ii) *If $\dim f(X) > a$ then $f^{-1}(\Delta_P)$ is connected.*

Remark 1. If in Theorem 1 we take $n = 0$ then X_{24} is a point and hence $a = \max\{m, \dim X_{13}\}$. Then $P \cong \mathbb{P}^m$, $f(X) \cong X_{13}$, and therefore the conclusion of Theorem 1 becomes:

i') If $\dim f(X) \geq m$ then $f^{-1}(\Delta_P) \neq \varnothing$, and
ii') If $\dim f(X) > m$ then $f^{-1}(\Delta_P)$ is connected.

In other words Theorem 1 for $n = 0$ yields exactly the Fulton–Hansen connectivity theorem.

Lemma 1. *Let $f: X \to P \times P$ be a morphism as in Theorem 1, with $P = \mathbb{P}^m(e) \times \mathbb{P}^n(h)$. Assume $m \geq n \geq 1$.*

i) *If $\dim f(X) \geq 2m + n$ then $\dim f(X) \geq a$.*
ii) *If $\dim f(X) > 2m + n$ then $\dim f(X) > a$.*

Proof. Since $X_{13} \subseteq \mathbb{P}^m(e) \times \mathbb{P}^m(e)$ and $X_{24} \subseteq \mathbb{P}^n(h) \times \mathbb{P}^n(h)$, $\dim X_{13} \leq 2m$ and $\dim X_{24} \leq 2n$. It follows that $a = \max\{m + \dim X_{24}, n + \dim X_{13}\} \leq \max\{m + 2n, n + 2m\} = 2m + n$. □

Via Lemma 1 we get the following Corollary of Theorem 1:

Corollary 1. *Let $f : X \to P \times P$ be a morphism as in Theorem 1, with X a complete irreducible variety and $P := \mathbb{P}^m(e) \times \mathbb{P}^n(h)$, $m \geq n \geq 1$. If $\dim f(X) > 2m + n$ then $f^{-1}(\Delta_P)$ is connected.*

Proof. Since $\dim X_{13} \leq 2m$, $\dim X_{24} \leq 2n$ and $m \geq n$, then $a \leq \max\{2m + n, 2n + m\} = 2m + n$, and the conclusion follows from Theorem 1 and Lemma 1. □

Definition 1 (Hironaka–Matsumura [17], or also [16], or also [2], Chap. 9). Let X be a complete irreducible variety over the field k, and let Y be a closed subvariety of X. Denote by $K(X)$ the field of rational functions of X, by $X_{/Y}$ the formal completion of X along Y, and by $K(X_{/Y})$ the ring of formal-rational functions of X along Y. According to Hironaka and Matsumura [17] we say that Y is $G3$ in X if the canonical injective map $\alpha_{X,Y} : K(X) \to K(X_{/Y})$ is an isomorphism of k-algebras. In other words, Y is $G3$ in X if every formal rational-function of X along Y extends to a rational function of X. We also say that Y is $G2$ in X if the natural injective map $\alpha_{X,Y} \colon K(X) \to K(X_{/Y})$ makes $K(X_{/Y})$ a finite field extension of $K(X)$.

Let $f \colon X' \to X$ be a proper surjective morphism of irreducible varieties, and let $Y \subset X$ and $Y' \subset X'$ be closed subvarieties such that $f(Y') \subseteq Y$. Then one can define a canonical map of k-algebras $\tilde{f}^* \colon K(X_{/Y}) \to K(X'_{/Y'})$ (pull back of formal-rational functions, see [17], or also [2], Corollary 9.8) rendering commutative the following diagram:

$$\begin{array}{ccc} K(X) & \xrightarrow{f^*} & K(X') \\ {\scriptstyle \alpha_{X,Y}} \downarrow & & \downarrow {\scriptstyle \alpha_{X',Y'}} \\ K(X_{/Y}) & \xrightarrow{\tilde{f}^*} & K(X'_{/Y'}) \end{array}$$

Proposition 1 (Hironaka–Matsumura [17], or also [2], Cor. 9.10). *Let X be an irreducible algebraic variety over k, and let Y be a closed subvariety of X. Let $u \colon \tilde{X} \to X$ be the (birational) normalization of X. Then $K(X_{/Y})$ is a field if and only if $u^{-1}(Y)$ is connected.*

Theorem 2 (Hironaka–Matsumura [17], or also [2], Thm. 9.11). *Let $f \colon X' \to X$ be a proper surjective morphism of irreducible varieties over k. Then for every closed subvariety Y of X there is a canonical isomorphism*

$$K(X'_{/f^{-1}(Y)}) \cong [K(X') \otimes_{K(X)} K(X_{/Y})]_0,$$

where $[A]_0$ denotes the total ring of fractions of a commutative unitary ring A.

Corollary 2. *Under the hypotheses of Theorem 2, assume that Y is G3 in X. Then $f^{-1}(Y)$ is G3 in X'.*

Theorem 3 (Bădescu–Schneider [5], or also [2], Cor. 9.22). *Let (X, Y) be a pair consisting of a complete irreducible variety X over k and a closed subvariety Y of X. The following conditions are equivalent:*

i) *For every proper surjective morphism $f: X' \to X$ from an irreducible variety X', $f^{-1}(Y)$ is connected.*
ii) *$K(X_{/Y})$ is a field and $K(X)$ is algebraically closed in $K(X_{/Y})$.*

Theorem 4 (Bădescu [3]). *Under the notation of Theorem 1 let $f: X \to P \times P$ be a morphism from a complete irreducible variety X, with $P = \mathbb{P}^m \times \mathbb{P}^n$ a product of the ordinary projective spaces $\mathbb{P}^m$ and $\mathbb{P}^n$ over k and let Δ_P be the diagonal of $P \times P$. Assume that* $\dim f(X) > m + n + 1$, $\dim X_{13} > m$ *and* $\dim X_{24} > n$. *Then $f^{-1}(\Delta_P)$ is G3 in X.*

Theorem 5 (Bădescu–Schneider [5], or also [2], Thm. 9.21). *Let $\zeta \in K(X_{/Y})$ be a formal-rational function of an irreducible variety X along a closed subvariety Y of X such that $K(X_{/Y})$ is a field. Then the following two conditions are equivalent:*

i) *ζ is algebraic over $K(X)$.*
ii) *There is a proper surjective morphism $f: X' \to X$ from an irreducible variety X' and a closed subvariety Y' of X' such that $f(Y') \subseteq Y$ and $\tilde{f}^*(\zeta) \in K(X')$ (more precisely, there exists a rational function $t \in K(X')$ such that $\tilde{f}^*(\zeta) = \alpha_{X',Y'}(t)$).*

3 Extending Formal-Rational Functions

Start with the following:

Lemma 2. *Under the above notation let $P = \mathbb{P}^m(e) \times \mathbb{P}^n(h)$ be the product of the weighted projective spaces $\mathbb{P}^m(e)$ and $\mathbb{P}^n(h)$ over k, let X be a closed irreducible subvariety of $P \times P$, and set* $a := \max\{m + \dim X_{24}, n + \dim X_{13}\}$.

i) *If* $\dim X > a$ *then* $\dim X > m + n + 1$, $\dim X_{13} > m$ *and* $\dim X_{24} > n$;
ii) *If* $\dim X \geq a$ *then* $\dim X \geq m + n$, $\dim X_{13} \geq m$ *and* $\dim X_{24} \geq n$.

Proof. By the hypothesis that $\dim X > a$ we get $\dim X > m + \dim X_{24}$ and $\dim X > n + \dim X_{13}$. Denote by $p: X \to X_{13}$ and $q: X \to X_{24}$ the two canonical (surjective) projections, and by F_p and F_q the general fibers of p and q respectively.

i) By way of contradiction assume for instance that $\dim X_{13} \leq m$. Then we get successively:

$$\begin{aligned} \dim X_{24} &< \dim X - m && \text{(by } \dim X > m + \dim X_{24}) \\ &\le \dim X - \dim X_{13} && \text{(by } \dim X_{13} \le m) \\ &= \dim F_p && \text{(by the theorem on dimension of fibers)} \\ &\le \dim X_{24} && \text{(the restriction } q|F_p\colon F_p \to X_{24} \text{ is injective).} \end{aligned}$$

Thus the assumption that $\dim X_{13} \le m$ leads to the contradiction that $\dim X_{24} < \dim X_{24}$. This proves that $\dim X_{13} > m$. In the same manner one proves that $\dim X_{24} > n$. Finally, from $\dim X > n + \dim X_{13}$ and $\dim X_{13} \ge m + 1$ we get $\dim X > m + n + 1$.

ii) If instead $\dim X \ge m + \dim X_{24}$ and $\dim X \ge n + \dim X_{13}$ we may again assume, by way of contradiction, that $\dim X_{13} < m$. Then we get successively:

$$\begin{aligned} \dim X_{24} &\le \dim X - m && \text{(by } \dim X \ge m + \dim X_{24}) \\ &< \dim X - \dim X_{13} && \text{(by } \dim X_{13} < m) \\ &= \dim F_p && \text{(by the theorem on dimension of fibers)} \\ &\le \dim X_{24} && \text{(the restriction } q|F_p\colon F_p \to X_{24} \text{ is injective).} \end{aligned}$$

Thus the assumption that $\dim X_{13} < m$ leads to the same contradiction as above. This proves that $\dim X_{13} \ge m$. In the same manner one proves that $\dim X_{24} \ge n$. Finally, from $\dim X \ge n + \dim X_{13}$ and $\dim X_{13} \ge m$ we get $\dim X \ge m + n$. □

Now we can strengthen part ii) of Theorem 1 above to get the main result of this paper:

Theorem 6. *Under the notation of the introduction, let $f\colon X \to P \times P$ be a morphism from a complete irreducible variety X over an algebraically closed field k of arbitrary characteristic, with $P = \mathbb{P}^m(e) \times \mathbb{P}^n(h)$, and let Δ_P be the diagonal of $P \times P$. If $\dim f(X) > a := \max\{m + \dim X_{24}, n + \dim X_{13}\}$ then $f^{-1}(\Delta_P)$ is $G3$ in X.*

Proof. By Corollary 2 applied to the proper surjective morphism $f\colon X \to f(X)$, it is enough to prove that $f(X) \cap \Delta_P$ if $G3$ in $f(X)$. In other words, replacing X by $f(X)$ we may assume that X is a closed subset of $P \times P$ of dimension $> a$ and then we have to prove that $X \cap \Delta_P$ is $G3$ in X.

Let $P' := \mathbb{P}^m \times \mathbb{P}^n$ be the product of two ordinary projective spaces of dimension m and n respectively. Then we have the canonical finite surjective morphisms $u_m(e)\colon \mathbb{P}^m \to \mathbb{P}^m(e)$ and $u_n(h)\colon \mathbb{P}^n \to \mathbb{P}^n(h)$. It follows that the morphism

$$u := u_m(e) \times u_n(h) \times u_m(e) \times u_n(h)\colon P' \times P' \to P \times P,$$

is finite and surjective. Choose an irreducible component X' of $u^{-1}(X)$ and denote by $v\colon X' \to X$ the restriction $u|X'$. Clearly, v is again a finite surjective morphism, and in particular, $\dim X' = \dim X$. Then it makes sense to define the irreducible subvarieties $X'_{13} \subseteq \mathbb{P}^m \times \mathbb{P}^m$ and $X'_{24} \subseteq \mathbb{P}^n \times \mathbb{P}^n$. Since the morphisms

$$(u_m(e) \times u_m(e))|X'_{13}: X'_{13} \to X_{13} \text{ and } (u_n(h) \times u_n(h))|X'_{24}: X'_{24} \to X_{24}$$

are finite and surjective we infer that $\dim X'_{13} = \dim X_{13}$ and $\dim X'_{24} = \dim X_{24}$, we get

$$\dim X' > a = \max\{m+\dim X_{24}, n+\dim X_{13}\} = \max\{m+\dim X'_{24}, n+\dim X'_{13}\}. \tag{1}$$

Then by Lemma 2, i), the inequality (1) yield the following inequalities

$$\dim X' > m + n + 1, \ \dim X'_{13} > m \text{ and } \dim X'_{24} > n. \tag{2}$$

Now, the inequalities (2) show that the hypotheses of Theorem 4 above are satisfied for the inclusion $X' \subset P' \times P'$. Therefore by Theorem 4 it follows that $X' \cap \Delta_{P'}$ is $G3$ in X'. Now the idea is to show that in our situation, this last fact implies that $X \cap \Delta_P$ is $G3$ in X as well.

Indeed, consider the following commutative diagram

$$\begin{array}{ccc} K(X) & \xrightarrow{\nu^*} & K(X') \\ {\scriptstyle \alpha_{X,X\cap\Delta_P}} \downarrow & & \downarrow {\scriptstyle \alpha_{X',X'\cap\Delta_{P'}}} \\ K(X_{/X\cap\Delta_P}) & \xrightarrow{\tilde{\nu}^*} & K(X'_{/X'\cap\Delta_{P'}}) \end{array} \tag{3}$$

in which the second vertical map is an isomorphism (because $X' \cap \Delta_{P'}$ is $G3$ in X') and the first horizontal map yields a finite field extension (because the morphism ν is finite). In particular, via the injective map $\alpha_{X',X'\cap\Delta_{P'}} \circ \nu^*$, $K(X'_{/X'\cap\Delta_{P'}})$ becomes a finite field extension of $K(X)$.

On the other hand, we claim that the ring $K(X_{/X\cap\Delta_P})$ of formal-rational functions of X along $X \cap \Delta_P$ is actually a field. Indeed by Proposition 1 we have to check that if $f: \tilde{X} \to X$ is the birational normalization of X, then $f^{-1}(X \cap \Delta_P)$ is connected. But the connectivity of $f^{-1}(X \cap \Delta_P)$ follows from Theorem 1. So $K(X_{/X\cap\Delta_P})$ field, and hence it can be identified with a subfield of $K(X'_{/X'\cap\Delta_{P'}})$ which contains $K(X)$. By the commutativity of diagram (3) we get

$$\tilde{\nu}^* \circ \alpha_{X,X\cap\Delta_P} = \alpha_{X',X'\cap\Delta_{P'}} \circ \nu^*,$$

so that the field extension $K(X_{/X\cap\Delta_P})|K(X)$ becomes a field subextension of the finite field extension $K(X'_{/X'\cap\Delta_{P'}})|K(X)$. It follows that the field extension $K(X_{/X\cap\Delta_P})|K(X)$ is finite, i.e., $X \cap \Delta_P$ is $G2$ in X.

It remains to see that the map $\alpha_{X,X\cap\Delta_P}$ is an isomorphism. Under our assumption that $\dim X > a = \max\{m + \dim X_{24}, n + \dim X_{13}\}$, we can apply Theorem 1, ii) to get that the condition i) of Theorem 3 is satisfied for the pair $(X, X \cap \Delta_P)$. Then by Theorem 3 above, this condition is equivalent to saying that the subfield $K(X)$ is algebraically closed in $K(X_{/X\cap\Delta_P})$. Recalling also that the field extension $K(X_{/X\cap\Delta_P})|K(X)$ is finite (and hence algebraic) we get that the map $\alpha_{X,X\cap\Delta_P}$ is an isomorphism, i.e., $X \cap \Delta_P$ is $G3$ in X. □

Corollary 3. *Let $f: X \to P \times P$ be a morphism from a complete irreducible variety X, with $P = \mathbb{P}^m(e) \times \mathbb{P}^n(h)$, such that $m \geq n \geq 1$ and $\operatorname{codim}_{P\times P} f(X) < n$. Then $f^{-1}(\Delta)$ is $G3$ in X.*

Proof. This follows from Theorem 6 and Corollary 1. □

Remark 2. In the special case when $P = \mathbb{P}^m \times \mathbb{P}^n$ is a product of two ordinary projective spaces over the field $\mathbb{C}$ of complex numbers, Corollary 1 also follows (via Corollary 2) from an old general result of Faltings (see [12], Satz 8, p. 161) proved in the case when P is a complex projective rational homogeneous space. In general $P = \mathbb{P}^m(e) \times \mathbb{P}^n(h)$ is singular, so that Corollary 1 (to our best knowledge) is new.

Corollary 4. *Let X and Y be two closed irreducible subvarieties of $P = \mathbb{P}^m(e) \times \mathbb{P}^n(h)$ such that $m \geq n \geq 1$ and $\dim X + \dim Y > 2m + n$. Then $X \cap Y$ is $G3$ in X and in Y.*

Proof. Let $p_1: X \times Y \to X$ be the first projection of $X \times Y$. Then we get the commutative diagram:

$$\begin{array}{ccc} (X\times Y)\cap\Delta_P & \xrightarrow{\subset} & X\times Y \\ \cong\downarrow & & \downarrow p_1 \\ X\cap Y & \xrightarrow{\subset} & X \end{array}$$

yields the following commutative diagram

$$\begin{array}{ccc} K(X) & \xrightarrow{p_1^*} & K(X\times Y) \\ \alpha_{X,X\cap Y}\downarrow & & \downarrow \alpha_{X\times Y,(X\times Y)\cap\Delta_P} \\ K(X_{/X\cap Y}) & \xrightarrow{\tilde{p}_1^*} & K(X\times Y_{/(X\times Y)\cap\Delta_P}) \end{array} \tag{4}$$

As $\dim X\times Y = \dim X + \dim Y > 2m+n$, by Corollary 3 we get that $(X\times Y)\cap\Delta_P$ is $G3$ in $X \times Y$, so that the second vertical arrow of diagram (4) is an isomorphism.

On the other hand, we claim that for every proper surjective morphism $f: Z \to X$, $f^{-1}(X\cap Y)$ is connected. Indeed, since the morphism $f\times \mathrm{id}_Y: Z\times Y \to X\times Y$ is proper and surjective (because $f: Z \to X$ is so), and since $(X \times Y) \cap \Delta_P$ is $G3$ in $X \times Y$, by Corollary 1, $(f \times \mathrm{id}_Y)^{-1}((X \times Y) \cap \Delta_P)$ is $G3$ in $Z \times Y$. It follows in particular that $(f \times \mathrm{id}_Y)^{-1}((X \times Y) \cap \Delta_P)$ is connected. As $(f \times \mathrm{id}_Y)^{-1}((X \times Y) \cap \Delta_P)$ is biregularly isomorphic to $f^{-1}(X \cap Y)$, the claim is proved.

The claim implies the following two things:

i) $K(X_{/X \cap Y})$ is a field (by Proposition 1), and
ii) $K(X)$ is algebraically closed in $K(X_{/X \cap Y})$ (by Theorem 3).

Now we can easily prove that $X \cap Y$ is $G3$ in X. Indeed, if not, there would exist a formal-rational function $\zeta \in K(X_{/X \cap Y})$ such that $\zeta \notin K(X)$. Then by diagram (4) (with the second vertical arrow isomorphism) and by Theorem 5 it would follow that in the field extension $K(X_{/X \cap Y})|K(X)$ the function $\zeta \in K(X_{/X \cap Y})$ would be an algebraic element over $K(X)$ non belonging to $K(X)$, and this would contradict i) and ii) above.

Similarly one proves that $X \cap Y$ is $G3$ in Y. □

Remark 3. Corollary 4 extends to the case when $P = \mathbb{P}^m(e) \times \mathbb{P}^n(h)$ an old result of Faltings [11] (see Corollary 3, p. 102) regarding the case when $P = \mathbb{P}^n$. Our proof (based on global arguments) is different from Faltings' proof (which uses local methods).

Corollary 5. *Under the hypotheses of Corollary* 4 *assume that* k *is the field of complex numbers. Then every meromorphic function defined on a complex connected open neighborhood* U *of* $X \cap Y$ *in* X *extends to a rational function in* X.

Proof. Let $\mathcal{M}(U)$ denote the field of meromorphic functions on U. By Bădescu [2], p. 117, $K(X) \subseteq \mathcal{M}(U) \subseteq K(X_{X \cap Y})$, and then apply Corollary 4. □

References

1. Bădescu, L.: Algebraic Barth-Lefschetz theorems. Nagoya Math. J. **142**, 17–38 (1996)
2. Bădescu, L.: Projective Geometry and Formal Geometry. Monografie Matematyczne, vol. 65. Birkhäuser, Boston (2004)
3. Bădescu, L.: On a connectedness theorem of Debarre. In: Bates, D., Besana, G.M., Di Rocco, S., Wampler, C. (eds.) Proceedings of the Conference on Interactions of Classical and Numerical Algebraic Geometry. Contemporary Mathematics, vol. 496, pp. 11–20. American Mathematical Society, Providence (2009)
4. Bădescu, L., Repetto, F.: A connectedness theorem for products of weighted projective spaces. Commun. Algebra **36**(8), 2958–2968 (2008)
5. Bădescu, L., Schneider, M.: Formal functions, connectivity and homogeneous spaces. In: Beltrametti, M.C., Catanese, F., Ciliberto, C., Lanteri, A., Pedrini, C. (eds.) Algebraic Geometry – A Volume in Memory of Paolo Francia, pp. 1–21. Walter de Gruyter, Berlin/New York (2002)
6. Beltrametti, M., Robbiano, L.: Introduction to the theory of weighted projective spaces. Expo. Math. **4**, 111–162 (1986)
7. Debarre, O.: Fulton-Hansen and Barth-Lefschetz theorems for subvarieties of abelian varieties. J. Reine Angew. Math. **467**, 187–197 (1995)
8. Debarre, O.: Théorèmes de connexité et variétés abéliennes. Am. J. Math. **117**, 787–805 (1995)

9. Debarre, O.: Théorèmes de connexité pour les produits d'espaces projectives et les Grassmanniennes. Am. J. Math. **118**, 1347–1367 (1996)
10. Dolgachev, I.: Weighted projective varieties. In: Lecture Notes in Mathematics, vol. 956, pp. 34–71. Springer, Berlin/Heidelberg/New York (1982)
11. Faltings, G.: A contribution to the theory of formal meromorphic functions. Nagoya Math. J. **77**, 99–106 (1980)
12. Faltings, G.: Formale Geometrie und homogene Räume. Invent. Math. **64**, 123–165 (1981)
13. Fulton, W., Hansen, J.: A connectedness theorem for proper varieties with applications to intersections and singularities. Ann. Math. **110**, 159–166 (1979)
14. Fulton, W., Lazarsfeld, R.: Connectivity and its applications in algebraic geometry. In: Lecture Notes in Mathematics, vol. 862, pp. 26–92. Springer, New York (1981)
15. Hansen, J.: A connectedness theorem for flagmanifolds and Grassmannians. Am. J. Math. **105**, 633–639 (1983)
16. Hartshorne, R.: Ample Subvarieties of Algebraic Varieties. In: Lecture Notes in Mathematics, vol. 156. Springer, New York (1970)
17. Hironaka, H., Matsumura, H.: Formal functions and formal embeddings. J. Math. Soc. Jpn. **20**, 52–82 (1968)

Hodge Invariants of Higher-Dimensional Analogues of Kodaira Surfaces

Vasile Brînzănescu

Abstract In the paper [17], Sankaran gives a construction of some complex analytic manifolds, which are higher-dimensional analogues of Inoue parabolic surfaces, by using methods of toric geometry (see also [9, 16]). Some higher-dimensional analogues of Kodaira surfaces are obtained as hypersurfaces in these Inoue manifolds. In this paper we construct another higher-dimensional analogues of primary Kodaira surfaces and we compute their invariants as the Hodge numbers.

Keywords Principal torus bundles • Kodaira manifolds • Hodge invariants • Holomorphic simplectic structure

1 Introduction

In the paper [17], Sankaran gives a construction of some complex analytic manifolds, which are higher-dimensional analogues of Inoue parabolic surfaces, by using methods of toric geometry (see also [9, 16]). Some higher-dimensional analogues of Kodaira surfaces are obtained as hypersurfaces in these Inoue manifolds.

In this paper we construct another higher-dimensional analogues of primary Kodaira surfaces (called here Kodaira manifolds), which are complex, non-kählerian principal torus bundles over an elliptic curve. Then, we compute their invariants as the Hodge numbers. Finally, we show that some of these Kodaira manifolds are complex (holomorphic) symplectic manifolds.

V. Brînzănescu (✉)
Institute of Mathematics of the Romanian Academy, P.O. Box 1-764, 70700 Bucharest, Romania
e-mail: Vasile.Brinzanescu@imar.ro

D. Ibadula, W. Veys (eds.), *Bridging Algebra, Geometry, and Topology*, Springer Proceedings in Mathematics & Statistics 96,
DOI 10.1007/978-3-319-09186-0_6

2 Complex Principal Torus Bundles

Let $T = V/\Lambda$ be an n-dimensional compact complex torus, defined by a lattice $\Lambda \subset V$ in the n-dimensional complex vector space V. Canonical notation concerning the torus T will be used:

$$T_0(T) = H^0(T, \Theta_T) = V \ , \ H^i(T, \Theta_T) = H^i(T, \mathcal{O}_T) \otimes V \ ,$$

$$H^0(T, \Omega^1_T) = H^0(T, \Theta_T)^* = V^* \ , \ \Lambda = H_1(T, \mathbf{Z}) \ , \ H^1(T, \mathbf{Z}) = \Lambda^* \ ,$$

$$H^{p,q}_T = H^{p,0}_T \otimes H^{0,q}_T \ .$$

If B is a compact complex manifold of dimension m, then $X \stackrel{\pi}{\to} B$ denotes a principal T-bundle over B. Let $\mathcal{O}_B(T)$ denote the sheaf of germs of locally holomorphic maps from B to T. The principal T-bundles are described by cohomology classes

$$\xi \in H^1(B, \mathcal{O}_B(T)) \ .$$

For a Čech 1-cocycle (ξ_{ij}) the function

$$\xi_{ij} : U_i \cap U_j \to T$$

identifies $(z, t) \in U_i \times T$ with $(z, t') = (z, \xi_{ij}(z) + t) \in U_j \times T$ for all $z \in U_i \cap U_j$, where (U_i) is an open covering of B.

In [12], Höfer defined a homomorphism

$$\Phi : H^1(B, \mathcal{O}^*_B) \otimes_{\mathbf{Z}} \Lambda \to H^1(B, \mathcal{O}_B(T)) \ ,$$

which can be used to describe some principal torus bundles by giving a combination of line bundles on B in $Pic(B) \otimes_{\mathbf{Z}} \Lambda$, $\sum \mathcal{L}_k \otimes \lambda_k$ (where $Pic(B) \cong H^1(B, \mathcal{O}^*_B)$). Indeed, by choosing a sufficiently fine open covering (U_i) of B, the transitions functions of each $\mathcal{L}_k$ are expressed by a cocycle $(f^{(k)}_{ij})$. Now, identify $(z, t_i) \in U_i \times T$ with $(z, t_j) \in U_j \times T$ if and only if

$$t_i = t_j + \left(\sum \frac{\lambda_k}{2\pi\sqrt{-1}} \log(f^{(k)}_{ij}) \right) \ ,$$

for all $z \in U_i \cap U_j$. This is exactly Höfer's morphism Φ, and the above construction is called a *generalized logarithmic transformation* applied to the trivial principal T-bundle $B \times T$ (see [7, 18]).

If B is a curve the homomorphism Φ is surjective (see [12, 7.2 and 7.4]). This allows us to construct any principal torus bundle over a curve or, more generally any torus quasi-bundle (see [7]), by means of logarithmic transformations (similar to the case of elliptic surfaces [14]) from the trivial principal torus bundle $B \times T$.

Let us describe this more geometric procedure (for more details, see [7, 19]): Any element $\eta \in H^1(B, \mathcal{O}_B^*) \otimes_{\mathbf{Z}} \Lambda$ has the form $\sum_{j=1}^{\ell} P_j \otimes \lambda_j$, where $P_j,\ j = 1, \ldots, \ell$, are points on the curve B and $\lambda_j \in \Lambda,\ \ j = 1, \ldots, \ell$. Denote $B' = B \setminus \{P_1, P_2, \ldots, P_\ell\}$. Let

$$D_j = \{z_j \in \mathbf{C} \mid |z_j| < \varepsilon\}$$

be a coordinate neighborhood of the point P_j. The mapping

$$\ell_{\lambda_j} : D_j^* \times T \to D_j^* \times T$$

$$(z_j, [t]) \to (z_j, [t - \frac{\lambda_j}{2\pi\sqrt{-1}} \log z_j])$$

is an isomorphism. We can patch together $B' \times T$ and $D_j \times T$'s by means of the isomorphisms ℓ_{λ_j}'s and obtain a principal T-bundle over B. It is denoted by

$$L_{P_1}(\lambda_1, 1) L_{P_2}(\lambda_2, 1) \ldots L_{P_\ell}(\lambda_\ell, 1)(B \times T)$$

and it is called a complex manifold obtained from $B \times T$ by means of logarithmic transformation.

Taking local sections of the constant sheaves

$$0 \to \Lambda \to V \to T \to 0$$

one gets an exact sequence of sheaves on the manifold B

$$0 \to \Lambda \to \mathcal{O}_B \otimes V \to \mathcal{O}_B(T) \to 0\,, \tag{1}$$

with the induced exact cohomology sequence

$$\ldots \to H^0(B, \mathcal{O}_B(T)) \to H^1(B, \Lambda) \to H^1(B, \mathcal{O}_B) \otimes V \to \tag{2}$$

$$\to H^1(B, \mathcal{O}_B(T)) \xrightarrow{c} H^2(B, \Lambda) \to H^2(B, \mathcal{O}_B) \otimes V \to \ldots\ .$$

The cohomology class ξ of the bundle in $H^1(B, \mathcal{O}_B(T))$ determines a characteristic class

$$c(\xi) \in H^2(B, \Lambda) \cong H^2(B, \mathbf{Z}) \otimes \Lambda.$$

By Blanchard's Theorem (see [2]) it follows that a principal T-bundle $X \xrightarrow{\pi} B$ is Kähler if and only if $c(\xi)$ is of finite order in $H^2(B, \Lambda)$.

The Höfer's map Φ is compatible with taking characteristic classes, i.e. if $\sum \mathscr{L}_k \otimes \lambda_k$ is a combination of line bundles in $Pic(B) \otimes_{\mathbf{Z}} \Lambda$ then the characteristic class of $\Phi(\sum \mathscr{L}_k \otimes \lambda_k)$ equals

$$\sum c_1(\mathscr{L}_k) \otimes \lambda_k \in H^2(B, \Lambda).$$

In other words, the following diagram is commutative:

$$\begin{array}{ccc} Pic(B) \otimes_{\mathbf{Z}} \Lambda & \xrightarrow{\Phi} & H^1(B, \mathscr{O}_B(T)) \\ c_1 \otimes id \downarrow & & \downarrow c \\ H^2(B, \mathbf{Z}) \otimes \Lambda & \xrightarrow{=} & H^2(B, \Lambda) \end{array}$$

(see [12, 7.1]). In the case B is a curve, the characteristic class of a principal T-bundle

$$L_{P_1}(\lambda_1, 1) L_{P_2}(\lambda_2, 1) \ldots L_{P_\ell}(\lambda_\ell, 1)(B \times T) \xrightarrow{\pi} B$$

is given by

$$\sum_{j=1}^{\ell} \lambda_j$$

if we identify $H^2(B, \Lambda)$ with Λ. Then the principal T-bundle $X \xrightarrow{\pi} B$ is Kähler if and only if

$$\sum_{j=1}^{\ell} \lambda_j = 0.$$

Because transition functions of the T-principal bundle $X \xrightarrow{\pi} B$ act trivially on the cohomology of fiber, we get natural identifications:

$$R^q \pi_* \mathbf{Z}_X = \mathbf{Z}_B \otimes_{\mathbf{Z}} H^q(T, \mathbf{Z}) \; ; \; R^q \pi_* \mathscr{O}_X = \mathscr{O}_B \otimes_{\mathbf{C}} H^q(T, \mathscr{O}_T) \, . \tag{3}$$

The transgression of the fiber bundle in integral cohomology is a map

$$\delta^{\mathbf{Z}} : H^1(T, \mathbf{Z}) \to H^2(B, \mathbf{Z}) \, .$$

Under the identification

$$H^1(T, \mathbf{Z}) = Hom(\Lambda, \mathbf{Z}) = \Lambda^* \, ,$$

the characteristic class

$$c(\xi) \in H^2(B, \mathbf{Z}) \otimes \Lambda$$

and the mapping

$$\delta^{\mathbf{Z}} : H^1(T, \mathbf{Z}) \to H^2(B, \mathbf{Z})$$

coincide (see [12, 6.1]).

The first possibly nontrivial d_2-homomorphism

$$H^0(B, R^1\pi_*\mathcal{O}_X) \to H^2(B, \pi_*\mathcal{O}_X)$$

in the Leray spectral sequence of $\mathcal{O}_X$ is denoted by

$$\varepsilon : H^1(T, \mathcal{O}_T) \to H^2(B, \mathcal{O}_B) \ .$$

We shall call a *Kodaira manifold* any non-Kähler principal T-bundle $X \xrightarrow{\pi} B$ over an elliptic curve B. If $\dim_{\mathbf{C}} T = 1$ (i.e. T is an elliptic curve), then we get a primary Kodaira surface. The homotopy exact sequence for the fibration $X \xrightarrow{\pi} B$ reduces to the exact sequence

$$0 \to \pi_1(T) \to \pi_1(X) \to \pi_1(B) \to 0 \ . \tag{4}$$

The characteristic class of the principal T-bundle $X \xrightarrow{\pi} B$ can be written in the following form

$$c(\xi) = m\lambda_0 \ ,$$

where m is an integer and $\lambda_0 \in \Lambda$ is a primitive element in the lattice $\Lambda \subset V \cong \mathbf{C}^n$. Then, by the invariant factors theorem (see [4]), we can choose a basis of Λ such that λ_0 is the first element of this basis.

The fundamental group of the fiber T is isomorphic to the lattice $\Lambda \subset \mathbf{C}^n$ generated by $\{\lambda_1 = \lambda_0, \dots, \lambda_{2n}\}$, the fundamental group of the basis B is isomorphic to a lattice $\Gamma \subset \mathbf{C}$ generated by $\{\beta_1, \beta_2\}$, and we have the following central extension, obtained from the exact sequence (4):

$$0 \to \Lambda \to G \to \Gamma \to 0 \ , \tag{5}$$

where G is the fundamental group $\pi_1(X)$. In fact, the group G has a presentation of the following form:

$$G = < \tilde{\lambda}_1, \dots, \tilde{\lambda}_{2n}, \tilde{\beta}_1, \tilde{\beta}_2 \mid [\tilde{\beta}_1, \tilde{\beta}_2] = \tilde{\lambda}_1^m, \ \tilde{\lambda}_1, \dots, \tilde{\lambda}_{2n} \ \text{central} > \ .$$

Thus, the Kodaira manifold $X \xrightarrow{\pi} B$ is a quotient of $\mathbf{C}^{n+1}$ by a group G of affine transformations; compare [6, 15] for the case of primary Kodaira surfaces.

Remark. In [7], it is computed the torsion of $H^2(X, \mathbf{Z})$ (Tors $H^2(X, \mathbf{Z}) \cong \mathbf{Z}_m$) and the Neron–Severi group of the principal T-bundle $X \xrightarrow{\pi} B$. For primary Kodaira surfaces, see [5].

3 Computation of Hodge Numbers

Let $X \xrightarrow{\pi} B$ be a principal T-bundle over a compact complex manifold B. Let $H_X^{p,q}$ denote as usual the cohomology spaces $H^q(X, \Omega_X^p)$ and let

$$h_X^{p,q} = \dim_{\mathbf{C}} H_X^{p,q}$$

be the Hodge numbers. The Leray spectral sequence for Ω_X^p converges to $H_X^{p,q}$ and has E_2-term

$$E_2^{i,j} = H^i(B, R^j \pi_* \Omega_X^p) .$$

Except for $p = 0$ the higher direct images sheaves are nontrivial and it is difficult to compute the Hodge numbers by using the Leray spectral sequences. We shall use the Borel's spectral sequences (see [3]) as in [12]; see, also [8].

The sheaf Ω_X^p can be resolved by the Dolbeault complex $(\mathscr{A}_X^{p,\bullet}, \overline{\partial})$ of $\mathscr{C}^\infty$ – $(p, \bullet)$-forms. The direct image complex $\pi_* \mathscr{A}_X^{p,\bullet}$ can be filtered by the base degree of the forms, i.e. $F^s \pi_* \mathscr{A}_X^{p,q}$ consists of those forms that in local bundle coordinates (z, t) can be written as a linear combination of

$$dz_I \wedge d\overline{z}_{\overline{I}} \wedge dt_J \wedge d\overline{t}_{\overline{J}}$$

with $|I| + |\overline{I}| \geq s$. Taking global sections

$$A_X^{p,q} := \Gamma(\mathscr{A}_X^{p,q})$$

we get a filtered complex $(A_X^{p,\bullet}, \overline{\partial})$ of modules and a corresponding spectral sequence

$${}^{p,q}E_r^{s,t} .$$

The usual spectral sequence graduation is given by (s, t), corresponding to the filtration. We also include (p, q) denoting the $(\partial, \overline{\partial})$-type but p is constant in each of the sequences and we have $p + q = s + t$. Note that p is not changed by the differential

$$ {}^{p,q}E_r^{s,t} \xrightarrow{d_r} {}^{p,q+1}E_r^{s+r,t-r+1} \,, $$

and q is determined by $p + q = s + t$, so in fact there is a single spectral sequence for each p, computing the cohomology of Ω_X^p. We get

$$ Gr(H_X^{p,q}) = \oplus_{s+t=p+q} \left({}^{p,q}E_\infty^{s,t} \right) \tag{6} $$

and

$$ {}^{p,q}E_2^{s,t} = \oplus_i \left(H_B^{i,s-i} \otimes H_T^{p-i,t-p+i} \right) $$

(see [3]).

For the determination of d_2 the basic maps are

$$ \varepsilon : \ {}^{0,1}E_2^{0,1} = H_T^{0,1} \xrightarrow{d_2} {}^{0,2}E_2^{2,0} = H_B^{0,2} $$

and

$$ \gamma : \ {}^{1,0}E_2^{0,1} = H_T^{1,0} \xrightarrow{d_2} {}^{1,1}E_2^{2,0} = H_B^{1,1} \,. $$

Höfer proved in [12, 4.3], that $d_r = 0$ for $r > 2$ (i.e., the Borel's spectral sequence degenerates at E_3-level) and the d_2-differential is completely determined by the two maps γ and ε. If B is a complex curve, then ε vanishes for dimensions reasons, so the d_2-differential is determined by γ.

We have:

Theorem 1. *Let $X \xrightarrow{\pi} B$ be a non-kählerian principal T-bundle over a complex curve B of genus $g \geq 1$. Then the Hodge numbers are given by*

$$ h_X^{p,q} = \binom{n-1}{p}\binom{n}{q} + (g-1)\binom{n}{p}\binom{n}{q-1} + g\binom{n}{p-1}\binom{n}{q} + \binom{n}{q-1}\left(\binom{n}{p-1} + \binom{n-1}{p}\right). $$

In particular, for a Kodaira manifold of dimension $n + 1$, we get

$$ h_X^{p,q} = \binom{n+1}{q}\left(\binom{n-1}{p} + \binom{n}{p-1}\right). $$

Proof: Let $X \xrightarrow{\pi} B$ be a principal T-bundle over a complex curve B of genus $g \geq 1$. Then $H^2(B, \mathbf{C})$ has a Hodge decomposition. By the Theorem 6.3 in [12], the transgression map

$$\delta : H^1(T, \mathbf{C}) \rightarrow H^2(B, \mathbf{C}) ,$$

(which is obtained from $\delta^{\mathbf{Z}}$ by scalar extension, i.e. $\delta = \delta^{\mathbf{Z}} \otimes id_{\mathbf{C}}$) determines γ and vice versa (since $\varepsilon = 0$).

Let $\{\lambda_1, \ldots, \lambda_{2n}\}$, be a basis of the lattice Λ, embedded in the complex vector space V with basis $\{e_1, \ldots, e_n\}$ and corresponding complex coordinates $(t_1, \ldots, t_n)$. Modulo an analytic isomorphism of the n-torus T, we can take Λ to be the lattice generated by the column vectors in $V \cong \mathbf{C}^n$ of the following period matrix:

$$\Omega = (I_n, A) .$$

We can suppose also that the principal T-bundle $X \xrightarrow{\pi} B$ is defined by a cohomology class

$$\xi \in H^1(B, \mathcal{O}_B(T))$$

such that

$$c(\xi) = m\lambda_1 \in H^2(B, \Lambda) \cong \Lambda$$

(see the end of the previous section).

Then

$$\delta(dt_1) = m\alpha, \ \delta(dt_2) = 0, \ldots, \delta(dt_n) = 0,$$

where $\{\alpha\}$ is a basis of $H^2(B, \mathbf{C})$; see [12, 6.5]. We get

$$\gamma(dt_j) = \delta(dt_j) .$$

Since

$$^{p,q}E_{\infty}^{s,t} = {}^{p,q}E_3^{s,t} ,$$

we have to compute $^{p,q}E_3^{s,t}$ from the sequences:

$$^{p,q-1}E_2^{s-2,t+1} \xrightarrow{d_2} {}^{p,q}E_2^{s,t} \xrightarrow{d_2} {}^{p,q+1}E_2^{s+2,t-1} .$$

The only possibly nonzero terms are obtained for $s = 0, 1, 2$.

For $s = 0$ we get the sequence

$$0 \xrightarrow{d_2} \left(H_B^{0,0} \otimes H_T^{p,0}\right) \otimes H_T^{0,q} \xrightarrow{\gamma^p \otimes id^q} \left(H_B^{1,1} \otimes H_T^{p-1,0}\right) \otimes H_T^{0,q} \ ,$$

where

$$\gamma^p : H_B^{0,0} \otimes H_T^{p,0} \longrightarrow H_B^{1,1} \otimes H_T^{p-1,0} \ ,$$

is the map induced by γ (we have the identity on the second factor since $\varepsilon = 0$; see [12, Proposition 4.3]).

A basis in $H_B^{0,0} \otimes H_T^{p,0}$ is given by

$$1 \otimes \left(dt_{i_1} \wedge \ldots \wedge dt_{i_p}\right) \ ,$$

where $1 \leq i_1 < i_2 < \ldots < i_p \leq n$. Because of our choice of the basis in lattice Λ, we get:

$$\gamma^p(1 \otimes (dt_{i_1} \wedge \ldots \wedge dt_{i_p})) = \begin{cases} m\alpha \otimes (dt_{i_1} \wedge \ldots \wedge dt_{i_p}) & \text{if } i_1 = 1 \\ 0 & \text{if } i_1 > 1 \end{cases} .$$

A basis of the kernel of the map γ^p is given by

$$1 \otimes (dt_{i_1} \wedge \ldots \wedge dt_{i_p}) \ ,$$

where $2 \leq i_1 < i_2 < \ldots < i_p \leq n$. It follows that the dimension of ${}^{p,q}E_\infty^{0,p+q}$ is given by

$$\binom{n-1}{p}\binom{n}{q} .$$

For $s = 1$ we get the sequence

$$0 \xrightarrow{d_2} \left(H_B^{0,1} \otimes H_T^{p,q-1}\right) \oplus \left(H_B^{1,0} \otimes H_T^{p-1,q}\right) \xrightarrow{d_2} 0 .$$

It follows that the dimension of ${}^{p,q}E_\infty^{1,p+q-1}$ is given by

$$g\left(\binom{n}{p}\binom{n}{q-1}+\binom{n}{p-1}\binom{n}{q}\right) .$$

For $s = 2$ we get the sequence

$$\left(H_B^{0,0} \otimes H_T^{p,0}\right) \otimes H_T^{0,q-1} \xrightarrow{\gamma^p \otimes id^{q-1}} \left(H_B^{1,1} \otimes H_T^{p-1,0}\right) \otimes H_T^{0,q-1} \xrightarrow{d_2} 0 \ ,$$

and the dimension of ${}^{p,q}E_{\infty}^{2,p+q-2}$ is given by

$$\binom{n}{q-1}\left(\binom{n}{p-1}+\binom{n-1}{p}-\binom{n}{p}\right).$$

From formula (6) we get the result.

Remark. If the principal T-bundle $X \xrightarrow{\pi} B$ is kählerian (i.e., $c(\xi) = 0$) then the Hodge numbers are the same as for the product $B \times T$, since $\gamma^p = 0$ in this case.

4 Structure of the Kodaira Manifolds

In this section we establish some facts about the Kodaira manifolds $X \xrightarrow{\pi} B$ introduced in Sect. 2. As we have seen a Kodaira manifold is a non-kählerian connected compact manifold.

Recall that a holomorphic symplectic structure on a complex manifold is a closed holomorphic two-form, which is non-degenerate in every point of the manifold, i.e., it has maximal rank; see [1, 13].

A holomorphic symplectic manifold X has an even complex dimension $2r$ and its canonical bundle K_X is trivial; indeed, if ω is a holomorphic symplectic structure on X, the form ω^r is a generator of K_X. For examples of holomorphic symplectic kählerian manifolds see [1]. Examples of compact holomorphic symplectic manifolds which are not Kähler are known since recently; see, for example [10, 11].

For Kodaira manifolds we have the following result:

Proposition 2. *A Kodaira manifold $X \xrightarrow{\pi} B$ has a holomorphic symplectic structure if and only if it has an even complex dimension.*

Proof: As we have seen at the end of Sect. 2, any Kodaira manifold is a quotient of $\mathbf{C}^{n+1}$ by a group G of affine transformations.

Let $n + 1 = 2k + 2$. By choosing as above the basis of the lattice $\Lambda \subset \mathbf{C}^n \cong V$, the complex torus T has the period matrix

$$(I_n, A)\ ,$$

where

$$A = (a_{ij}) \in GL(n, \mathbf{C})\ .$$

The group G has generators

$$\{\tilde{\lambda}_1, \ldots, \tilde{\lambda}_{2n}, \tilde{\beta}_1, \tilde{\beta}_2\}$$

and they act on $\mathbf{C}^{n+1}$ by

$$\tilde{\lambda}_j(z_1, \ldots, z_n; w) = (z_1, \ldots, z_j + 1, \ldots, z_n; w) ,$$

for $1 \leq j \leq n$,

$$\tilde{\lambda}_j(z_1, \ldots, z_n; w) = (z_1 + a_{1j}, \ldots, z_n + a_{nj}; w) ,$$

for $n + 1 \leq j \leq 2n$, and

$$\tilde{\beta}_1(z_1, \ldots, z_n; w) = (z_1 + \mu_1 w + \mu_3, z_2 + \gamma_2, \ldots, z_n + \gamma_n; w + \alpha_1) ,$$

$$\tilde{\beta}_2(z_1, \ldots, z_n; w) = (z_1 + \mu_2 w + \mu_4, z_2 + \theta_2, \ldots, z_n + \theta_n; w + \alpha_2) ,$$

where $(\alpha_1 \alpha_2)$ is a period matrix for the elliptic curve B. From the relation

$$[\tilde{\beta}_1, \tilde{\beta}_2] = \tilde{\lambda}_1^m ,$$

we get the condition

$$\gamma_1 \alpha_2 - \gamma_2 \alpha_1 = m .$$

Consider on $\mathbf{C}^{n+1}$ $(n + 1 = 2k + 2)$ the standard non-degenerate, closed holomorphic symplectic two-form

$$\omega = dz_1 \wedge dw + dz_2 \wedge dz_3 + \ldots + dz_{2k} \wedge dz_{2k+1} .$$

Clearly, this two-form is invariant by the action of the group G, so it defines a closed, holomorphic symplectic two-form $\tilde{\omega}$ on $X = \mathbf{C}^{n+1}/G$, which is non-degenerate in every point of X.

Acknowledgements This work was supported by a grant from the Ministry of National Education CNCS-UEFISCDI, project number PN-II-ID-PCE-2012-4-0156.

References

1. Beauville, A.: Variétés kähleriennes dont la première classe de Chern est nulle. J. Differ. Geom. **18**, 755–782 (1983)
2. Blanchard, A.: Espaces fibrés kählériens compacts. C.R. Acad. Sci. **238**, 2281–2283 (1954)
3. Borel, A.: A spectral sequence for complex analytic bundles. In: Hirzebruch, F.: Topological Methods in Algebraic Geometry. 3rd edn. (Appendix II). Springer, Berlin/Heidelberg/New York (1966)
4. Bourbaki, N.: Algèbre II. Hermann, Paris (1964)
5. Brînzănescu, V.: The Picard group of a primary Kodaira surface. Math. Ann. **296**, 725–738 (1993)

6. Brînzănescu, V.: Holomorphic Vector Bundles over Compact Complex Surfaces. Lecture Notes in Mathematics, vol. 1624. Springer, Berlin/Heidelberg/New York (1996)
7. Brînzănescu, V., Ueno, K.: Neron-Severi group of torus quasi bundles over curves. In: Maruyama, M. (ed.) Moduli of Vector Bundles. Lecture Notes in Pure and Applied Mathematics, vol. 179, pp. 11–32. Marcel Dekker, New York/Basel/Hong Kong (1996)
8. Brinzanescu, V., Halanay, A.D., Trautmann, G.: Vector bundles on non-Kähler elliptic principal bundles. Ann. Inst. Fourier (Grenoble) **63**(3), 1033–1054 (2013)
9. Friedman, R., Shepherd-Barron, N.I.: Degenerations of Kodaira surfaces. In: Friedman, R., Morrison, D.R. (eds.) The Birational Geometry of Degenerations. Progress in Mathematics, vol. 29, pp. 261–276. Birkhäuser, Boston/Basel/Stuttgart (1984)
10. Guan, D.: Examples of compact holomorphic symplectic manifolds which are not Kählerian II. Invent. Math. **121**, 135–145 (1995)
11. Guan, D.: Examples of compact holomorphic symplectic manifolds which are not Kählerian III. Intern. J. Math. **6**–5, 709–718 (1995)
12. Höfer, T.: Remarks on torus principal bundles. J. Math. Kyoto Univ. (JMKYAZ) **33**(1), 227–259 (1993)
13. Kobayashi, S.: Differential Geometry of Complex Vector Bundles. Iwanami Shoten/Princeton University Press, Tokyo/Princeton (1987)
14. Kodaira, K.: On compact analytic surfaces III. Ann. Math. **78**, 1–40 (1963)
15. Kodaira, K.: On the structure of compact analytic surfaces I. Am. J. Math. **86**, 751–798 (1964)
16. Oda, T.: On Torus Embeddings and Applications. Tata Institute Lecture Notes. Tata Institute of Fundamental Research, Bombay (1978)
17. Sankaran, G.K.: Higher-dimensional analogues of Inoue and Kodaira surfaces. Math. Nachr. **166**, 83–94 (1994)
18. Ueno, K.: Degeneration of Elliptic Surfaces and Certain Non-Kähler Manifolds. Progress in Mathematics, vol. 39, pp. 545–566. Birkhäuser, Boston/Basel/Stuttgart (1983)
19. Ueno, K.: Geometry of torus quasi bundles over curves. In: Proc. Conf. Geometry from the Pacific Rim (Singapore, 1994), pp. 377–394. de Gruyter, Berlin (1997)

An Invitation to Quasihomogeneous Rigid Geometric Structures

Sorin Dumitrescu

Dedicated to Alex Dimca and Ştefan Papadima

Abstract This is a survey paper dealing with quasihomogeneous geometric structures, in the sense that they are locally homogeneous on a nontrivial open set, but not on all of the manifold. Our motivation comes from Gromov's open-dense orbit theorem which asserts that if the pseudogroup of local automorphisms of a *rigid geometric structure* acts with a dense orbit, then this orbit is open. Fisher conjectured that the maximal open set of local homogeneity is all of the manifold as soon as the following three conditions are fulfilled: the automorphism group of the manifold acts with a dense orbit, the geometric structure is a G-structure (meaning that it is locally homogeneous at the first order) and the manifold is compact. In a recent joint work, with Adolfo Guillot, we succeeded to prove Fisher's conjecture for real analytic torsion free affine connections on surfaces: we construct and classify those connections which are quasihomogeneous; their automorphism group never act with a dense orbit.

Keywords Quasihomogeneous affine connections • Rigid geometric structures • Normal forms • Transitive Killing Lie algebra • Complex geometry

1 Introduction

Following the work of Lie [28] and then Ehresmann [19], the study of *locally homogeneous rigid* geometric structures (think of pseudo-Riemannian metrics or connections) became a classical subject in differential geometry. Recall that these locally homogeneous geometric structures are closely related to manifolds *locally modeled* on homogeneous spaces in the following way.

S. Dumitrescu (✉)
Université Nice-Sophia Antipolis, Laboratoire J.-A. Dieudonné, UMR 7351 CNRS, Parc Valrose, 06108 Nice Cedex 2, France
e-mail: dumitres@unice.fr

D. Ibadula, W. Veys (eds.), *Bridging Algebra, Geometry, and Topology*, Springer Proceedings in Mathematics & Statistics 96,
DOI 10.1007/978-3-319-09186-0_7

Let G be a connected Lie group and I a closed subgroup of G. A manifold M is said to be *locally modeled* on the homogeneous space G/I if it admits an atlas with open sets diffeomorphic to open sets in G/I, such that the transition maps are given by restrictions of elements in G. Any G-invariant geometric structure $\tilde{\phi}$ on G/I uniquely defines a locally homogeneous geometric structure ϕ on M which is locally isomorphic to $\tilde{\phi}$.

In [30], Mostow gave modern proofs for Lie's classification of surfaces locally modeled on homogeneous spaces (see also the more recent [32]). It is interesting to notice that the dimension of the Lie group G is unbounded. Locally homogeneous connections on surfaces were studied in [2, 27, 33]. In [7], answering a question of Lie, the authors classified Riemannian metrics on surfaces whose underlying Levi-Civita connections are projectively locally homogeneous.

In dimension three, the locally homogeneous Riemannian metrics formed the context on Thurston's geometrization program [37].

The celebrated open-dense orbit theorem of Gromov [10, 24] (see also [3, 8, 20]) asserts that *a rigid geometric structure admitting a pseudogroup of local automorphisms which acts with a dense orbit is locally homogeneous on an open dense set.*

This maximal locally homogeneous open-dense set appears to be mysterious and it might very well happen that it coincides with all of the (connected) manifold in many interesting geometric backgrounds. This was proved, for instance, for Anosov flows with differentiable stable and unstable foliations and transverse contact structure [4] and for three dimensional compact Lorentz manifolds admitting a nonproper one parameter group acting by automorphisms [38]. In these papers, the authors make use of the nontrivial dynamics of the automorphism group of the geometric structure, to prove that the geometric structure needs to be locally homogeneous on all of the manifold.

Surprisingly, the extension of a locally homogeneous open dense subset to all of the (connected) manifold might stand *even without assuming the existence of a big automorphism group*. This is known to be true in the Riemannian setting [34], as a consequence of the fact that all scalar invariants are constant. This was also proved in the frame of three dimensional real analytic Lorentz metrics [13], for real analytic unimodular affine connections on surfaces [15] and for complete real analytic pseudo-Riemannian metrics admitting a semi-simple Killing Lie algebra in [29].

In [5], the authors deal with this question and their results indicate ways in which some rigid geometric structures cannot degenerate off the open dense set. In [21], Fisher conjectured that the maximal open set of local homogeneity is all of the manifold as soon as the following three conditions are fulfilled: the automorphism group of the manifold acts with a dense orbit, the geometric structure is a G-structure (meaning that it is locally homogeneous at the first order) and the manifold is compact.

In a recent joint work, with Adolfo Guillot [16], we succeeded to prove Fisher's conjecture for real analytic torsion free affine connections on surfaces. We classify torsion free real-analytic affine connections on compact oriented real-analytic surfaces which are *quasihomogeneous*, in the sense that they are locally homogeneous on a nontrivial open set, without being locally homogeneous on all of the surface. In particular, we prove that such connections exist, but their automorphism group never acts with a dense orbit, which gives a positive answer to Fisher's conjecture for analytic connections on surfaces.

This classification relies in a local result that gives normal forms for germs of torsion free real-analytic affine connections on a neighborhood of the origin in the plane which are *quasihomogeneous*, in the sense that they are locally homogeneous on an open set containing the origin in its closure, but not locally homogeneous in the neighborhood of the origin.

Another motivation to study quasihomogeneous rigid geometric structures, is the result proved in [11] (see also [6]) which asserts, in particular, that *holomorphic rigid geometric structures on compact complex manifolds of algebraic dimension zero are locally homogeneous on an open dense set.* As we will see in the last section of this survey, the holomorphic rigidity implies that many rigid holomorphic geometric structures on compact complex manifolds are locally homogeneous, as it was shown in [11, 12, 14].

2 Rigid Geometric Structures and Killing Fields

We give here the definitions in the framework of smooth real manifolds and smooth geometric structures. The definitions are similar in the holomorphic category.

Consider an n-manifold M and, for all integers $r \geq 1$, consider the associated bundle $R^r(M)$ of r-frames, which is a $D^r(\mathbf{R}^n)$-principal bundle over M, with $D^r(\mathbf{R}^n)$ the real algebraic group of r-jets at the origin of local diffeomorphisms of $\mathbf{R}^n$ fixing 0.

Let us consider, as in [10, 24], the following

Definition 1. A *geometric structure* (of order r and of algebraic type) ϕ on a manifold M is a $D^r(\mathbf{R}^n)$-equivariant smooth map from $R^r(M)$ to a real algebraic variety Z endowed with an algebraic action of $D^r(\mathbf{R}^n)$.

Riemannian and pseudo-Riemannian metrics, affine and projective connections and the most encountered geometric objects in differential geometry are known to verify the previous definition [3, 8, 10, 20, 24]. For instance, if the image of ϕ in Z is exactly one orbit, this orbit identifies with a homogeneous space $D^r(\mathbf{R}^n)/G$, where G is the stabilizer of a chosen point in the image of ϕ. We get then a reduction of the structure group of $R^r(M)$ to the subgroup G. This is exactly the classical definition of a G-structure (of order r): the case $r = 1$ and $G = O(n, \mathbf{R})$ corresponds to a Riemannian metric and that of $r = 2$ and $G = GL(n, \mathbf{R})$ gives a torsion free affine connection.

Definition 2. A (local) Killing field of ϕ is a (local) vector field on M whose canonical lift to $R^r(M)$ preserves ϕ.

For all $k \in \mathbf{N}$, the k-th prolongation of ϕ is a geometric structure of order $r+k$ (and of algebraic type), given by an equivariant smooth map $\phi^{(k)} : R^{(r+k)}(M) \to J^{n,k}(Z)$, where $J^{n,k}(Z)$ is the bundle of k-jets at 0 of smooth maps from $\mathbf{R}^n$ to Z. The real algebraic variety $J^{n,k}(Z)$ admits an algebraic action of $D^{(r+k)}(\mathbf{R}^n)$ which is obtained in a canonical way from the $D^r(\mathbf{R}^n)$-action on Z (see, for example, [20]).

For all $m \in M$, we denote by $Is^{(r+k)}(m)$ the group of $(r+k)$-isometric jets. They are $(r+k)$-jets at m of local diffeomorphisms f of M fixing m such that for any $\xi \in R^{(r+k)}(M)$ above m, we have $\phi^{(r+k)}(f \cdot \xi) = \phi^{(r+k)}(\xi)$. Remark that the previous relation only depends on the $(r+k)$-jet of f at m.

For every point m in M, there exists a natural homomorphism

$$Is^{(r+k+1)}(m) \to Is^{(r+k)}(m).$$

Following Gromov [10, 24] we define rigidity as:

Definition 3. The geometric structure ϕ (of order r) is rigid at order $(r+k)$, if for every point $m \in M$, the homomorphism $Is^{(r+k+1)}(m) \to Is^{(r+k)}(m)$ is injective.

A consequence of the previous definition is that the Lie algebra of Killing fields of a rigid geometric structure is finite dimensional in the neighborhood of every point in M [20].

Recall that (pseudo)-Riemannian metrics, as well as affine and projective connections, or conformal structures in dimension ≥ 3 are known to be rigid [3,8,10,20,24].

Definition 4. The geometric structure ϕ is said to be locally homogeneous on the open subset $U \subset M$ if for every $u \in U$ and every tangent vector $V \in T_u U$ there exists a local Killing field X of ϕ such that $X(u) = V$.

The Lie algebra of Killing fields is the same at the neighborhood of any point of a locally homogeneous geometric structure ϕ. This still holds for any real analytic rigid geometric structure (not necessarily locally homogeneous). In this case, one need to make use of an extendibility result for local Killing fields proved first for Nomizu in the Riemannian setting [31] and generalized then for rigid geometric structures by Amores and Gromov [1, 24] (see also [8, 10, 20]). This phenomenon states roughly that a local Killing field of a *rigid analytic* geometric structure can be extended along any curve in M. We then get a multivalued Killing field defined on all of M or, equivalently, a global Killing field defined on the universal cover. In particular, the Killing algebra in the neighborhood of any point is the same (as long as M is connected). It will be simply called *the Killing algebra of* ϕ.

Notice also that Nomizu's extension phenomenon doesn't imply that the extension of a family of linearly independent Killing fields, stays linearly independent. In general, the extension of a locally transitive Killing algebra fails to be transitive on a nowhere dense analytic subset S in M. This is exactly what happens for quasihomogeneous geometric structures (see later our examples of quasihomogeneous connections on surfaces). Moreover, this explains also that *a real analytic rigid geometric structure which is locally homogeneous on some nontrivial open set, is also locally homogeneous away from an analytic subset S of positive codimension* (S might be empty).

We recall that there exists locally homogeneous Riemannian metrics on five-dimensional manifolds which are not locally isometric to a invariant Riemannian metric on a homogeneous space. However this phenomenon cannot happen in lower dimension:

Theorem 1. *Let M be a manifold of dimension ≤ 4 bearing a locally homogeneous rigid geometric structure ϕ with Killing algebra $\mathfrak{g}$. Then M is locally modeled on a homogeneous space G/I, where G is a connected Lie group with Lie algebra $\mathfrak{g}$ and I is a closed subgroup of G. Moreover, (M, ϕ) is locally isomorphic to a G-invariant geometric structure on G/I.*

Proof. Let $\mathfrak{g}$ be the Killing algebra of ϕ. Denote by $\mathfrak{I}$ the (isotropy) subalgebra of $\mathfrak{g}$ composed by Killing fields vanishing at a given point in M. Let G be the unique connected simply connected Lie group with Lie algebra $\mathfrak{g}$. Since $\mathfrak{I}$ is of codimension ≤ 4 in $\mathfrak{g}$, a result of Mostow [30] (Chap. 5, p. 614) shows that the Lie subgroup I in G associated with $\mathfrak{I}$ is *closed*. Then ϕ induces a G-invariant geometric structure $\tilde{\phi}$ on G/I locally isomorphic to it. Moreover, M is locally modeled on G/I.

Remark 1. By the previous construction, G is simply connected and I is connected (and closed), which implies that G/I is simply connected (see [30, p. 617], Corollary 1). In general, the G-action on G/I admits a nontrivial discrete kernel. We can assume that this action is effective considering the quotient of G and I by the maximal normal subgroup of G contained in I (see proposition 3.1 in [35]).

We give now a last definition:

Definition 5. A geometric structure ϕ on M is said to be of *Riemannian type* if there exists a Riemannian metric on M preserved by all Killing fields of ϕ.

Roughly speaking a locally homogeneous geometric structure is of Riemannian type if it is constructed by putting together a Riemannian metric and any other geometric structure (e.g., a vector field). Since Riemannian metrics are rigid, a geometric structure of Riemannian type is automatically rigid.

With this terminology we have the following corollary of Theorem 1.

Theorem 2. *Let M be a compact manifold of dimension ≤ 4 equipped with a locally homogeneous geometric structure ϕ of Riemannian type. Then M is isomorphic to a quotient of a homogeneous space G/I, endowed with a G-invariant geometric structure, by a lattice in G.*

Proof. By Theorem 1, M is locally modeled on a homogeneous space G/I. Since ϕ is of Riemannian type, G/I admits a G-invariant Riemannian metric. This implies that the isotropy I is compact.

On the other hand, compact manifolds locally modeled on homogeneous space G/I with compact isotropy group I are classically known to be complete (meaning exactly that M is isomorphic to a quotient of G/I by a lattice in G): this is a consequence of the Hopf–Rinow's geodesical completeness [35].

Remark 2. A G-invariant geometric structure on G/I is of Riemannian type if and only if I is compact.

Recall that a homogeneous space G/I is said to be *imprimitive* if the canonical G-action preserves a non trivial foliation.

Proposition 1. *If M is a compact surface locally modeled on an imprimitive homogeneous space, then M is a torus.*

Proof. The G-invariant one-dimensional foliation on G/H descends on M to a nonsingular foliation. Hopf–Poincaré's theorem implies then that the genus of M equals one: M is a torus.

Note that the results of [2, 27] imply in particular:

Theorem 3. *A locally homogeneous affine connection on a surface which is neither torsion free and flat, nor of Riemannian type, is locally modeled either on an imprimitive homogeneous space or on $SL(2,\mathbf{R})/I$, where I the diagonal one parameter subgroup.*

Indeed, Arias-Marco and Kowalski study in [2] all possible local normal forms for locally homogeneous affine connections on surfaces with the corresponding Killing algebra. Their results are summarized in a nice table (see [2, pp. 3–5]). In all cases, except for the Killing algebra of the (standard) torsion free affine connection and for Levi-Civita connections of Riemannian metrics of constant sectional curvature, either there exists at least one Killing field non contained in the isotropy algebra which is normalized by the Killing algebra, or the Lie algebra is the standard $sl(2,\mathbf{R})$. In the first case, the normalized Killing field direction defines a G-invariant line field on G/I; in the second case the corresponding homogeneous space is $SL(2,\mathbf{R})/I$, where I is the diagonal one parameter subgroup.

The previous result combined with proposition 1 imply the main result in [33]:

Theorem 4. *A compact surface M bearing a locally homogeneous affine connection of non-Riemannian type is a torus.*

Recall first that a well-known result of J. Milnor shows that a compact surface bearing a flat torsion free affine connection is a torus.

Proof. In the case of a non-flat connection, Theorem 3 shows that M is locally modeled on an imprimitive homogeneous space or on $SL(2,\mathbf{R})/I$. Proposition 1 finishes the proof in the first case. In the second case, M inherits of a flat torsion free affine connection and we conclude using Milnor's theorem.

As an application of the Nomizu's phenomenon we give (compare with Theorem 1):

Theorem 5. *Let M be a compact simply connected real analytic manifold admitting a real analytic locally homogeneous rigid geometric structure. Then M is isomorphic to a homogeneous space G/I endowed with a G-invariant geometric structure.*

Proof. Since ϕ is locally homogeneous and M is simply connected and compact, the local transitive action of the Killing algebra extends to a global action of the associated simply connected Lie group G (we need compactness to ensure that vector fields on M are complete). All orbits have to be open, so there is only one orbit: the action is transitive and M is a homogeneous space.

3 Quasihomogeneous Real-Analytic Connections

We present now the recent results obtained in collaboration with Adolfo Guillot [16].

The main results in [16] are stated in Theorem 7 (the local classification) and in Theorem 8 (the global classification). In particular, we show that such (strictly) quasihomogeneous connections exist.

Motivated by these results, Theorem 8 constructs and characterizes torsion free real-analytic affine connections on compact surfaces which are quasihomogeneous (but not homogeneous). Even if we cannot say that a real analytic connection on a compact surface which is locally homogeneous somewhere is locally homogeneous everywhere, we fully understand the connections that do not satisfy this property.

Let us first give the following example of a quasihomogeneous (unimodular) real analytic affine connection.

Proposition 2. *Let ∇ be a connection on $\mathbf{R}^2$ such that $\nabla_{\frac{\partial}{\partial x}}\frac{\partial}{\partial x} = \mu y^3 \frac{\partial}{\partial y}$, with μ a nonzero real constant, and all other Christoffel symbols vanish. Then the (torsion free) unimodular affine connection determined by ∇ and the volume form vol $= dx \wedge dy$ is locally homogeneous on $y > 0$ and on $y < 0$, but not on all of $\mathbf{R}^2$.*

Proof. We check easily that ∇ and the volume form are invariant by the flows of $\frac{\partial}{\partial x}$ and of $x\frac{\partial}{\partial x} - y\frac{\partial}{\partial y}$.

Remark that the invariant vector fields of the previous action are $A = y\frac{\partial}{\partial y}$ and $B = \frac{1}{y}\frac{\partial}{\partial x}$. They are of constant volume and ∇ has constant Christoffel symbols with respect to A and B:

$$\nabla_A A = A, \nabla_B B = \mu A, \nabla_B A = 0, \nabla_A B = [A, B] = -B.$$

Thus the unimodular affine connection is locally homogeneous on the open sets $y > 0$ and $y < 0$, where A and B are pointwise linearly independent.

The only nonzero component of the curvature tensor is

$$R(\frac{\partial}{\partial x}, \frac{\partial}{\partial y})\frac{\partial}{\partial x} = \nabla_{\frac{\partial}{\partial x}}\nabla_{\frac{\partial}{\partial y}}\frac{\partial}{\partial x} - \nabla_{\frac{\partial}{\partial y}}\nabla_{\frac{\partial}{\partial x}}\frac{\partial}{\partial x} = -3\mu y^2 \frac{\partial}{\partial y}.$$

The curvature tensor vanishes exactly on $y = 0$. Thus ∇ is not locally homogeneous on all of $\mathbf{R}^2$.

The main theorem in [15] shows that the previous germ of connection cannot be extended on a compact surface:

Theorem 6. *Let ∇ be a real analytic unimodular affine connection on a compact connected real analytic surface M. If ∇ is locally homogeneous on a nontrivial open set in M, then ∇ is locally homogeneous on all of M.*

The main local ingredient of Theorem 8 is the following classification of (all) germs of torsion free real-analytic affine connections which are quasihomogeneous. It is, in some sense, the quasihomogeneous analogue to the local results in [27].

Theorem 7. *Let ∇ be a torsion free real-analytic affine connection in a neighborhood of the origin in $\mathbf{R}^2$. Suppose that the maximal open set where $\mathfrak{K}(\nabla)$, the Lie algebra of Killing vector fields of ∇, is transitive, contains the origin in its closure, but does not contain the origin. Then, up to an analytic change of coordinates, the germ of ∇ at the origin is one of the following:*

[Type I(n), $n \in \frac{1}{2}\mathbf{Z}$, $n \geq \frac{1}{2}$*] The germ at* $(0,0)$ *of*
$\nabla_{\frac{\partial}{\partial x}}\frac{\partial}{\partial x} = 0$, $\nabla_{\frac{\partial}{\partial x}}\frac{\partial}{\partial y} = -\gamma x^n \frac{\partial}{\partial x}$, $\nabla_{\frac{\partial}{\partial y}}\frac{\partial}{\partial y} = -\frac{1}{n}\epsilon x^{2n+1}\frac{\partial}{\partial x} - \phi x^n \frac{\partial}{\partial y}$,
with $\phi = 0$ *and* $\gamma = 0$, *if* $n \notin \mathbf{Z}$ *and* $(n, \phi, \epsilon) \neq (1, -\gamma, -\gamma^2)$.
For these, $\mathfrak{K}(\nabla) = \langle x\partial/\partial x - ny\partial/\partial y, \partial/\partial y\rangle$.

[Type II(n), $n \in \frac{1}{2}\mathbf{Z}$, $n \geq \frac{5}{2}$*] The germ at* $(0,0)$*—Type* II$^0(n)$*—or the germ at* $(0,1)$*—Type* II$^1(n)$*—of*
$\nabla_{\frac{\partial}{\partial x}}\frac{\partial}{\partial x} = \left(-\frac{1}{n}\epsilon x^{2n-3}y^2 + 2\gamma x^{n-2}y\right)\frac{\partial}{\partial x} + \left(-\frac{1}{n}\epsilon x^{2n-4}y^3 + [2\gamma - \phi]x^{n-3}y^2\right)\frac{\partial}{\partial y}$, $\nabla_{\frac{\partial}{\partial x}}\frac{\partial}{\partial y} = \left(\frac{1}{n}\epsilon x^{2n-2}y - \gamma x^{n-1}\right)\frac{\partial}{\partial x} + \left(\frac{1}{n}\epsilon x^{2n-3}y^2 + [\phi - \gamma]x^{n-2}y\right)\frac{\partial}{\partial y}$, $\nabla_{\frac{\partial}{\partial y}}\frac{\partial}{\partial y} = -\frac{1}{n}\epsilon x^{2n-1}\frac{\partial}{\partial x} - \left(\frac{1}{n}\epsilon x^{2n-2}y + \phi x^{n-1}\right)\frac{\partial}{\partial y}$,
with $\phi = 0$ *and* $\gamma = 0$, *if* $n \notin \mathbf{Z}$. *For these,* $\mathfrak{K}(\nabla) = \langle x\partial/\partial x + (1-n)y\partial/\partial y, x\partial/\partial y\rangle$.

[Type III*] The germ at* $(0,0)$ *of*
$\nabla_{\frac{\partial}{\partial x}}\frac{\partial}{\partial x} = \left(-\frac{1}{2}\epsilon xy^2 + 2\gamma y\right)\frac{\partial}{\partial x} - \frac{1}{2}\epsilon y^3\frac{\partial}{\partial y}$, $\nabla_{\frac{\partial}{\partial x}}\frac{\partial}{\partial y} = \left(\frac{1}{2}\epsilon x^2 y - \gamma x\right)\frac{\partial}{\partial x} + \left(\frac{1}{2}\epsilon xy^2 + \gamma y\right)\frac{\partial}{\partial y}$,
$\nabla_{\frac{\partial}{\partial y}}\frac{\partial}{\partial y} = -\frac{1}{2}\epsilon x^3\frac{\partial}{\partial x} - \left(\frac{1}{2}\epsilon x^2 y + 2\gamma x\right)\frac{\partial}{\partial y}$,
for which $\mathfrak{K}(\nabla)$ *is the Lie algebra of divergence-free linear vector fields in* $\mathbf{R}^2$.

In Types I *and* II, $(\gamma, \phi, \epsilon) \in \mathbf{R}^3 \setminus \{(0,0,0)\}$ *and the connection with parameters* (γ, ϕ, ϵ) *is equivalent to the one with parameters* $(\mu\gamma, \mu\phi, \mu^2\epsilon)$, *for* $\mu > 0$ *in the case of Type* II^1 *and, for* $\mu \in \mathbf{R}^*$, *in the other cases. In Type* III, $(\gamma, \epsilon) \in \mathbf{R}^2 \setminus \{(0,0)\}$ *and the connection with parameters* (γ, ϵ) *is equivalent to the one with parameters* $(\mu\gamma, \mu^2\epsilon)$ *for* $\mu \in \mathbf{R}^*$. *Apart from this, all the above connections are inequivalent.*

Remark 3. For the connections of Type III, $\mathfrak{K}(\nabla) \approx \mathfrak{sl}(2, \mathbf{R})$ and there is one two-dimensional orbit of the Killing algebra (the complement of the origin). In the other cases, $\mathfrak{K}(\nabla) \approx \mathfrak{aff}(\mathbf{R})$, the Lie algebra of the affine group of the real line, and the components of the complement of the geodesic $\{x = 0\}$ in $\mathbf{R}^2$ are the two-dimensional orbits of the Killing algebra. In particular, all these germs admit nontrivial open sets on which ∇ is locally isomorphic to a translation invariant connection on the connected component of the affine group of the real line.

The closed set where ∇ is not locally homogeneous is either a geodesic, or a point. Moreover, in all cases, every vector field in the Killing algebra $\mathfrak{K}(\nabla)$ is an affine one and thus these Killing Lie algebras also preserve a flat torsion free affine connection.

The previous theorem admits the following corollary:

Corollary 1. *If a surface S admits a quasihomogeneous real analytic torsion free affine connection ∇, then S also admits a flat torsion free affine connection preserved by the Killing algebra of ∇. In particular, if S is compact, then S is diffeomorphic to a torus.*

The idea for the proof of Theorem 7 is the following. We prove that for a germ of quasihomogeneous real-analytic affine connection, we can always find a two-dimensional subalgebra of the Killing algebra which is transitive on a nontrivial open set, but not at the origin. Consequently, there exists a nontrivial open set where the connection is locally isomorphic either to a translation-invariant connection on $\mathbf{R}^2$, or to a left-invariant connection on the affine group. Then we show that a quasihomogeneous connection cannot be locally isomorphic (on a nontrivial open set) to a translation invariant connection on $\mathbf{R}^2$ (without being locally homogeneous everywhere). In order to deal with the affine case, we will have to study the invariant affine connections and their Killing algebras in the affine group. The method consists in considering normal forms at the origin of left invariant vector fields X, Y on the affine group, with respect to which we compute Christoffel coefficients (in general X and Y are meromorphic at the origin).

Our global classification is the following one:

Theorem 8. *1. For integers n_1, n_2, with $n_2 \geq n_1 \geq 2$, there exists a unique (up to automorphism) real-analytic torsion free affine connection ∇_{n_1,n_2} on $\mathbf{R}^2$ such that*

a. ∇_{n_1,n_2} is locally homogeneous on a nontrivial open set, but not on all of $\mathbf{R}^2$. For $i = 1, 2$, there exists a point $p_i \in \mathbf{R}^2$ such that ∇_{n_1,n_2} is, in a neighborhood of p_i, given by a normal form of type $\mathrm{II}^1(n_i)$, if n_i is odd and by a normal form of type $\mathrm{I}(n_i)$, if n_i is even (in particular, the Killing Lie algebra of ∇_{n_1,n_2} is isomorphic to that of the affine group of the real line).

b. There exist groups of automorphisms of ∇_{n_1,n_2} acting freely, properly discontinuously and cocompactly on $\mathbf{R}^2$.

2. Let S be a compact orientable analytic surface endowed with a real-analytic torsion free affine connection that is locally homogeneous on some nontrivial open set, but not on all of S. Then (S, ∇) is isomorphic to a quotient of $(\mathbf{R}^2, \nabla_{n_1,n_2})$.

3. The moduli space of compact quotients of $(\mathbf{R}^2, \nabla_{n_1,n_2})$ is $\Xi = \mathbf{N} \times \mathbf{R} \times \mathbf{R}/\mathbf{Z}$. Every compact quotient of $(\mathbf{R}^2, \nabla_{n_1,n_2})$ is a torus. For the torus T corresponding to $(k, \tau, \theta) \in \Xi$ we have:

a. The open set of local homogeneity is dense and is a union of $2k$ (if $n_1 \neq n_2$) or k (if $n_1 = n_2$) cylinders bounded by simple closed geodesics.

b. *There exists a globally defined Killing field A on T, unique up to multiplication by a constant. When normalized such that there exists a Killing vector field B defined in some open subset such that $[A, B] = B$, A is periodic with period τ.*
c. *If γ_1 and γ_2 are generators of the fundamental group of T and γ_1 is homotopic to an orbit of A, then the analytic continuations of B along γ_1 and γ_2 are, respectively, $e^{\tau} B$ and $e^{(\theta+k)\tau} B$.*

Example 1. Let us describe explicitly the quotients of $\nabla_{2,2}$ with parameters $(1, \tau, [\theta]) \in \Xi$. This will also give a self-contained proof of the fact that tori admit quasihomogeneous connections. Consider, in $\mathbf{R}^2$, the torsion free affine connection ∇ such that

$$\nabla_{\frac{\partial}{\partial x}} \frac{\partial}{\partial x} = 0,\ \nabla_{\frac{\partial}{\partial x}} \frac{\partial}{\partial y} = \frac{3}{4}x^2 \frac{\partial}{\partial x},\ \nabla_{\frac{\partial}{\partial y}} \frac{\partial}{\partial y} = \frac{3}{8}x^5 \frac{\partial}{\partial x} + \frac{3}{4}x^2 \frac{\partial}{\partial y}$$

(it is a connection of type I for $n = 2$ and $\epsilon = \phi = \gamma = -\frac{3}{4}$). Remark that $B = \partial/\partial y$ is a Killing vector field for ∇. Consider the commuting meromorphic vector fields

$$A = \frac{1}{2}x\frac{\partial}{\partial x} - y\frac{\partial}{\partial y},\ Z = \frac{1}{2}x\frac{\partial}{\partial x} + x^{-2}\frac{\partial}{\partial y}.$$

If we let $h(x, y) = x^2 y$, we have

$$\nabla_A Z = \frac{3}{4}A - Z,\ \nabla_Z Z = -\frac{1}{4}Z,\ \nabla_A A = \frac{3}{4}(h^2 + 2h)A + \left(\frac{1}{4} - \frac{3}{8}[h^2 + 2h]\right) Z. \tag{1}$$

Notice that since $[A, Z] = 0$ and h is a first integral of A, the vector field A is a Killing field of ∇. The Lie algebra of Killing vector fields of ∇ contains the subalgebra generated by A and by B. The rank of this Lie algebra of vector fields is one in $\{x = 0\}$ and two in its complement. A direct computation shows that the curvature tensor of ∇ vanishes exactly on $\{x = 0\}$. The connection is thus locally homogeneous in the half planes $\{x < 0\}$ and $\{x > 0\}$, but not on all of $\mathbf{R}^2$.

Consider the orientation-preserving birational involution $\sigma(x, y) = (-x, -y - 2x^{-2})$. It preserves the vector fields A and Z. Moreover, $h \circ \sigma = -2 - h$ and hence $(h^2 + 2h) \circ \sigma = h^2 + 2h$. This implies, by (1), that ∇ is preserved by σ. Let $\Omega = \{(x, y); y < 0, h > -2\}$, $U^+ = \Omega \cap \{x > 0\}$, $U^- = \Omega \cap \{x < 0\}$. Notice that $\sigma|_{U^+} : U^+ \to U^-$ is an analytic diffeomorphism. Let $\phi : \Omega \to \Omega$ be the diffeomorphism generated by the flow of A in time τ. The diffeomorphism ϕ preserves ∇ and commutes with σ. The quotient of Ω under the group generated by ϕ is a cylinder containing a simple closed curve coming from $\{x = 0\}$, whose complement is the union of the two cylinders $U^+/\langle\phi\rangle$ and $U^-/\langle\phi\rangle$. Let $K : U^+/\langle\phi\rangle \to U^-/\langle\phi\rangle$ be given by, first, the restriction to $U^+/\langle\phi\rangle$ of the flow of A in time $\theta\tau$ and then composing with σ (notice that adding an integer to θ yields the same result). By identifying $U^+/\langle\phi\rangle$ and $U^-/\langle\phi\rangle$ (as open subsets of $\Omega/\langle\phi\rangle$), via K, we obtain a torus S, naturally endowed with a connection ∇_s, coming from ∇, a globally defined Killing vector field for ∇_s, induced by A, and a multivalued one, induced by B. There is one simple closed curve in S coming from $\{x = 0\}/\langle\phi\rangle$. The rank of the Killing algebra of vector fields of ∇_s is one along this curve and two in the complement: the connection is not locally homogeneous everywhere (since the curvature tensor vanishes exactly on $\{x = 0\}$), but is locally homogeneous in a dense open subset.

Note that in the previous example the connected component of the automorphism group of ∇_s is the flow generated by A, all of whose orbits are closed. The proof of Theorem 8 shows that, in general, the automorphism group of a quasihomogeneous connection is, up to a finite group, the flow of a Killing field, all of whose orbits are simple closed curves. Hence, the automorphism group doesn't admit dense orbits (our quasihomogeneous connections are not counter-examples to Fisher's conjecture, but they give a positive answer).

Another framework where Fisher's conjecture hold is real analytic Lorentz geometry in dimension three. In this context we proved in [13]:

Theorem 9. *Let g be a real analytic Lorentz metric on a compact connected real analytic threefold M. If g is locally homogeneous on a nontrivial open set in M, then g is locally homogeneous on all of M.*

Later, in a common work with Zeghib [18], we completely classified those locally homogeneous compact Lorentz threefolds. All of them have to be quotients of Lorentz homogeneous spaces (completeness). We also proved that there are exactly four essential and maximal Lorentz homogeneous spaces which admit compact quotients: the flat Lorentz space, the anti-de Sitter space (constant negative curvature), a left invariant Lorentz metric on the Heisenberg group and a left invariant Lorentz metric on the Sol group. This is a Lorentz version of Thurston's classification of eight maximal Riemannian geometries in dimension three and leads to the following uniformization result:

Corollary 2. *If a compact manifold of dimension three is locally modeled on a Lorentz homogeneous space (of non-Riemannian kind), then it admits, on a finite cover, a Lorentz metric of constant (non-positive) sectional curvature.*

4 Holomorphic Geometric Structures

Recall that a complex manifold M is of algebraic dimension zero if there are no nonconstant meromorphic functions on M. In this context the following theorem was proved in [11].

Theorem 10. *Let M be a connected complex manifold of algebraic dimension zero, endowed with a meromorphic rigid geometric structure ϕ. Then ϕ is locally homogeneous on an open dense subset in M.*

Moreover, if ϕ is unimodular and holomorphic and its Killing Lie algebra $\mathcal{G}$ is unimodular and simply transitive, then ϕ is locally homogeneous on all of M.

Proof. Let us denote by n the complex dimension of M and by $Is^{loc}(\phi)$ the pseudogroup of local isometries of ϕ.

Let ϕ be of order r, given by a map $\phi : R^r(M) \to Z$. For each positive integer s we consider the s-jet $\phi^{(s)}$ of ϕ. This is a $D^{(r+s)}(\mathbf{C}^n)$-equivariant meromorphic map $R^{(r+s)}(M) \to Z^{(s)}$, where $Z^{(s)}$ is the algebraic variety of the s-jets at the origin of holomorphic maps from $\mathbf{C}^n$ to Z. One can find the expression of the (algebraic) $D^{(r+s)}(\mathbf{C}^n)$-action on $Z^{(s)}$ in [8, 20].

Since ϕ is rigid, there exists a nowhere dense analytic subset S' in M, containing the poles of ϕ, and a positive integer s such that two points m, m' in $M \setminus S'$ are related by local automorphisms if and only if $\phi^{(s)}$ sends the fibers of $R^{(r+s)}(M)$ above m and m' on the same $D^{(r+s)}(\mathbf{C}^n)$-orbit in $Z^{(s)}$ [10, 24].

Rosenlicht's theorem shows that there exists a $D^{(r+s)}(\mathbf{C}^n)$-invariant stratification

$$Z^{(s)} = Z_0 \supset \ldots \supset Z_l,$$

such that Z_{i+1} is Zariski closed in Z_i, the quotient of $Z_i \setminus Z_{i+1}$ by $D^{(r+s)}(\mathbf{C}^n)$ is a complex manifold and rational $D^{(r+s)}(\mathbf{C}^n)$-invariant functions on Z_i separate orbits in $Z_i \setminus Z_{i+1}$.

Consider the open dense $Is^{loc}(\phi)$-invariant subset U in $M \setminus S'$, such that $\phi^{(s)}$ is of constant rank above U and the image of $R^{(r+s)}(M)|_U$ through $\phi^{(s)}$ lies in $Z_i \setminus Z_{i+1}$, but not in Z_{i+1}. Then the orbits of $Is^{loc}(\phi)$ in U are the fibers of a fibration of constant rank (on the quotient of $Z_i \setminus Z_{i+1}$ by $D^{(r+s)}(\mathbf{C}^n)$). Obviously, $U = M \setminus S$, where S is a nowhere dense analytic subset in M.

Assume now, by contradiction, that m and m' are two points in U which are not in the same $Is^{loc}(\phi)$-orbit, then the corresponding fibers of $R^{(r+s)}(M)|_U$ are sent by $\phi^{(s)}$ on two distinct $D^{(r+s)}(\mathbf{C}^n)$-orbits in $Z_i \setminus Z_{i+1}$. By Rosenlicht's theorem there exists a $D^{(r+s)}(\mathbf{C}^n)$-invariant rational function $F : Z_i \setminus Z_{i+1} \to \mathbf{C}$, which takes distinct values at these two orbits.

The meromorphic function $F \circ \phi^{(s)} : R^{(r+s)}(M) \to \mathbf{C}$ is $D^{(r+s)}(\mathbf{C}^n)$-invariant and descends in a $Is^{loc}(\phi)$-invariant meromorphic function on M which takes distinct values at m and at m': a contradiction.

We proved that the Killing Lie algebra $\mathcal{G}$ of ϕ is transitive on a maximal open dense subset U in M. Suppose now that ϕ is unimodular and also that $\mathcal{G}$ is unimodular and simply transitive.

Pick up a point m in M. We want to show that m is in U. The point m admits an open neighborhood U_m in M such that any local holomorphic Killing field of ϕ defined on a connected open subset in U_m extends on all of U_m [1, 31].

Since $\mathcal{G}$ acts transitively on U, choose local linearly independent Killing fields $X_1, \ldots, X_n$ on a connected open set included in $U \cap U_m$. As ϕ is unimodular, it determines a holomorphic volume form vol on U_m (if necessary we restrict to a smaller U_m). But $Is^{loc}(\phi)$ acts transitively on $U \cap U_m$ and $\mathcal{G}$ is supposed to be unimodular. This implies that the function $vol(X_1, \ldots, X_n)$ is $\mathcal{G}$-invariant and, consequently, a (nonzero) constant on $U \cap U_m$.

On the other hand $X_1, \ldots, X_n$ extend in some holomorphic Killing fields $\tilde{X}_1, \ldots, \tilde{X}_n$ defined on all of U_m. The holomorphic function $vol(\tilde{X}_1, \ldots, \tilde{X}_n)$ is a nonzero constant on U_m: in particular, $\tilde{X}_1(m), \ldots, \tilde{X}_n(m)$ are linearly independent. We proved that $\mathcal{G}$ acts transitively in the neighborhood of m and thus $m \in U$.

As a consequence of Theorem 10, we proved in [14] the following result:

Theorem 11. *Any holomorphic geometric structure τ on a Inoue surface is locally homogeneous.*

Indeed, it is known that Inoue surfaces admit holomorphic affine structures (locally modeled on $(GL(2, \mathbf{C}) \ltimes \mathbf{C}^2, \mathbf{C}^2)$). We put together this holomorphic affine structure and any other holomorphic geometric structure τ in some extra *rigid* holomorphic geometric structure ϕ, for which Theorem 10 applies (since Inoue surfaces have algebraic dimension zero). It remains to prove that ϕ is locally homogeneous on all of the surface. The next step of the proof is to use the fact that Inoue surfaces don't admit nontrivial complex submanifolds and to show that ϕ is locally homogeneous at least in the complementary of a finite number of points. The last step of the proof uses the fact that nontrivial holomorphic vector fields on Inoue surfaces don't vanish to conclude that ϕ is locally homogeneous. In particular, τ is also locally homogeneous.

Let us give an example where the open set of local homogeneity is not all of M.

Example 2. Consider the Hopf surface M of algebraic dimension zero, which is the quotient of $\mathbf{C}^2 \setminus \{0\}$ by the group $\mathbf{Z}$ generated by $T(z_1, z_2) = (\frac{1}{2}z_1, \frac{1}{3}z_2)$. The only curves on M are the two elliptic curves obtained as projections of the lines $\{z_1 = 0\}$ and $\{z_2 = 0\}$.

The Hopf surface M inherits of the standard affine structure of $\mathbf{C}^2$ and also of the vector fields $X_1 = z_1 \frac{\partial}{\partial z_1}$ et $X_2 = z_2 \frac{\partial}{\partial z_2}$ (they are T-invariant). Let us denote by ϕ the holomorphic rigid geometric structure which is the juxtaposition of the affine structure and of the vector fields X_1 and X_2.

The open dense set U of local homogeneity is the complement of the two elliptic curves in M. This is exactly the open set where the vector fields X_1 and X_2, which are Killing fields of ϕ (since they are linear and commute), are linearly independent. On the other hand, since X_1 and X_2 are part of the geometric structure, ϕ is not locally homogeneous in the neighborhood of points where X_1 and X_2 are collinear (on each elliptic curve, either X_1 or X_2 vanishes).

Let us also remind that Inoue, Kobayashi, and Ochiai classified in [25] all compact complex surfaces admitting holomorphic affine connections. In particular, they proved the following:

Theorem 12. *If a complex compact surface M admits a holomorphic affine connection, then it also admits a flat holomorphic affine connection (locally modeled on $(GL(2, \mathbf{C}) \ltimes \mathbf{C}^2, \mathbf{C}^2)$).*

Moreover, if M is Kaehler then M admits a finite cover which is a complex torus.

Following their work, in [14] we determined the local structure of all holomorphic torsion free affine connections on compact complex surfaces. It turns out that the Killing Lie algebra of those connections is either of dimension one, or it is isomorphic to $\mathbf{C}^2$ acting transitively (these connections are locally homogeneous). In particular, we conclude that *there are no quasihomogeneous torsion free holomorphic affine connections on compact complex surfaces*. Precisely, the classification is given by the following:

Theorem 13. *Let (M, ∇) be a compact complex surface endowed with a torsion free holomorphic affine connection.*

(i) If M is not biholomorphic to a principal elliptic bundle over a Riemann surface of genus $g \geq 2$, with odd first Betti number, then ∇ is locally isomorphic to a translation invariant connection on $\mathbf{C}^2$. In particular, ∇ is locally homogeneous.

(ii) In the other case, ∇ is invariant by the fundamental vector field of the elliptic fibration. If ∇ is non flat, then the corresponding Killing algebra is of dimension one.

Corollary 3. *Normal holomorphic projective connections on compact complex surfaces are flat (locally modeled on $(PGL(3, \mathbf{C}), P^2(\mathbf{C}))$.*

Let us now define a new holomorphic rigid geometric structure which is the complex analogous of a pseudo-Riemannian metric.

Definition 6. A holomorphic Riemannian metric on M is a holomorphic section q of the bundle $S^2(T^*M)$ such that at any point m in M, the complex quadratic form $q(m)$ is non degenerated.

For this geometric structure, we proved in [12] the following:

Theorem 14. *A holomorphic Riemannian metric on a compact complex connected threefold is locally homogeneous.*

Later, in a joint work with Zeghib [17], we proved a holomorphic analogous of our Lorentz classification in dimension three:

Theorem 15. *Let M be a compact complex connected threefold endowed with a holomorphic Riemannian metric g.*

(i) If the Killing Lie algebra of g admits a nontrivial semi-simple part, then it preserves a holomorphic Riemannian metric of constant sectional curvature on M.

(ii) If the Killing Lie algebra of g is solvable, then M admits a finite cover which is a quotient of the complex Heisenberg group or of the complex SOL group by a lattice.

Corollary 4. *M possesses a finite cover which admits a holomorphic Riemannian metric of constant sectional curvature.*

Contrary to the situation in Lorentz geometry, the classification of the compact complex three-manifolds endowed with a holomorphic Riemannian metric of constant sectional curvature is still an open problem. Two different situations could appear:

Flat Case. In this case M is locally modeled on $(O(3,\mathbb{C}) \ltimes \mathbb{C}^3, \mathbb{C}^3)$. The challenge remains:

1) *Markus conjecture*: Is M complete (i.e., is M a quotient of the model)?
2) *Auslander conjecture*: Assuming M as above, is Γ solvable?

Note that these questions are settled in the setting of (real) flat Lorentz manifolds [9, 22], but unsolved for general (real) pseudo-Riemannian metrics. The real part of the holomorphic Riemannian metric is a (real) pseudo-Riemannian metric of signature $(3, 3)$ for which both previous conjectures are still open.

Non Flat Case. In this case, M is locally modeled on G/I, with $G = SL(2, \mathbf{C}) \times SL(2, \mathbf{C})$ and $I = SL(2, \mathbf{C})$ diagonally embedded in G. The completeness of this geometry on compact complex manifolds is still an open problem (i.e., whether our manifold M is isomorphic to a quotient of the model G/I, or not), despite a positive local result of Ghys [23].

Nevertheless recent results of Tholozan [36] together with a theorem of Kassel [26] show that the space of complete structures is a union of connected components (in the space of group homomorphisms from the fundamental group of M into $SL(2, \mathbf{C}) \times SL(2, \mathbf{C})$). The question now is to find out if some connected component formed by (exotic) non complete structures does exist.

References

1. Amores, A.M.: Vector fields of a finite type G-structure. J. Differ. Geom. **14**(1), 1–6 (1979/1980)
2. Arias-Marco, T., Kowalski, O.: Classification of locally homogeneous affine connections with arbitrary torsion on 2-dimensional manifolds. Monatsh. Math. **153**(1), 1–18 (2008)
3. Benoist, Y.: Orbites des structures rigides (d'après M. Gromov). In: Integrable systems and foliations/Feuilletages et systèmes intégrables (Montpellier, 1995). Progr. Math., vol. 145, pp. 1–17. Birkhäuser, Boston (1997)
4. Benoist, Y., Foulon, P., Labourie, F.: Flots d'Anosov à distributions stable et instable différentiables. J. Am. Math. Soc. **5**(1), 33–74 (1992)
5. Benveniste, E.J., Fisher, D.: Nonexistence of invariant rigid structures and invariant almost rigid structures. Comm. Anal. Geom. **13**(1), 89–111 (2005)
6. Bogomolov, F.: Holomorphic tensors and vector bundles on projective varieties. Math. USSR Izvestija **13**(3), 499–555 (1979)
7. Bryant, R.L., Manno, G., Matveev, V.S.: A solution of a problem of Sophus Lie: normal forms of two-dimensional metrics admitting two projective vector fields. Math. Ann. **340**(2), 437–463 (2008)
8. Candel, A., Quiroga-Barranco, R.: Gromov's centralizer theorem. Geom. Dedicata **100**, 123–155 (2003)

9. Carrière, Y.: Autour de la conjecture de l. markus sur les variétés affines. Inv. Math. **95**, 615–628 (1989)
10. D'Ambra, G., Gromov, M.: Lectures on transformation groups: geometry and dynamics. In: Surveys in Differential Geometry (Cambridge, 1990), pp. 19–111. Lehigh University, Bethlehem (1991)
11. Dumitrescu, S.: Structures géométriques holomorphes sur les variétés complexes compactes. Ann. ENS **34**, 557–571 (2001)
12. Dumitrescu, S.: Homogénéité locale pour les métriques riemanniennes holomorphes en dimension 3. Ann. Inst. Fourier **57**(3), 739–773 (2007)
13. Dumitrescu, S.: Dynamique du pseudo-groupe des isométries locales sur une variété lorentzienne analytique de dimension 3. Ergodic Theory Dyn. Syst. **28**(4), 1091–1116 (2008)
14. Dumitrescu, S.: Connexions affines et projectives sur les surfaces complexes compactes. Math. Z. **264**(2), 301–316 (2010)
15. Dumitrescu, S.: Locally homogeneous rigid geometric structures on surfaces. Geom. Dedicata **160**, 71–90 (2012)
16. Dumitrescu, S., Guillot, A.: Quasihomogeneous real analytic connections on surfaces. J. Topol Anal. **5**(4), 491–532 (2013)
17. Dumitrescu, S., Zeghib, A.: Global rigidity of holomorphic riemannian metrics on compact complex 3-manifolds. Math. Ann. **345**(1), 53–81 (2009)
18. Dumitrescu, S., Zeghib, A.: Géométries lorentziennes de dimension trois : classification et complétude. Geom. Dedicata **149**, 243–273 (2010)
19. Ehresmann, C.: Sur les espaces localement homogènes. Enseign. Math. **35**, 317–333 (1936)
20. Feres, R.: Rigid geometric structures and actions of semisimple Lie groups. In: Rigidité, groupe fondamental et dynamique. Panor. Synthèses, vol. 13, pp. 121–167. Soc. Math. France, Paris (2002)
21. Fisher, D.: Groups acting on manifolds: around the Zimmer program. In: Geometry, Rigidity, and Group Actions. Chicago Lectures in Mathematics, pp. 72–157. University of Chicago Press, Chicago (2011)
22. Fried, D., Goldman, W.: Three-dimensional affine crystallographic groups. Adv. Math. **47**(1), 1–49 (1983)
23. Ghys, E.: Déformations des structures complexes sur les espaces homogènes de sl(2,c). J. Reine Angew. Math. **468**, 113–138 (1995)
24. Gromov, M.: Rigid transformations groups. In: Géométrie différentielle (Paris, 1986). Travaux en Cours, vol. 33, pp. 65–139. Hermann, Paris (1988)
25. Inoue, M., Kobayashi, S., Ochiai, T.: Holomorphic affine connections on compact complex surfaces. J. Fac. Sci. Univ. Tokyo **27**(2), 247–264 (1980)
26. Kassel, F.: Proper actions on corank-one proper reductive homogeneous spaces. J. Lie Theory **4**, 961–978 (2008)
27. Kowalski, O., Opozda, B., Vlášek, Z.: A classification of locally homogeneous connections on 2-dimensional manifolds via group-theoretical approach. Cent. Eur. J. Math. **2**(1), 87–102 (2004, electronic)
28. Lie, S.: Theorie der Transformationsgruppen I. Math. Ann. **16**(4), 441–528 (1880)
29. Melnick, K.: Compact Lorentz manifolds with local symmetry. J. Diff. Geom. **81**(2), 355–390 (2009)
30. Mostow, G.D.: The extensibility of local Lie groups of transformations and groups on surfaces. Ann. Math. (2) **52**, 606–636 (1950)
31. Nomizu, K.: On local and global existence of Killing vector fields. Ann. Math. (2) **72**, 105–120 (1960)
32. Olver, P.J.: Equivalence, Invariants, and Symmetry. Cambridge University Press, Cambridge (1995)
33. Opozda, B.: Locally homogeneous affine connections on compact surfaces. Proc. Am. Math. Soc. **132**(9), 2713–2721 (2004, electronic)
34. Prüfer, F., Tricerri, F., Vanhecke, L.: Curvature invariants, differential operators and local homogeneity. Trans. Am. Math. Soc. **348**(11), 4643–4652 (1996)

35. Sharpe, R.W.: Differential Geometry. Graduate Texts in Mathematics, vol. 166. Springer, New York (1997) [Cartan's generalization of Klein's Erlangen program, With a foreword by S. S. Chern]
36. Tholozan, N.: Sur la complétude des certaines variétés pseudo-riemanniennes localement symétriques. Arxiv (2013)
37. Thurston, W.P.: In: Levy, S. (ed.) Three-Dimensional Geometry and Topology, Vol. 1. Princeton Mathematical Series, vol. 35. Princeton University Press, Princeton (1997)
38. Zeghib, A.: Killing fields in compact Lorentz 3-manifolds. J. Diff. Geom. **43**(4), 859–894 (1996)

Koszul Binomial Edge Ideals

Viviana Ene, Jürgen Herzog, and Takayuki Hibi

Abstract It is shown that if the binomial edge ideal of a graph G defines a Koszul algebra, then G must be chordal and claw free. A converse of this statement is proved for a class of chordal and claw-free graphs.

Keywords Koszul algebra • Binomial edge ideals

Subject Classifications:13C13, 13A30, 13F99, 05E40

1 Introduction

A Koszul algebra in our context will be a standard graded (commutative) K-algebra whose graded maximal ideal has a linear resolution. This class of K-algebras occurs quite frequently among toric rings and other K-algebras arising in combinatorial commutative algebra and algebraic geometry. It is known and easily seen that a Koszul algebra is defined by quadrics. This statement has a partial converse,

---

V. Ene (✉)
Faculty of Mathematics and Computer Science, Ovidius University, Bd. Mamaia 124, 900527 Constanta, Romania

Simion Stoilow Institute of Mathematics of Romanian Academy, Research group of the project ID-PCE-2011-1023, P.O.Box 1-764, Bucharest 014700, Romania
e-mail: vivian@univ-ovidius.ro

J. Herzog
Fachbereich Mathematik, Universität Duisburg-Essen, Campus Essen, 45117 Essen, Germany
e-mail: juergen.herzog@uni-essen.de

T. Hibi
Department of Pure and Applied Mathematics, Graduate School of Information Science and Technology, Osaka University, Toyonaka, Osaka 560-0043, Japan
e-mail: hibi@math.sci.osaka-u.ac.jp

D. Ibadula, W. Veys (eds.), *Bridging Algebra, Geometry, and Topology*, Springer Proceedings in Mathematics & Statistics 96,
DOI 10.1007/978-3-319-09186-0_8

which says that a K-algebra is Koszul if its defining ideal admits a reduced Gröbner basis of quadrics. The proof of these statements can for example be found in [9]. For a nice survey on commutative Koszul algebras we refer to [4].

In the present paper we consider K-algebras defined by binomial edge ideals. Given a finite simple graph G on the vertex set $[n] = \{1, 2 \dots, n\}$, one defines the binomial edge ideal J_G associated with G as the ideal generated by the quadrics $f_{ij} = x_i y_j - x_j y_i$ in $S = K[x_1, \dots, x_n, y_1, \dots, y_n]$ with $\{i, j\}$ an edge of G.

This class of ideals was introduced in [13, 19]. Part of the motivation to consider such ideals arises from algebraic statistic as explicated in [13], see also [9]. In recent years several papers appeared [6, 10, 11, 14–16] attempting to describe algebraic and homological properties of binomial edge ideals in terms of the underlying graph. Since by its definition J_G is generated by quadrics it is natural to ask for which graphs G the K-algebra S/J_G is Koszul. If this happens to be the case we call G Koszul with respect to K. As noted above, G will be Koszul if J_G has a quadratic Gröbner basis. This is the case with respect to the lexicographic order induced by $x_1 > \cdots > x_n > y_1 > \cdots > y_n$ if and only if G is a closed graph with respect to the given labeling, in other words, if G satisfies the following condition: whenever $\{i, j\}$and $\{i, k\}$ are edges of G and either $i < j, i < k$ or $i > j, i > k$ then $\{j, k\}$ is also an edge of G. One calls a graph G closed if it is closed with respect to some labeling of its vertices. It was observed in [13] that a closed graph must be chordal and claw free. However the class of closed graphs is much smaller than that of the chordal and claw-free graphs. Interesting combinatorial characterizations of closed graphs are given in [5, 11].

By what we said so far it follows that all closed graphs are Koszul. On the other hand, it is not hard to find non-closed graphs that are Koszul. Thus the problem arises to classify all Koszul graphs. In Sect. 2 we show that Koszul graphs must be closed and claw free. Thus we have the implications

$$\text{Closed graph} \Rightarrow \text{Koszul graph} \Rightarrow \text{Chordal and claw-free graph.}$$

The first implication cannot be reversed. In Sect. 3 we give an example of a graph which is chordal and claw free but not Koszul. Thus the second implication cannot be reversed as well. The results that we have so far allow a classification of all Koszul graphs whose cliques are of dimension at most 2.

2 Koszul Graphs Are Chordal and Claw Free

The goal of this section is to prove the statement made in the section title. We first recall some concepts from graph theory. Let G be a finite simple graph, that is, a graph with no loops or multiple edges. We denote by $V(G)$ the set of vertices and by $E(G)$ the set of edges of G. A *cycle* C of G of length n is a subgraph of G whose vertices $V(C) = \{v_1, \dots, v_n\}$ can be labeled such that the edges of C are $\{v_i, v_{i+1}\}$ for $i = 1, \dots, n-1$ and $\{v_1, v_n\}$. A graph H is called an *induced subgraph* of G if there exists a subset $W \subset V(G)$ with $V(H) = W$ and $E(H) = \{\{u, v\} \in$

$E(G)$: $u, v \in W\}$. The graph G is called *chordal* if any cycle C of G has a chord, where a chord of C is defined to be an edge $\{u, v\}$ of G with $u, v \in V(C)$ but $\{u, v\} \notin E(C)$. Finally, the graph Cl with $V(Cl) = \{v_1, v_2, v_3, v_4\}$ and $E(Cl) = \{\{v_1, v_2\}, \{v_1, v_3\}, \{v_1, v_4\}\}$ is called a *claw*, and G is called *claw free* if G does not contain an induced subgraph which is isomorphic to Cl.

Now we are in the position to formulate the main result of this section.

Theorem 2.1. *Let G be a Koszul graph. Then G is chordal and claw free.*

For the proof of this theorem we shall need the following lemma which provides a necessary condition for Koszulness.

Lemma 2.2. *Let $S = K[x_1, \dots, x_n]$ be the polynomial ring over the field K in the variables $x_1, \dots, x_n$, and let $I \subset S$ be a graded ideal of S generated by quadrics. Denote the graded Betti numbers of S/I by $\beta_{ij}^S(S/I)$ and suppose that $\beta_{2j}^S(S/I) \neq 0$ for some $j > 4$. Then S/I is not Koszul.*

This lemma is an immediate consequence of Formula (2) given in the introduction of [2], where, as a consequence of results in that paper, it is stated that if S/I is Koszul, then $t_{i+1}(S/I) \leq t_i(S/I) + 2$ for $i \leq \operatorname{codim} S/I + 1$. Here $t_i(S/I) = \max\{j\colon \beta_{ij}^S(S/I) \neq 0\}$ for $i = 0, \dots, \operatorname{proj\,dim} S/I$.

For the convenience of the reader we give a direct proof of the lemma: let $(R, \mathfrak{m}, K)$ be a (Noetherian) local ring or a standard graded K-algebra (in which case we assume that $\mathfrak{m}$ is the graded maximal ideal of R). Tate in his famous paper [22] constructed an R-free resolution

$$X\colon \cdots \longrightarrow X_i \longrightarrow \cdots \longrightarrow X_2 \longrightarrow X_1 \longrightarrow X_0 \longrightarrow 0,$$

of the residue class field $R/\mathfrak{m} = K$, that is, an acyclic complex of finitely generated free R-modules X_i with $H_0(X) = K$, admitting an additional structure, namely the structure of a differential graded R-algebra. It was Gulliksen [12] who proved that if Tate's construction is minimally done, as explained below, then X is indeed a minimal free R-resolution of K. For details we refer to the original paper of Tate and to a modern treatment of the theory as given in [1].

Here we sketch Tate's construction as much as is needed to prove the lemma. In Tate's theory X is a DG-algebra, that is, a graded skew-symmetric R-algebra with free R-modules X_i as graded components and $X_0 = R$, equipped with a differential d of degree -1 such that

$$d(ab) = d(a)b + (-1)^i a d(b) \qquad (1)$$

for $a \in X_i$ and $b \in X$. Moreover, (X, d) is an acyclic complex with $H_0(X) = K$.

The algebra X is constructed by adjunction of variables: given any DG-algebra Y and a cycle $z \in Y_i$, then the DG-algebra $Y' = Y\langle T\colon dT = z\rangle$ is obtained by adjoining the variable T of degree $i + 1$ to Y in order to kill the cycle z.

If i is even we let

$$Y'_j = Y_j \oplus Y_{j-i-1}T \quad \text{with } T^2 = 0 \text{ and } d(T) = z.$$

If i is odd we let

$$Y'_j = X_j \oplus X_{j-(i+1)}T^{(1)} \oplus X_{j-2(i+1)}T^{(2)} \oplus \cdots$$

with $T^{(0)} = 1, T^{(1)} = T, T^{(i)}T^{(j)} = ((i+j)!/i!j!)T^{(i+j)}$ and $d(T^{(i)}) = zT^{(i-1)}$. The $T^{(j)}$ are called the divided powers of T. The degree of $T^{(j)}$ is defined to be $j \deg T$.

The construction of X proceeds as follows: Say, $\mathfrak{m}$ is minimally generated by $x_1, \ldots, x_n$. Then we adjoin to R (which is a DG-algebra concentrated in homological degree 0) the variables $T_{11}, \ldots, T_{1n}$ of degree 1 with $d(T_{1i}) = x_i$. The DG-algebra $X^{(1)} = R\langle T_{11}, \ldots, T_{1n}\rangle$ so obtained is nothing but the Koszul complex of the sequence $x_1, \ldots, x_n$ with values in R. If $X^{(1)}$ is acyclic, then R is regular and $X = X^{(1)}$ is the Tate resolution of K. Otherwise $H_1(X^{(1)}) \neq 0$ and we choose cycles $z_1, \ldots, z_m$ whose homology classes form a K-basis of $H_1(X^{(1)})$, and we adjoin variables $T_{21}, \ldots, T_{2m}$ of degree 2 to $X^{(1)}$ with $d(T_{2i}) = z_i$ to obtain $X^{(2)}$. It is then clear that $H_j(X^{(2)}) = 0$ for $j = 1$. Suppose $X^{(k)}$ has been already constructed with $H_j(X^{(k)}) = 0$ for $j = 1, \ldots, k-1$. We first observe that $H_k(X^{(k)})$ is annihilated by $\mathfrak{m}$. Indeed, let z be a cycle of $X^{(k)}$, then $x_i z = d(T_{1i}z)$, due to the product rule (1). Now one chooses a K-basis of cycles representing the homology classes of $H_k(X^{(k)})$ and adjoins variables in degree $k+1$ to kill these cycles, thereby obtaining $X^{(k+1)}$. In this way one obtains a chain of DG-algebras

$$R = X^{(0)} \subset X^{(1)} \subset X^{(2)} \subset \cdots \subset X^{(2)} \subset \cdots$$

which in the limit yields the Tate resolution X of K. It is clear that if R is standard graded then in each step the representing cycles that need to be killed can be chosen to be homogeneous, so that X becomes a graded minimal free R-resolution of K if we assign to the variables T_{ij} inductively the degree of the cycles they do kill and apply the following rule: denote the internal degree (different from the homological degree) of a homogeneous element a of X by $\mathrm{Deg}(a)$. Then we require that $\mathrm{Deg}\, T^{(i)} = i\, \mathrm{Deg}\, T$ for any variable of even homological degree and furthermore $\mathrm{Deg}(ab) = \mathrm{Deg}(a) + \mathrm{Deg}(b)$ for any two homogeneous elements in X.

Now we are ready to prove Lemma 2.2: the Koszul complex $X^{(1)}$ as a DG-algebra over S/I is generated by the variable T_{1i} with $d(T_{1i}) = x_i$ for $i = 1, \ldots, n$. Thus $\mathrm{Deg}\, T_{1i} = 1$ for all i. Let $f_1, \ldots, f_m$ be quadrics which minimally generate I, and write $f_i = \sum_{j=1}^m f_{ij} x_j$ with suitable linear forms f_{ij}. Then $H_1(X^{(1)})$ is minimally generated by the homology classes of the cycles $z_i = \sum_{j=1}^m f_{ij} T_{1j}$. Let $T_{2i} \in X^{(2)}$ be the variables of homological degree 2 with $d(T_{2i}) = z_i$ for $i = 1, \ldots, m$. Then $\mathrm{Deg}\, T_{2i} = \mathrm{Deg}\, z_i = 2$ for all i. To proceed in the construction of X we have to kill the cycles $w_1, \ldots, w_r$ whose homology classes form a K-basis of $H_2(X^{(2)})$. Since $\mathrm{Tor}_i(K, S/I) \cong H_i(X^{(1)})$, our hypothesis implies that there is a cycle $z \in (X^{(1)})_2$ with $\mathrm{Deg}\, z = j > 4$ which is not a boundary. Of course z is also a cycle in $X^{(2)}$ because $X^{(1)}$ is a subcomplex of $X^{(2)}$. We claim that z is not a boundary in $X^{(2)}$. To see this we consider the exact sequence of complexes

$$0 \longrightarrow X^{(1)} \longrightarrow X^{(2)} \longrightarrow X^{(2)}/X^{(1)} \longrightarrow 0,$$

which induces the long exact sequence

$$\cdots \longrightarrow H_3(X^{(2)}/X^{(1)}) \xrightarrow{\delta} H_2(X^{(1)}) \longrightarrow H_2(X^{(2)}) \longrightarrow \cdots$$

Thus it suffices to show that the homology class $[z]$ of the cycle z is not in the image of δ. Notice that the elements $T_{1i}T_{2j}$ form a basis of the free S/I-module $(X^{(2)}/X^{(1)})_3$ and that the differential on $X^{(2)}/X^{(1)}$ maps $T_{1i}T_{2j}$ to $x_i T_{2j}$, so that $w \in (X^{(2)}/X^{(1)})_3$ is a cycle if and only if $w = \sum_{j=1}^{m} w_j T_{2j}$ where each $w_j \in X_1^{(1)}$ is a cycle. Now the connecting homomorphism δ maps $[w]$ to $[-\sum_{j=1}^{m} w_j z_j]$. It follows that $\mathrm{Im}\,\delta = H_1(X^{(1)})^2$. Since $H_1(X^{(1)})$ is generated in degree 2 we conclude that the subspace $H_1(X^{(1)})^2$ of $H_2(X^{(1)})$ is generated in degree 4. Hence our element $[z] \in H_2(X^{(1)})$ which is of degree > 4 cannot be in the image of δ, as desired.

Thus the homology class of z, viewed as an element of $H_2(X^{(2)})$ has to be killed by adjoining a variable of degree $j > 4$. This shows that $\beta_{3j}^{S/I}(S/\mathfrak{m}) \neq 0$, and hence S/I is not Koszul.

Proof (Proof of Theorem 2.1). We may assume that $[n]$ is the vertex set of G. Let H by any induced subgraph of G. We may further assume that $V(H) = [k]$. Let $S = K[x_1, \ldots, x_n, y_1, \ldots, y_n]$ and $T = K[x_1, \ldots, x_k, y_1, \ldots, y_k]$. Then T/J_H is an algebra retract of S/J_G. Indeed, let $L = (x_{k+1}, \ldots, x_n, y_{k+1}, \ldots, y_n)$. Then the composition $T/J_H \to S/J_G \to S/(J_G, L) \cong T/J_H$ of the natural K-algebra homomorphisms is an isomorphism. It follows therefore from [18, Corollary 2.6] that any induced subgraph of a G is again Koszul.

Suppose that G is not claw free. Then there exists an induced subgraph H of G which is isomorphic to a claw. We may assume that $V(H) = \{1, 2, 3, 4\}$, and let $R = K[x_1, \ldots, x_4, y_1, \ldots, y_4]$. A computation with Singular [7] shows that $\beta_{3,5}^{R/J_H}(K) \neq 0$. Thus H is not Koszul, a contradiction.

Suppose that G is not chordal. Then there exist a cycle C of length ≥ 4 which has no chord. Then C is an induced subgraph and hence should be Koszul. We may assume that $V(C) = \{1, 2, \ldots, m\}$ with edges $\{i, i+1\}$ for $i = 1, \ldots, m-1$ and edge $\{1, m\}$ and set $T = K[x_1, \ldots, x_m, y_1, \ldots, y_m]$. We claim that $\beta_{2,m}^{T}(T/J_C) \neq 0$. For $m > 4$ this will imply that C is not Koszul. That a four-cycle is not Koszul can again be directly checked with Singular [7].

In order to prove the claim we let $F = \bigoplus_{i=1}^{m} Te_i$ and consider the free presentation

$$\varepsilon\colon F \to J_C \longrightarrow 0, \quad e_i \mapsto f_{i,i+1} \text{ for } i = 1, \dots, m$$

For simplicity, here and in the following, we read $m+1$ as 1.

Obviously, $g = \sum_{i=1}^{m} (\prod_{j=1}^{m} x_j)/(x_i x_{i+1}) e_i \in \operatorname{Ker} \varepsilon$. We will show that g is a minimal generator of $\operatorname{Ker}\varepsilon$. Indeed, let $g' = \sum_{i=1} g_i e_i \in \operatorname{Ker}\varepsilon$ be an arbitrary relation, and suppose that some $g_j = 0$. Since the $f_{i,i+1}$ for $i \neq j$ form a regular sequence, it then follows that all the other g_i belong to J_C. However, since the coefficients of g do not belong to J_C, we conclude that g cannot be written as a linear combination of relations for which one of its coefficients is zero.

Now assume that all $g_i \neq 0$. Let ε_i denotes the ith canonical unit vector of $\mathbb{Z}^n$. Since J_C is a $\mathbb{Z}^n$-graded ideal with $\deg_{\mathbb{Z}^n} x_i = \deg_{\mathbb{Z}^n} y_i = \varepsilon_i$, we may assume that $g' = \sum_{i=1} g_i e_i$ is a homogeneous relation where $\deg_{\mathbb{Z}^n} e_i = \deg f_{i,i+1} = \varepsilon_i + \varepsilon_{i+1}$ and g_i is homogeneous satisfying $\deg_{\mathbb{Z}^n} g' = \deg_{\mathbb{Z}^n} g_i + \varepsilon_i + \varepsilon_{i+1}$ for all i. This is only possible if $\deg_{\mathbb{Z}^n} g' \geq \sum_{i=1}^{m} \varepsilon_i$, coefficientwise. In particular it follows that $\deg g' \geq m$, where $\deg g'$ denotes the total degree of g'. Thus g cannot be a linear combination of relations of lower (total) degree and hence is a minimal generator of $\operatorname{Ker}\varepsilon$. Since $\deg g = m$, we conclude that $\beta_{2,m}^{T}(T/J_C) \neq 0$.

3 Gluing of Koszul Graphs Along a Vertex

In this section we first show that Koszulness is preserved under the operation of gluing two graphs along a vertex in the sense that we are going to explain below. We begin with two general statements about Koszul algebras which may be found in [3]. For the convenience of the reader, we include their proofs here.

Proposition 3.1. *Let $R = K[x_1, \dots, x_n]/I$ and $S = K[x_{n+1}, \dots, x_m]/J$ be two standard graded K-algebras. Then $R \otimes_K S$ is Koszul if and only if R and S are Koszul.*

Proof. Let $\mathfrak{m}$ and $\mathfrak{n}$ be the maximal ideals of R and, respectively, S. Let $\mathbb{F} \to R/\mathfrak{m} \to 0$ and $\mathbb{G} \to S/\mathfrak{n} \to 0$ be the minimal graded free resolutions of $R/\mathfrak{m}$ over R and, respectively, of $S/\mathfrak{n}$ over S. Then the total complex of $\mathbb{F} \otimes \mathbb{G}$ is the minimal graded free resolution over $R \otimes S$ of the maximal graded ideal of $R \otimes S$. If $F_i = \bigoplus_k R(-k)^{\beta_{ik}}$ for all i and $G_j = \bigoplus_\ell S(-\ell)^{\beta'_{j\ell}}$ for all j, then

$$F_i \otimes G_j \cong \bigoplus_{k,\ell} R \otimes S(-k-\ell)^{\beta_{ik}\beta'_{j\ell}}.$$

This immediately implies the desired conclusion if $\mathbb{F}$ and $\mathbb{G}$ are linear. For the converse, we note that $\operatorname{Tor}_p^{R\otimes S}(K,K) \cong \bigoplus_{i+j=p} \bigoplus_{k,\ell} K(-k-\ell)^{\beta_{ik}\beta'_{j\ell}}$. Then we must have $k + \ell = p$ for all i, j with $i + j = p$. As $k \geq i$ and $\ell \geq j$, it follows that $k = i$ and $\ell = j$ for all i, j. Therefore, $\mathbb{F}$ and $\mathbb{G}$ are linear resolutions as well.

The above proposition shows, in particular, that it is enough to study the Koszul property for connected graphs.

Corollary 3.2. *Let G be a graph with connected components $G_1, \ldots, G_r$. Then G is Koszul if and only if G_i is Koszul for $1 \leq i \leq r$.*

Proof. Let $V(G) = [n]$ and $S = K[x_1, \ldots, x_n, y_1, \ldots, y_n]$. Then $S/J_G \cong \otimes_{i=1}^r S_i/J_{G_i}$ where $S_i = K[\{x_j, y_j : j \in V(G_i)\}]$ for $1 \leq i \leq r$. The claim follows by applying Proposition 3.1.

Proposition 3.3. *Let R be a standard graded K-algebra with maximal graded ideal $\mathfrak{m}$ and $f_1, \ldots, f_m \in \mathfrak{m} \setminus \mathfrak{m}^2$ a regular sequence of homogeneous elements in R. Then R is Koszul if and only if $R/(f_1, \ldots, f_m)$ is Koszul.*

Proof. By induction on m, it is sufficient to prove the claim for $m = 1$. Let then $f \in R$ be a form of degree 1. We have to show that R is Koszul if and only if $R/(f)$ is Koszul or, equivalently, K has a linear resolution over R if and only if it has a linear resolution over $R/(f)$. But this is a direct consequence of [1, Theorem 2.2.3].

Now we come to the main subject of this section. By Corollary 3.2, in the sequel we may assume that all the graphs are connected.

Let G be a graph. A *clique* of G is a complete subgraph of G. The cliques of G form a simplicial complex $\Delta(G)$ which is called the *clique complex* of G. The facets of $\Delta(G)$ are the maximal cliques of G with respect to inclusion. A *free* vertex of $\Delta(G)$ or, simply, of G is a vertex of G which belongs only to one facet of $\Delta(G)$.

Let G_1, G_2 be to graphs such that $V(G_1) \cap V(G_2) = \{v\}$ and v is a free vertex in G_1 and G_2. Let $G = G_1 \cup G_2$ with $V(G) = V(G_1) \cup V(G_2)$ and $E(G) = E(G_1) \cup E(G_2)$. We say that G is obtained by *gluing G_1 and G_2 along the vertex v*.

Theorem 3.4. *Let G be a graph obtained by gluing the graphs G_1 and G_2 along a vertex. Then G is Koszul if and only if G_1 and G_2 are Koszul.*

Proof. Let $V(G) = [n]$ and assume that G_1 and G_2 are glued along the vertex $v \in [n]$. Let v' be a vertex which does not belong to $V(G)$ and let G_2' be the graph with $V(G_2') = (V(G_2) \setminus \{v\}) \cup \{v'\}$ whose edge set is $E(G_2') = E(G_2 \setminus \{v\}) \cup \{\{i, v'\} : \{i, v\} \in E(G_2)\}$. We set $S = K[x_1, \ldots, x_n, y_1, \ldots, y_n]$ and $S' = S[x_{v'}, y_{v'}]$. Let $\ell_x = x_v - x_{v'}$ and $\ell_y = y_v - y_{v'}$. By the proof of [20, Theorem 2.7], we know that ℓ_x, ℓ_y is a regular sequence on $S'/J_{G'}$, where G' is the graph whose connected components are G_1 and G_2'. Moreover, we obviously have

$$S'/(J_G', \ell_x, \ell_y) \cong S/J_G.$$

By Proposition 3.3, it follows that G is Koszul if and only if G' is Koszul. Next, by Corollary 3.2, it follows that G' is Koszul if and only its connected components, namely G_1 and G_2', are Koszul. Finally, we observe that G_2' is Koszul if and only if G_2 is so.

Let G be a graph. By Dirac's theorem [8], G is chordal if and only if the facets of $\Delta(G)$ can be ordered as $F_1, \ldots, F_r$ such that for all $i > 1$, F_i is a leaf of the simplicial complex $\langle F_1, \ldots, F_{i-1} \rangle$. This means that there exists a facet F_j with $j < i$ which intersects F_i maximally, that is, for each $\ell < i$, $F_\ell \cap F_i \subset F_j \cap F_i$. F_j is called a *branch* of F_i.

The following corollary gives a class of chordal and claw-free graphs which are Koszul.

Corollary 3.5. *Let G be a chordal and claw-free graph with the property that $\Delta(G)$ admits a leaf order $F_1, \ldots, F_r$ such that for all $i > 1$, the facet F_i intersects any of its branches in one vertex. Then G is Koszul.*

Proof. We proceed by induction on r. If $r = 1$, there is nothing to prove since any clique is Koszul. Let $r > 1$ and assume that the graph G' with $\Delta(G') = \langle F_1, \ldots, F_{r-1} \rangle$ is Koszul. We may assume that F_{r-1} is a branch of F_r and let $\{v\} = F_r \cap F_{r-1}$. The desired statement follows by applying Theorem 3.4 for G' and the clique F_r, once we show that v is a free vertex of G'.

Let us assume that v is not free in G' and choose a maximal clique F_j with $j \leq r - 2$ such that $v \in F_j$. We may find three vertices $a, b, c \in V(G)$ such that $a \in F_r \setminus (F_{r-1} \cup F_j)$, $b \in F_{r-1} \setminus (F_r \cup F_j)$, and $c \in F_j \setminus (F_r \cup F_{r-1})$. If $\{a, b\} \in E(G)$, then there exists a maximal clique F_k with $k \leq r - 1$ such that $a, b \in F_k$. This implies that $a \in F_k \cap F_r \subset \{v\}$, contradiction. Therefore, $\{a, b\}$ is not an edge of G. Similarly, one proves that $\{a, c\} \notin E(G)$. Let us now assume that $\{b, c\} \in E(G)$. The clique on the vertices v, b, c is contained in some maximal clique F_k. We have $k \leq r - 2$ since $F_k \neq F_{r-1}$. Then it follows that $|F_k \cap F_{r-1}| \geq 2$ which is a contradiction to our hypothesis on G. Consequently, we have proved that $\{a, b\}, \{b, c\}, \{a, c\} \notin E(G)$. Hence, G contains a claw as an induced subgraph, contradiction. Therefore, v is a free vertex of G'.

In Fig. 1 is shown a graph which satisfies the conditions of Corollary 3.5 and is not closed.

Figure 2 displays a chordal and claw-free graph G which is not Koszul. That G is not Koszul can be seen as follows: we first observe that the graph G' restricted to the vertex set [4] is Koszul by Corollary 3.5, and that $B = K[x_1, \ldots, x_4, y_1, \ldots, y_4]/J_{G'}$ is an algebra retract of $A = K[x_1, \ldots, x_6, y_1, \ldots, y_6]/J_G$ with retraction map $A \to A/(x_5, x_6, y_5, y_6) \cong B$.

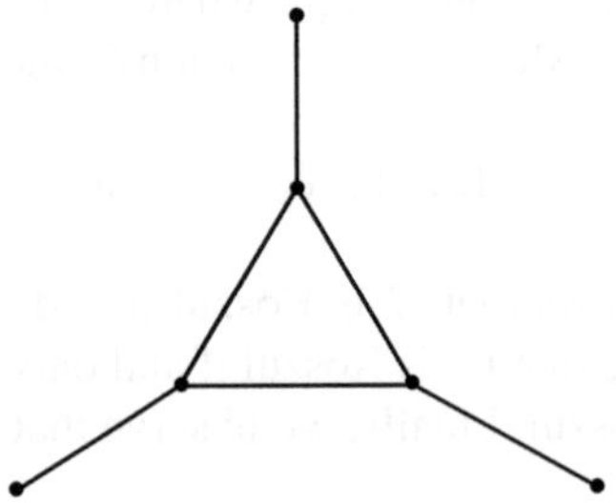

Fig. 1 Koszul non-closed graph

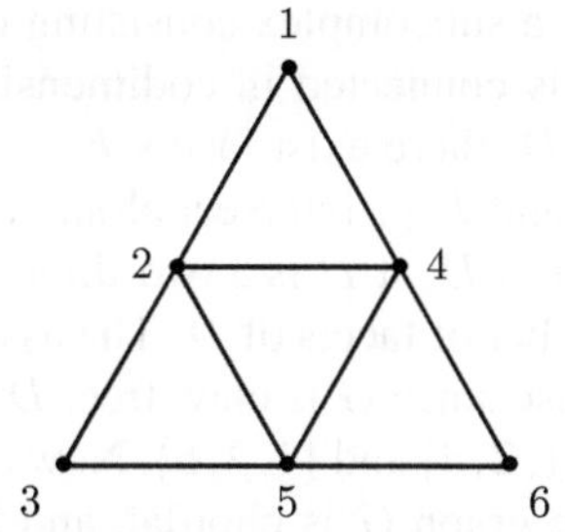

Fig. 2 Chordal and claw-free graph which is not Koszul

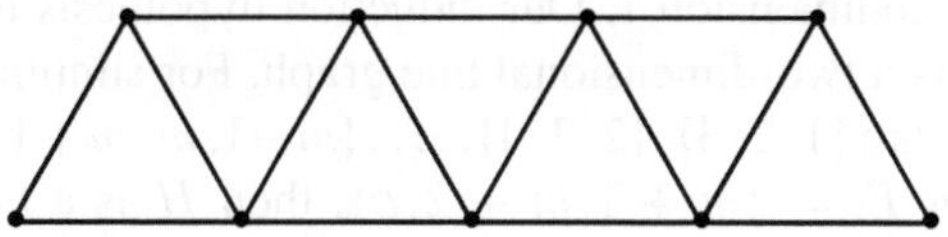

Fig. 3 Two-dimensional line graph

Thus if A would be Koszul, the ideal (x_5, x_6, y_5, y_6) would have to have an A-linear resolution, see [18, Proposition 1.4]. It can be verified with Singular [7] that this is not the case.

A *line graph* of length m is a graph which is isomorphic to the graph with edges $\{1,2\}, \{2,3\}, \ldots, \{m-1, m\}$. A *two-dimensional line graph* is a graph whose cliques are two-dimensional cliques composed as shown in Fig. 3.

To be precise, a two-dimensional line graph of length m is a graph whose clique complex is isomorphic to the simplicial complex with facets

$$\{1,2,3\}, \{2,3,4\}, \ldots, \{m-1, m, m+1\}, \{m, m+1, m+2\}.$$

From what we have shown so far it is not too difficult to obtain the following classification result, which roughly says that any connected Koszul graph whose clique complex is of dimension ≤ 2 is obtained by gluing one-dimensional and two-dimensional line graphs.

Theorem 3.6. *Let G be a connected graph whose clique complex is of dimension ≤ 2. The following conditions are equivalent:*

(a) G is Koszul;

(b) There exists a tree T whose vertices have order at most 3 such that G is obtained from T as follows:

(i) each vertex v of T is replaced by a one-dimensional or two-dimensional line graph G_v;

(ii) if $\{v, w\}$ is an edge of T then G_v and G_w are glued via a free vertex of G_v and G_w;

(iii) if v is a vertex of order 3 and w_1, w_2, w_3 are the neighbors of v, then G_v is a simplex and each G_{w_i} is glued to a different vertex of G_v.

Proof. (a) $\Rightarrow$ (b): Let D be a subcomplex consisting of two-dimensional facets of $\Delta(G)$, and assume that D is connected in codimension 1. By that we mean that for any two facets $F, F' \in D$, there exist facets $F_1, \ldots, F_r$ such that $F = F_1$ and $F' = F_r$, and such that F_i and F_{i+1} intersect along an edge for $i = 1, \ldots, r-1$. We claim that the one-skeleton H of D is a two-dimensional line graph, and prove this by induction on the number of facets of D. The assertion is trivial if the number of facets of D is ≤ 3, because since G is claw free, D cannot be isomorphic to the graph with edges $\{1,2,3\}, \{1,2,4\}$ and $\{1,2,5\}$. Now assume that D has $m+1 > 3$ facets. By Theorem 2.1, the graph G is chordal, and hence H is chordal as well. Applying Dirac's theorem [8] we conclude that D admits a leaf F. Let D' be the subcomplex of D which is obtained from D by removing the leaf F. Then D' is again connected in codimension 1. Our induction hypothesis implies that the one-skeleton H' of D' is a two-dimensional line graph. For simplicity we may assume that the facets of D' are $\{1,2,3\}, \{2,3,4\}, \ldots, \{m-1,m,m+1\}, \{m,m+1,m+2\}$. If $F = \{a,1,2\}$ or $F = \{m+1, m+2, b\}$, then H is a two-dimensional line graph, and we are done. Otherwise, $F = \{i, i+1, c\}$ or $F = \{i, i+2, c\}$ for some $i \in [m]$ and some vertex c of D. The first case cannot happen, since G is claw free. In the second case, if $1 < i < m$, then D, and consequently, $\Delta(G)$ contains an induced subgraph isomorphic to the graph in Fig. 2. Thus G is not Koszul, a contradiction. On the other hand, if $i = 1$, then H is not claw free, because then edges $\{3,c\}, \{3,2\}, \{3,5\}$ form a claw which is an induced subgraph of G, contradiction to the fact that G must be claw free.

Let $D_1, \ldots, D_r$ be the maximal two-dimensional subcomplexes of $\Delta(G)$ which are connected in codimension 1, and $L_1, \ldots, L_s$ be the maximal one-dimensional connected subcomplexes of $\Delta(G)$. Each L_i is a one-dimensional line graph, otherwise G would not be claw free, and the D_i are all two-dimensional line graphs, as we have seen above. The maximality of the L_i implies that $V(L_i) \cap V(L_j) = \emptyset$ for $i \neq j$, and the maximality of D_i implies that each facet of D_i intersects any facet of $\Delta(G)$ not belonging to D_i in at most one vertex.

Now we let T be the graph whose vertices are $D_1, \ldots, D_r, L_1, \ldots, L_s$. The edge set $E(T)$ consists of the edges $\{D_i, D_j\}$ if $V(D_i) \cap V(D_j) \neq \emptyset$ and $\{D_i, L_j\}$ if $V(D_i) \cap V(L_j) \neq \emptyset$.

If $\{D_i, D_j\} \in E(T)$ and v is a common vertex of D_i and D_j, then v must be a free vertex of D_i and of D_j, because otherwise G would not be claw free. Moreover, $|V(D_i) \cap V(D_j)| \leq 1$, because otherwise G contains a cycle of length > 3 without chord, contradicting the fact that G is chordal. By the same reason we have that $|V(D_i) \cap V(L_j)| \leq 1$ for all i and j.

Next observe that the intersection of any three of the sets

$$V(D_1), \ldots, V(D_r), V(L_1), \ldots, V(L_s)$$

is the empty set, which follows from the fact that G is claw free. Thus the order of the vertices of T can be at most the number of free vertices of an D_i or L_j, and hence is at most 3, where the maximal number 3 can be reached only if one of the D_i is a two-simplex. Finally T must be a tree, because otherwise G would not be chordal.

(b) ⇒ (a): We proceed by induction on $V(T)$. If $V(T) = 1$, then G is a one-dimensional or two-dimensional line graph. In both cases G is closed and hence has a quadratic Gröbner basis. This implies that G is Koszul. Now let $V(T) > 1$, and choose a free vertex $v \in V(T)$. Then $T' = T \setminus v$ is again a tree satisfying conditions (b). Let $W = \bigcup_{w \in T'} V(G_w)$ and G' the restriction of G to W. Then our induction hypothesis implies that G' is Koszul. Since G_v is Koszul and since G is obtained from G' and G_v by gluing along a common free vertex, the desired conclusion follows from Theorem 3.4.

Remark 3.7. Let G be a connected Koszul graph whose clique complex is of dimension ≤ 2, and let T be its "intersection tree" as described in Theorem 3.6(b). If $\Delta(G)$ does not contain a subcomplex as given in Fig. 1, then T is a line graph. In this case G is obtained by gluing successively one-dimensional and two-dimensional line graphs. Thus, it follows from [11, Theorem 2.2] that G is closed and hence J_G has a quadratic Gröbner basis.

On the other hand, if T contains a subcomplex as shown in Fig. 1, then G is not closed, and hence by a result of Crupi and Rinaldo [6, Theorem 3.4] it follows that J_G has no quadratic Gröbner basis for any monomial order. Thus those graphs provide an infinite family of binomial ideals which define a Koszul algebra but do not have a quadratic Gröbner basis.

Examples of toric rings which are Koszul but do not have a quadratic Gröbner basis were found independently by Roos and Sturmfels [21] and by Ohsugi and Hibi [17].

Acknowledgements We would like to thank the referee for the careful reading of our paper. Viviana Ene was supported by the grant UEFISCDI, PN-II-ID-PCE- 2011-3-1023.

References

1. Avramov, L.L.: Infinite free resolutions. In: Elias, J., et al. (eds.) Six Lectures on Commutative Algebra. Prog. Math., vol. 166, pp. 1–118. Birkhäuser, Basel (1998)
2. Avramov, L.L., Conca, A., Iyengar, S.: Subadditivity of syzygies of Koszul algebras. Math. Ann. doi: 10.1007/s00208-014-1060-4 (2014)
3. Backelin, J., Fröberg, R.: Koszul algebras, Veronese subrings and rings with linear resolutions. Rev. Roumaine Math. Pures Appl. **30**(2), 85–97 (1985)
4. Conca, A., De Negri, E., Rossi, M.E.: Koszul algebras and regularity. In: Peeva, I. (ed.) Commutative Algebra, Expository Papers Dedicated to David Eisenbud on the Occasion of His 65th Birthday, pp. 285–315. Springer, New York (2013)
5. Cox, D.A., Erskine, A.: On closed graphs. arXiv:1306.5149 (2013)
6. Crupi, M., Rinaldo, G.: Binomial edge ideals with quadratic Gröbner bases. Electr. J. Comb. **18**, P211 (2011)
7. Decker, W., Greuel, G.-M., Pfister, G., Schönemann, H.: A computer algebra system for polynomial computations. http://www.singular.uni-kl.de (2011)

8. Dirac, G.A.: On rigid circuit graphs. Abh. Math. Sem. Univ. Hamburg **38**, 71–76 (1961)
9. Ene, V., Herzog, J.: Gröbner Bases in Commutative Algebra. Graduate Studies in Mathematics, vol. 130. American Mathematical Society, Providence (2011)
10. Ene, V., Zarojanu, A.: On the regularity of binomial edge ideals. Math. Nachr. doi: 10.1002/mana.201300186 (2014)
11. Ene, V., Herzog, J., Hibi, T.: Cohen-Macaulay binomial edge ideals. Nagoya Math. J. **204**, 57–68 (2011)
12. Gulliksen, T.H.: A proof of the existence of minimal R-algebra resolutions. Acta Math. **120**, 53–58 (1968)
13. Herzog, J., Hibi, T., Hreinsdotir, F., Kahle, T., Rauh, J.: Binomial edge ideals and conditional independence statements. Adv. Appl. Math **45**, 317–333 (2010)
14. Saeedi Madani, S., Kiani, D.: Binomial edge ideals of graphs. Electr. J. Comb. **19**, ♯ P44 (2012)
15. Saeedi Madani, S., Kiani, D.: On the binomial edge ideal of a pair of graphs. Electr. J. Comb. **20**, ♯ P48 (2013)
16. Matsuda, K., Murai, S.: Regularity bounds for binomial edge ideals. J. Comm. Algebra **5**, 141–149 (2013)
17. Ohsugi, H., Hibi, T.: Toric ideals generated by quadratic binomials. J. Algebra **218**, 509–527 (1999)
18. Ohsugi, H., Herzog, J., Hibi, T.: Combinatorial pure subrings. Osaka J. Math. **37**, 745–757 (2000)
19. Ohtani, M.: Graphs and ideals generated by some 2-minors. Commun. Algebra **39**(3), 905–917 (2011)
20. Rauf, A., Rinaldo, G.: Construction of Cohen-Macaulay binomial edge ideals. Commun. Algebra **42**(1), 238–252 (2014)
21. Roos, J.E., Sturmfels, B.: A toric ring with irrational Poincaré-Betti series. C. R. Acad. Sci. Paris Sér. I Math. **326**, 141–146 (1998)
22. Tate, J.: Homology of noetherian rings and local rings. Illinois J. Math. **1**, 14–25 (1957)

On the Fundamental Groups of Non-generic $\mathbb{R}$-Join-Type Curves

Christophe Eyral and Mutsuo Oka

Dedicated to S. Papadima and A. Dimca for their 60th birthday

Abstract An $\mathbb{R}$-*join-type curve* is a curve in $\mathbb{C}^2$ defined by an equation of the form $a \cdot \prod_{j=1}^{\ell}(y - \beta_j)^{\nu_j} = b \cdot \prod_{i=1}^{m}(x - \alpha_i)^{\lambda_i}$, where the coefficients a, b, α_i and β_j are *real* numbers. For generic values of a and b, the singular locus of the curve consists of the points (α_i, β_j) with $\lambda_i, \nu_j \geq 2$ (so-called *inner* singularities). In the non-generic case, the inner singularities are not the only ones: the curve may also have "*outer*" singularities. The fundamental groups of (the complements of) curves having only inner singularities are considered in Oka (J Math Soc Jpn 30:579–597, 1978). In the present paper, we investigate the fundamental groups of a special class of curves possessing outer singularities.

Keywords Plane curves • Fundamental group • Bifurcation graph • Monodromy • Zariski–van Kampen's pencil method

MSC 2010: 14H30 (14H20, 14H45, 14H50)

1 Introduction

Let $\nu_1, \ldots, \nu_\ell, \lambda_1, \ldots, \lambda_m$ be positive integers. Denote by ν_0 (respectively, λ_0) the greatest common divisor of $\nu_1, \ldots, \nu_\ell$ (respectively, of $\lambda_1, \ldots, \lambda_m$). Set $d := \sum_{j=1}^{\ell} \nu_j$ and $d' := \sum_{i=1}^{m} \lambda_i$. A curve C in $\mathbb{C}^2$ is called a *join-type curve*

C. Eyral (✉)
Institute of Mathematics, Polish Academy of Sciences, ul. Śniadeckich 8, 00-656 Warsaw, Poland
e-mail: eyralchr@yahoo.com

M. Oka
Department of Mathematics, Tokyo University of Science, 1-3 Kagurazaka, Shinjuku-ku, Tokyo 162-8601, Japan
e-mail: oka@rs.kagu.tus.ac.jp

D. Ibadula, W. Veys (eds.), *Bridging Algebra, Geometry, and Topology*, Springer Proceedings in Mathematics & Statistics 96,
DOI 10.1007/978-3-319-09186-0_9

with exponents $(\nu_1,\dots,\nu_\ell;\lambda_1,\dots,\lambda_m)$ if it is defined by an equation of the form $f(y)=g(x)$, where

$$f(y) := a\cdot\prod_{j=1}^{\ell}(y-\beta_j)^{\nu_j} \quad\text{and}\quad g(x) := b\cdot\prod_{i=1}^{m}(x-\alpha_i)^{\lambda_i}. \tag{1}$$

Here, a and b are nonzero complex numbers, and $\beta_1,\dots,\beta_\ell$ (respectively, $\alpha_1,\dots,\alpha_m$) are mutually distinct complex numbers. We say that C is an $\mathbb{R}$-join-type curve if the coefficients a, b, α_i $(1\le i\le m)$ and β_j $(1\le j\le \ell)$ are *real* numbers.

The singular points of C (i.e., the points (x,y) satisfying $f(y)=g(x)$ and $f'(y)=g'(x)=0$) divide into two categories: the points (x,y) which also satisfy the equations $f(y)=g(x)=0$, and those for which $f(y)\neq 0$ and $g(x)\neq 0$. Clearly, the singular points contained in the intersection of lines $f(y)=g(x)=0$ are the points (α_i,β_j) with $\lambda_i,\nu_j\ge 2$. Hereafter, such singular points will be called *inner* singularities, while the singular points (x,y) with $f(y)\neq 0$ and $g(x)\neq 0$ will be called *outer* or *exceptional* singularities. It is easy to see that the singular points of a join-type curve are Brieskorn–Pham singularities $\mathbf{B}_{\nu,\lambda}$ (normal form $y^\nu - x^\lambda$). For example, inner singularities are of type $\mathbf{B}_{\nu_j,\lambda_i}$. In the case of $\mathbb{R}$-join-type curves, we shall see, more specifically, that outer singularities can be only node singularities (i.e., Brieskorn–Pham singularities of type $\mathbf{B}_{2,2}$).

Clearly, for generic values of a and b, under any fixed choice of the coefficients α_i $(1\le i\le m)$ and β_j $(1\le j\le\ell)$, the curve C has only inner singularities. In this case, it is shown in [2] that the fundamental group $\pi_1(\mathbb{C}^2\setminus C)$ is isomorphic to the group $G(\nu_0;\lambda_0)$ obtained by taking $p=\nu_0$ and $q=\lambda_0$ in the presentation (2) below. (In [2] it is assumed that $d=d'$ but the same proof works for $\pi_1(\mathbb{C}^2\setminus C)$ and $d\neq d'$.) For example, if C has only inner singularities and if λ_0 or ν_0 is equal to 1, then $\pi_1(\mathbb{C}^2\setminus C)\simeq\mathbb{Z}$.

In the present paper, we prove that the result of [2] extends to certain $\mathbb{R}$-join-type curves possessing *outer* singularities.[1] These curves are defined as follows. Let C be an $\mathbb{R}$-join-type curve. Then, without loss of generality, we can assume that the real numbers α_i $(1\le i\le m)$ and β_j $(1\le j\le\ell)$ are indexed so that $\alpha_1<\dots<\alpha_m$ and $\beta_1<\dots<\beta_\ell$. Then, by considering the restriction of the function $g(x)$ to the real numbers, we see easily that the equation $g'(x)=0$ has at least one real root γ_i in the open interval (α_i,α_{i+1}) for each $i=1,\dots,m-1$. Since the degree of

$$g'(x)\Big/\prod_{i=1}^{m}(x-\alpha_i)^{\lambda_i-1}$$

[1]Note that if C is a join-type curve with *non-real* coefficients and with only inner singularities, then it can always be deformed to an $\mathbb{R}$-join-type curve C_1 by a deformation $\{C_t\}_{0\le t\le 1}$ such that $C_0=C$ and C_t is a join-type curve with only inner singularities and with the same exponents as C (cf. [2]). (In particular, the topological type of C_t (respectively, $\mathbb{C}^2\setminus C_t$) is independent of t.) For curves possessing outer singularities, this is no longer true in general.

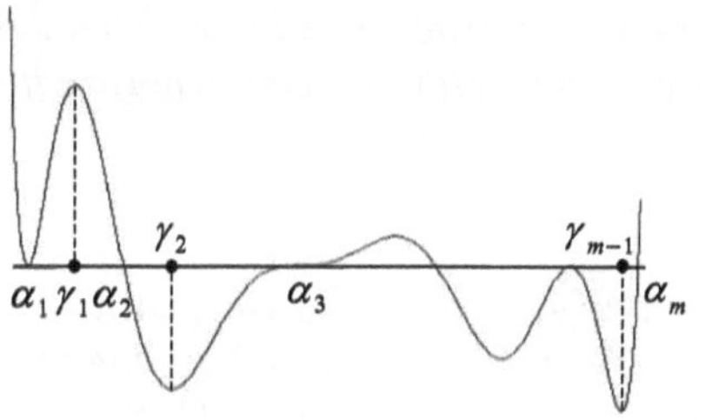

Fig. 1 Real graph of g

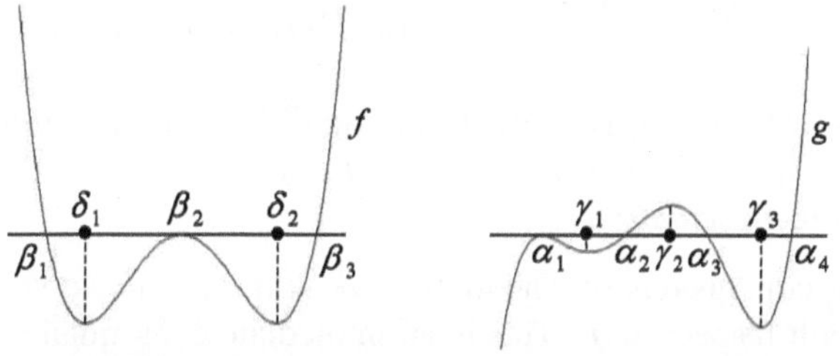

Fig. 2 Semi-genericity with respect to g ($i_0 = 1$ or 2)

is $m-1$, it follows that the roots of $g'(x) = 0$ are exactly $\gamma_1, \dots, \gamma_{m-1}$ and the coefficients α_i with $\lambda_i \geq 2$ (cf. Fig. 1). In particular, this shows that $\gamma_1, \dots, \gamma_{m-1}$ are *simple* roots of $g'(x) = 0$. Similarly, the equation $f'(y) = 0$ has $\ell - 1$ simple roots $\delta_1, \dots, \delta_{\ell-1}$ such that $\beta_j < \delta_j < \beta_{j+1}$ for each $j = 1, \dots, \ell - 1$. The other roots of $f'(y) = 0$ are the coefficients β_j with $\nu_j \geq 2$. (They are simple for $\nu_j = 2$.)

We fix the following terminology.

Definition 1. 1. We say that the curve C is *generic* if it has only inner singularities. In other words, C is generic if and only if, for any $1 \leq i \leq m-1$, $g(\gamma_i)$ is a regular value for f (i.e., $g(\gamma_i) \neq f(\delta_j)$ for any $1 \leq j \leq \ell - 1$). (Of course, this is also equivalent to the condition that for any $1 \leq j \leq \ell - 1$, $f(\delta_j)$ is a regular value for g.)

2. We say that C is *semi-generic with respect to* g if there exists an integer i_0 ($1 \leq i_0 \leq m$) such that $g(\gamma_{i_0-1})$ and $g(\gamma_{i_0})$ are regular values for f. (For $i_0 = 1$, this condition reduces to $g(\gamma_1) \notin \mathscr{V}_{\text{crit}}(f)$, and for $i_0 = m$, it reduces to $g(\gamma_{m-1}) \notin \mathscr{V}_{\text{crit}}(f)$, where $\mathscr{V}_{\text{crit}}(f)$ is the set of critical values of f.) The semi-genericity *with respect to* f is defined similarly by exchanging the roles of f and g.

Remark 1. It is obvious that a generic curve is also semi-generic with respect to both g and f, while the converse is not true. Also, note that C can be semi-generic with respect to g without being semi-generic with respect to f. For example, consider the curve defined by the polynomials f and g given in Fig. 2, where $f(\delta_1) = f(\delta_2) = g(\gamma_3)$.

Here is our main result.

Theorem 1. *Let C be an $\mathbb{R}$-join-type curve in $\mathbb{C}^2$ defined by the equation $f(y) = g(x)$, where f and g are as in (1). If C is semi-generic with respect to g, then*

$$\pi_1(\mathbb{C}^2 \setminus C) \simeq G(\nu_0; \lambda_0),$$

where, as above, $\nu_0 := \gcd(\nu_1, \dots, \nu_\ell)$, $\lambda_0 := \gcd(\lambda_1, \dots, \lambda_m)$ and $G(\nu_0; \lambda_0)$ is the group obtained by taking $p = \nu_0$ and $q = \lambda_0$ in the presentation (2).

Furthermore, if $\tilde{C}$ is the projective closure of C, then

$$\pi_1(\mathbb{P}^2 \setminus \tilde{C}) \simeq \begin{cases} G(\nu_0; \lambda_0; d/\nu_0) & \text{if } d \geq d', \\ G(\lambda_0; \nu_0; d'/\lambda_0) & \text{if } d' \geq d, \end{cases}$$

where $G(\nu_0; \lambda_0; d/\nu_0)$ (respectively, $G(\lambda_0; \nu_0; d'/\lambda_0)$) is the group obtained by taking $p = \nu_0$, $q = \lambda_0$ and $r = d/\nu_0$ (respectively, $p = \lambda_0$, $q = \nu_0$ and $r = d'/\lambda_0$) in the presentation (3).

Remark 2. The conclusions of Theorem 1 are still valid if we suppose that C is semi-generic with respect to f. This is an immediate consequence of the theorem itself and Proposition 1 below.

Example 1. With the same hypotheses as in Theorem 1, if d is a prime number and $\ell \geq 2$, then $\nu_0 = 1$, and hence $\pi_1(\mathbb{C}^2 \setminus C)$ is isomorphic to $G(1; \lambda_0) \simeq \mathbb{Z}$ while $\pi_1(\mathbb{P}^2 \setminus \tilde{C})$ is isomorphic to $\mathbb{Z}_d$ or $\mathbb{Z}_{d'}$ depending on whether $d \geq d'$ or $d' \geq d$. (Of course, if d' is a prime number and $m \geq 2$, then $\lambda_0 = 1$, and we get the same conclusion.)

2 The Groups $G(p; q)$ and $G(p; q; r)$

Let p, q, r be positive integers. In this section, we recall the definitions and collect the basic properties of the groups $G(p; q)$ and $G(p; q; r)$ introduced in [2] and which appear in Theorem 1 as the fundamental groups of the affine and the projective semi-generic $\mathbb{R}$-join-type curves, respectively.

The group $G(p; q)$ is defined by the presentation

$$\langle\, \omega,\, a_k\ (k \in \mathbb{Z}) \mid \omega = a_{p-1} a_{p-2} \dots a_0,\ \mathcal{R}_{q,k},\ \mathcal{R}'_{p,k}\ (k \in \mathbb{Z}) \,\rangle, \tag{2}$$

where

$$\begin{aligned} \mathcal{R}_{q,k}&\colon a_{k+q} = a_k \text{ (periodicity relation)};\\ \mathcal{R}'_{p,k}&\colon a_{k+p} = \omega a_k \omega^{-1} \text{ (conjugacy relation)}. \end{aligned}$$

The following proposition is used to show that the conclusions of Theorem 1 still hold if we suppose that C is semi-generic with respect to f (cf. Remark 2).

Proposition 1. *The groups $G(p;q)$ and $G(q;p)$ are isomorphic.*

Proof. From a purely algebraic point of view, this proposition is not obvious. However, by [2], we know that if C is the generic join-type curve $y^p = x^q$, then $\pi_1(\mathbb{C}^2 \setminus C) \simeq G(p;q)$. Now, by exchanging the roles of y and x, we also have $\pi_1(\mathbb{C}^2 \setminus C) \simeq G(q;p)$.

The following proposition will be useful to prove Theorem 1.

Proposition 2 (cf. [2]). *The relations $\mathscr{R}'_{p,k}$ $(k \in \mathbb{Z})$ and $\omega = a_{p-1}a_{p-2}\dots a_0$ imply the following new relation for any $k \in \mathbb{Z}$:*

$$\omega = a_k a_{k-1}\dots a_{k-p+1}.$$

It follows from this proposition that for any $n \in \mathbb{Z}$, we can reorder the generators as $b_k := a_{k+n}$ without changing the relations. That is, we have $\omega = b_{p-1}b_{p-2}\dots b_0$, $b_{k+q} = b_k$ and $b_{k+p} = \omega b_k \omega^{-1}$ $(k \in \mathbb{Z})$.

Now, let $q_1,\dots,q_n$ be positive integers, and $G(p;\{q_1,\dots,q_n\})$ the group defined by the presentation

$$\langle\, \omega,\, a_k\ (k \in \mathbb{Z}) \mid \omega = a_{p-1}a_{p-2}\dots a_0,\ \mathscr{R}_{q_i,k},\ \mathscr{R}'_{p,k}\ (1 \le i \le n,\ k \in \mathbb{Z})\,\rangle,$$

where

$$\mathscr{R}_{q_i,k}\colon a_{k+q_i} = a_k.$$

We shall also use the next result in the proof of Theorem 1.

Proposition 3 (cf. [2]). *The group $G(p;\{q_1,\dots,q_n\})$ is isomorphic to the group $G(p;q_0)$, where $q_0 := \gcd(q_1,\dots,q_n)$.*

The next proposition gives necessary and sufficient conditions for the group $G(p;q)$ to be abelian. Thus, it can be used to test the commutativity of the group $\pi_1(\mathbb{C}^2 \setminus C)$ which appears in Theorem 1.

Proposition 4 (cf. [2]). *The group $G(p;q)$ is abelian if and only if $q = 1$ or $p = 1$ or $p = q = 2$. More precisely,*

$$G(p;q) \simeq \begin{cases} \mathbb{Z} & \text{if } q = 1 \text{ or } p = 1; \\ \mathbb{Z}\times\mathbb{Z} & \text{if } p = q = 2. \end{cases}$$

The group $G(p;q;r)$ is defined by the presentation

$$\langle\, \omega,\, a_k\ (k \in \mathbb{Z}) \mid \omega = a_{p-1}a_{p-2}\dots a_0,\ \omega^r = e,\ \mathscr{R}_{q,k},\ \mathscr{R}'_{p,k}\ (k \in \mathbb{Z})\,\rangle, \tag{3}$$

where e is the unit element. In other words, $G(p;q;r)$ is the quotient of $G(p;q)$ by the normal subgroup generated by ω^r.

The next proposition is an interesting special case.

Proposition 5 (cf. [2]). *If* $\gcd(p,q) = 1$ *and* $r = q$, *then* $G(p;q;q)$ *is isomorphic to the free product* $\mathbb{Z}_p * \mathbb{Z}_q$.

Finally, we conclude this section with the following proposition which gives necessary and sufficient conditions for the group $G(p;q;r)$ to be abelian. This proposition can be used to test the commutativity of the group $\pi_1(\mathbb{P}^2 \setminus \tilde{C})$ which appears in Theorem 1.

Proposition 6 (cf. [2]). *The group* $G(p;q;r)$ *is abelian if and only if one of the following conditions is satisfied:*

1. $\gcd(p,q) = \gcd(q,r) = 1$;
2. $p = 1$;
3. $\gcd(p,q) = 2$, $\gcd(q/2,r) = 1$ *and* $p = 2$.

More precisely,

$$G(p;q;r) \simeq \begin{cases} \mathbb{Z}_{pr} & \text{if } \gcd(p,q) = \gcd(q,r) = 1; \\ \mathbb{Z}_r & \text{if } p = 1; \\ \mathbb{Z} \times \mathbb{Z}_r & \text{if } \gcd(p,q) = 2,\ \gcd(q/2,r) = 1 \text{ and } p = 2. \end{cases}$$

3 Special Pencil Lines

To compute the fundamental group $\pi_1(\mathbb{C}^2 \setminus C)$ in Theorem 1, we use the Zariski–van Kampen theorem with the pencil $\mathscr{P}$ given by the vertical lines $L_\gamma : x = \gamma$, where $\gamma \in \mathbb{C}$ (cf. [1, 4, 7]).[2] This theorem says that

$$\pi_1(\mathbb{C}^2 \setminus C) \simeq \pi_1(L_{\gamma_0} \setminus C)/\mathscr{M},$$

where L_{γ_0} is a generic line of $\mathscr{P}$ and $\mathscr{M}$ is the normal subgroup of $\pi_1(L_{\gamma_0} \setminus C)$ generated by the monodromy relations associated with the "special" lines of $\mathscr{P}$. Here, a line L_γ of $\mathscr{P}$ is called *special* if it meets the curve C at a point (γ, δ) with intersection multiplicity at least 2. This happens if and only if $f(\delta) = g(\gamma)$ and $f'(\delta) = 0$.

Let $\gamma_{j,1}, \ldots, \gamma_{j,d'}$ be the roots of $g(x) = f(\delta_j)$ for $1 \le j \le \ell - 1$, where δ_j is defined as in Sect. 1. If $g'(\gamma_{j,k}) \neq 0$, then $(\gamma_{j,k}, \delta_j)$ is a simple point of C. In a small neighborhood of this point, C is topologically described by

$$(y - \delta_j)^2 = c(x - \gamma_{j,k}), \tag{4}$$

where $c \neq 0$, and the line $x = \gamma_{j,k}$ is tangent to the curve at $(\gamma_{j,k}, \delta_j)$ with intersection multiplicity 2. (We recall that δ_j is a *simple* root of $f'(y) = 0$.) This

[2] Note that this pencil is "admissible" in the sense of [4].

is the case if $\gamma_{j,k} \in \mathbb{C} \setminus \mathbb{R}$, as $g'(x) = 0$ has only real roots. If $g'(\gamma_{j,k}) = 0$, then $(\gamma_{j,k}, \delta_j)$ is an outer singularity of type $\mathbf{B}_{2,2}$. (Note that $\gamma_{j,k}$ is a simple root of $g'(x) = 0$.) Near this point, the curve is topologically equivalent to

$$(y - \delta_j)^2 = c(x - \gamma_{j,k})^2. \tag{5}$$

For each β_j with $\nu_j \geq 2$, the roots of $g(x) = f(\beta_j)$ are $\alpha_1, \ldots, \alpha_m$. If $\lambda_i = 1$, then (α_i, β_j) is a simple point of C. In a small neighborhood of it, C is topologically given by

$$(y - \beta_j)^{\nu_j} = c(x - \alpha_i), \tag{6}$$

and the line $x = \alpha_i$ is tangent to C at (α_i, β_j) with intersection multiplicity ν_j. If $\lambda_i \geq 2$, then the point (α_i, β_j) is an inner singularity of type $\mathbf{B}_{\nu_j,\lambda_i}$, and in a small neighborhood of it, the curve is topologically equivalent to

$$(y - \beta_j)^{\nu_j} = c(x - \alpha_i)^{\lambda_i}. \tag{7}$$

4 Bifurcation Graph

The special lines of the pencil $\mathscr{P}$ correspond to certain vertices of a graph called the "bifurcation graph." This graph is defined as follows. Let $\mathscr{V}_{\text{crit}}(f)$ (respectively, $\mathscr{V}_{\text{crit}}(g)$) be the set of critical values of f (respectively, g), and let $\mathscr{V}_{\text{crit}} := \mathscr{V}_{\text{crit}}(f) \cup \mathscr{V}_{\text{crit}}(g)$. Thus, $\mathscr{V}_{\text{crit}} = \{0, g(\gamma_1), \ldots, g(\gamma_{m-1}), f(\delta_1), \ldots, f(\delta_{\ell-1})\}$ if there exists i_0 or j_0 such that $\lambda_{i_0} \geq 2$ or $\nu_{j_0} \geq 2$; otherwise, $\mathscr{V}_{\text{crit}} = \{g(\gamma_1), \ldots, g(\gamma_{m-1}), f(\delta_1), \ldots, f(\delta_{\ell-1})\}$. Denote by Σ the bamboo-shaped graph (embedded in the real axis) whose vertices are the points of $\mathscr{V}_{\text{crit}} \cup \{0\}$ (cf. Fig. 3). This graph can be decomposed into two connected subgraphs Σ_+ and Σ_-, where Σ_+ (respectively, Σ_-) is the subgraph whose vertices are ≥ 0 (respectively, ≤ 0). Hereafter, we shall denote by $v_+ := \sup\{v \mid v \in \mathscr{V}_{\text{crit}}\}$ and $v_- := \inf\{v \mid v \in \mathscr{V}_{\text{crit}}\}$. The pull-back graph $\Gamma := g^{-1}(\Sigma)$ of Σ by g is called the *bifurcation graph (or "dessin d'enfants") associated with the curve C with respect to g*. Its vertices are the points of the set $g^{-1}(\mathscr{V}_{\text{crit}} \cup \{0\})$.

Lemma 1. *The special lines $x = \gamma$ of the pencil $\mathscr{P}$ are given by the vertices γ of Γ such that $g(\gamma) \in \mathscr{V}_{\text{crit}}(f)$.*

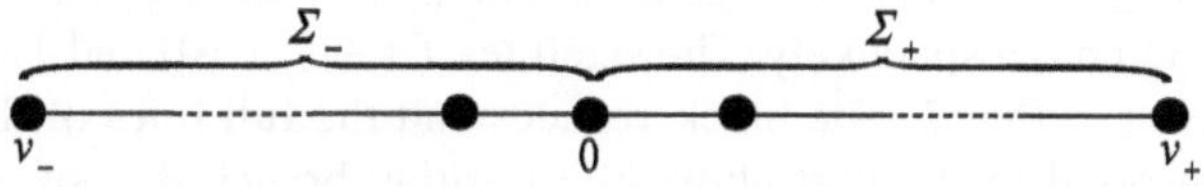

Fig. 3 Graph Σ

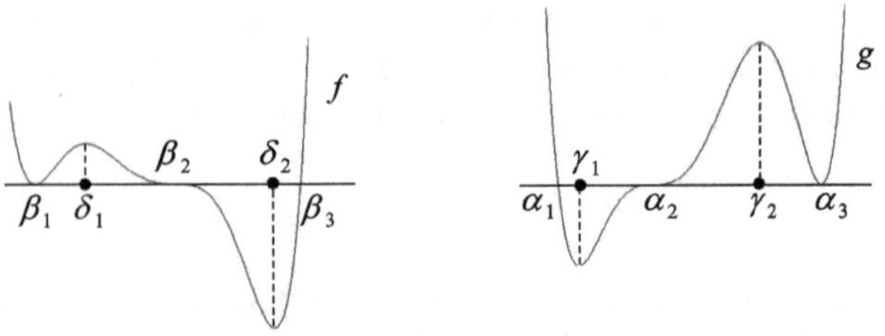

Fig. 4 Real graphs of f and g (Example 2)

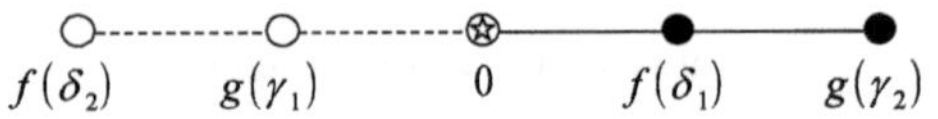

Fig. 5 Graph Σ (Example 2)

The bifurcation graph uniquely decomposes as the union of connected subgraphs $\Gamma(\alpha_1), \ldots, \Gamma(\alpha_m)$ such that for $1 \leq i \leq m$, the following properties are satisfied:

1. $\Gamma(\alpha_i)$ is a star-shaped graph with "center" α_i, and with $2\lambda_i$ branches (respectively, λ_i branches) if $v_+ > 0$ and $v_- < 0$ (respectively, if v_+ or v_- is zero);
2. the restriction of g to $\Gamma(\alpha_i)$ is an λ_i-fold branched covering onto Σ, whose branched locus is $\{0\}$, and $g^{-1}(0) \cap \Gamma(\alpha_i) = \{\alpha_i\}$;
3. for $i \neq m$, $\Gamma(\alpha_i) \cap \Gamma(\alpha_{i+1}) = \{\gamma_i\}$, and if $g(\gamma_i) \notin \{v_-, v_+\}$, then the branch of $\Gamma(\alpha_i)$ (respectively, $\Gamma(\alpha_{i+1})$) with γ_i as a vertex goes vertically downward (respectively, vertically upward) at γ_i.

Definition 2. The subgraphs $\Gamma(\alpha_i)$ $(1 \leq i \leq m)$ are called the *satellite graphs* of Γ. We say that a satellite $\Gamma(\alpha_i)$ is *regular* if $g(\gamma_{i-1})$ and $g(\gamma_i)$ are regular values for f. (For $i = 1$, this condition reduces to $g(\gamma_1) \notin \mathscr{V}_{\mathrm{crit}}(f)$, and for $i = m$, it reduces to $g(\gamma_{m-1}) \notin \mathscr{V}_{\mathrm{crit}}(f)$.)

Clearly, the curve C is generic if and only if all the satellites subgraphs of Γ are regular. The curve is semi-generic with respect to g if and only if Γ has at least one regular satellite.

Example 2. Consider the $\mathbb{R}$-join-type curve C defined by the polynomials $f(y) = (y+1)^2 y^3 (y-2)$ and $g(x) = 2(x+1)x^3(x-1)^2$. Then, f has four critical points $\beta_1 = -1$, $\delta_1 = (1-\sqrt{5})/2$, $\beta_2 = 0$ and $\delta_2 = (1+\sqrt{5})/2$. The polynomial g also has four critical points $\gamma_1 = -(1+\sqrt{73})/12$, $\alpha_2 = 0$, $\gamma_2 = -(1-\sqrt{73})/12$ and $\alpha_3 = 1$. See Fig. 4. (In the figure, the numerical scale is not respected; however, the order $f(\delta_2) < g(\gamma_1) < f(\delta_1) < g(\gamma_2)$ is rigorously respected.) As $g(\gamma_i) \neq f(\delta_j)$ for any $1 \leq i, j \leq 2$, the curve C is generic. The corresponding graphs Σ and Γ are given in Figs. 5 and 6 respectively. The satellites $\Gamma(-1)$, $\Gamma(0)$ and $\Gamma(1)$ associated with Γ are given in Fig. 7. The black vertices and the full lines of the bifurcation graph Γ correspond to the part above the positive branch Σ_+ of Σ. The white vertices and the dotted lines correspond to the part above the negative branch Σ_-. The star-style vertices represent the points $\alpha_1 = -1$, $\alpha_2 = 0$ and $\alpha_3 = 1$ which are the centers of the satellites. All the vertices (black, white, and star-style) surrounded

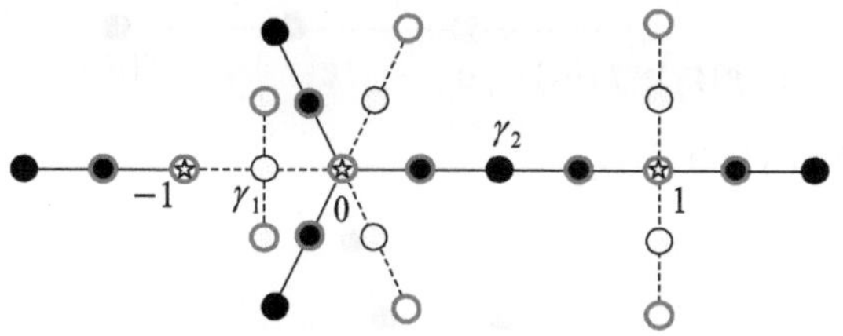

Fig. 6 Bifurcation graph Γ (Example 2)

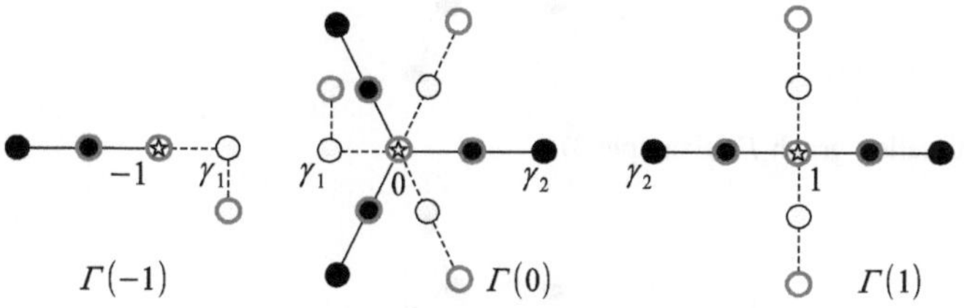

Fig. 7 Satellites $\Gamma(-1)$, $\Gamma(0)$ and $\Gamma(1)$ (Example 2)

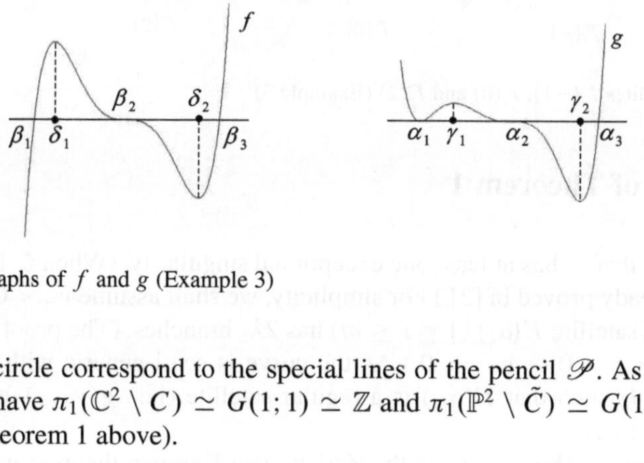

Fig. 8 Real graphs of f and g (Example 3)

with a gray circle correspond to the special lines of the pencil $\mathcal{P}$. As the curve is generic, we have $\pi_1(\mathbb{C}^2 \setminus C) \simeq G(1;1) \simeq \mathbb{Z}$ and $\pi_1(\mathbb{P}^2 \setminus \tilde{C}) \simeq G(1;1;6) \simeq \mathbb{Z}_6$ (by [2] or Theorem 1 above).

Now, let us give an example with a semi-generic curve which is not generic.

Example 3. Consider the $\mathbb{R}$-join-type curve C defined by the polynomials $f(y) = c(y+1)y^3(y-1)$ and $g(x) = (x+1)^2x^3(x-2)$, where the coefficient c is positive. We check easily that f has three critical points $\delta_1 < \beta_2 = 0 < \delta_2$, while the polynomial g has four critical points $\alpha_1 = -1 < \gamma_1 < \alpha_2 = 0 < \gamma_2$. We choose the coefficient c so that $f(\delta_2) = g(\gamma_2)$ (in particular, C is not generic). See Fig. 8. (Again, the figure is not numerically correct but the order $f(\delta_2) = g(\gamma_2) < g(\gamma_1) < f(\delta_1)$ is respected.) Now, as $g(\gamma_1)$ is a regular value for f, the satellite $\Gamma(\alpha_1)$ is regular, and therefore the curve C is semi-generic with respect to g. The corresponding graphs Σ and Γ are given in Figs. 9 and 10 respectively. The satellites $\Gamma(-1)$, $\Gamma(0)$ and $\Gamma(2)$ associated with Γ are given in Fig. 11. (The significance of the colors is as above.) As the curve is semi-generic with respect to g, Theorem 1 says that $\pi_1(\mathbb{C}^2 \setminus C) \simeq G(1;1) \simeq \mathbb{Z}$ and $\pi_1(\mathbb{P}^2 \setminus \tilde{C}) \simeq G(1;1;6) \simeq \mathbb{Z}_6$.

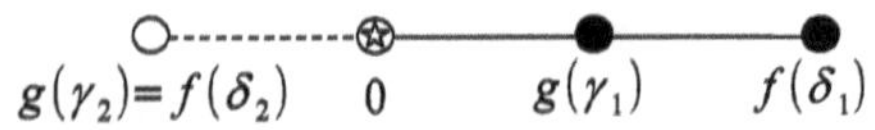

Fig. 9 Graph Σ (Example 3)

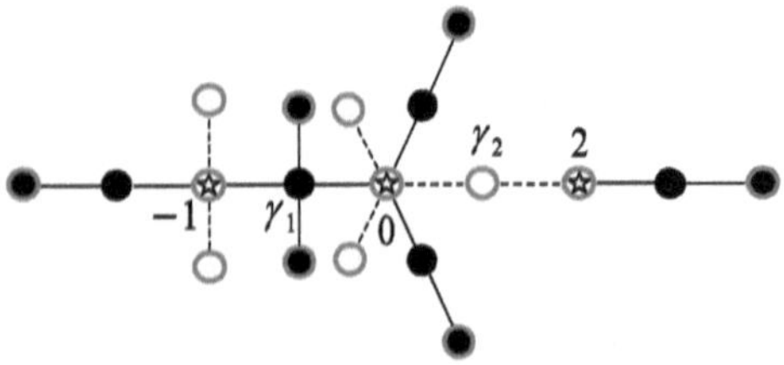

Fig. 10 Bifurcation graph Γ (Example 3)

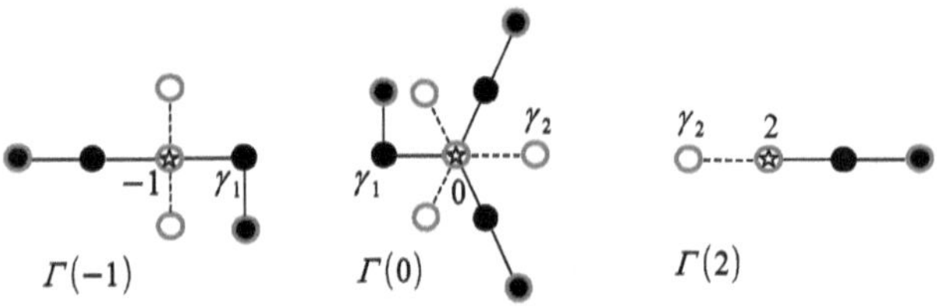

Fig. 11 Satellites $\Gamma(-1)$, $\Gamma(0)$ and $\Gamma(2)$ (Example 3)

5 Proof of Theorem 1

We suppose that C has at least one exceptional singularity. (When C is generic, the result is already proved in [2].) For simplicity, we shall assume $\nu_- < 0$ and $\nu_+ > 0$, so that each satellite $\Gamma(\alpha_i)$ $(1 \le i \le m)$ has $2\lambda_i$ branches. (The proof can be easily adapted if $\nu_- = 0$ or $\nu_+ = 0$.) As the curve is semi-generic with respect to g, there exists i_0 such that $\Gamma(\alpha_{i_0})$ is a regular satellite (i.e., $g(\gamma_{i_0-1}) \notin \mathscr{V}_{\text{crit}}(f)$ and $g(\gamma_{i_0}) \notin \mathscr{V}_{\text{crit}}(f)$).

As mentioned above, we use the Zariski–van Kampen theorem with the pencil $\mathscr{P}$ given by the vertical lines $L_\gamma : x = \gamma$, where $\gamma \in \mathbb{C}$. We take a sufficiently small positive number ε, and for any real number η, we write $\eta^- := \eta - \varepsilon$ and $\eta^+ := \eta + \varepsilon$. We consider the generic line $L_{\alpha_{i_0}^+}$, and we choose generators

$$\xi_{1,0}, \ldots, \xi_{1,\nu_1-1}, \ldots, \xi_{\ell,0}, \ldots, \xi_{\ell,\nu_\ell-1}$$

of the fundamental group $\pi_1(L_{\alpha_{i_0}^+} \setminus C)$ as in Fig. 12. (In the figure, we do not respect the numerical scale; we even zoom on the part that collapses to β_j when $\varepsilon \to 0$.) Here, the loops ξ_{j,r_j} $(1 \le j \le \ell, 0 \le r_j \le \nu_j - 1)$ are counterclockwise-oriented *lassos* around the intersection points of $L_{\alpha_{i_0}^+}$ with C. We shall refer to these generators as *geometric* generators.

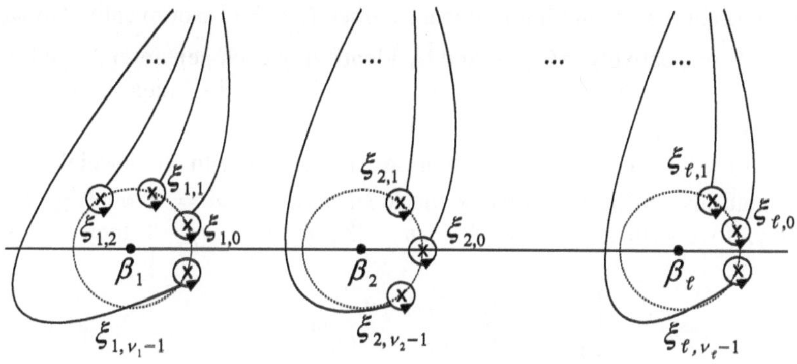

Fig. 12 Generators of $\pi_1(L_{\alpha_{i_0}^+} \setminus C)$

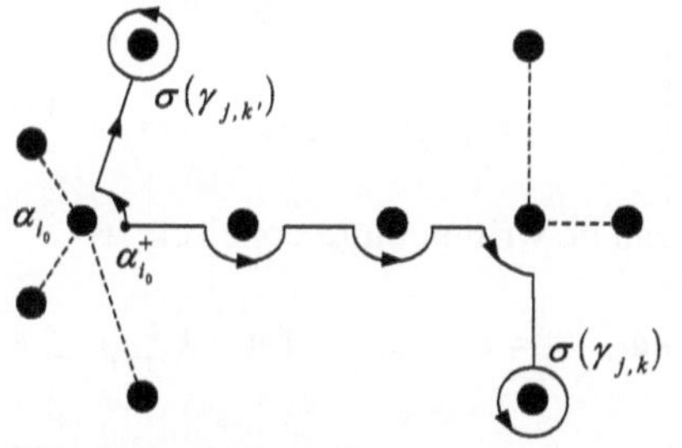

Fig. 13 Example of standard generators of $\pi_1(\mathbb{C} \setminus \mathscr{S})$

For $1 \leq j \leq \ell$, $0 \leq r_j \leq \nu_j - 1$ and $n \in \mathbb{Z}$, let

$$\omega_j := \xi_{j,\nu_j-1} \cdots \xi_{j,0} \quad \text{and} \quad \xi_{j,n\nu_j+r_j} := \omega_j^n \cdot \xi_{j,r_j} \cdot \omega_j^{-n}.$$

These relations define elements $\xi_{j,k}$ for any $1 \leq j \leq \ell$ and any $k \in \mathbb{Z}$. (Indeed, any $k \in \mathbb{Z}$ can be written as $k = n\nu_j + r_j$, with $n \in \mathbb{Z}$ and $0 \leq r_j \leq \nu_j - 1$.) It is easy to see that

$$\xi_{j,n\nu_j+r} = \omega_j^n \cdot \xi_{j,r} \cdot \omega_j^{-n} \quad \text{for} \quad 1 \leq j \leq \ell \text{ and } n, r \in \mathbb{Z}. \tag{8}$$

As usual, to find the monodromy relations associated with the special lines of the pencil $\mathscr{P}$, we consider a "standard" system of counterclockwise-oriented generators of the fundamental group $\pi_1(\mathbb{C} \setminus \mathscr{S})$, where $\mathscr{S}$ is the set consisting of the vertices α_i $(1 \leq i \leq m)$ and $\gamma_{j,k}$ $(1 \leq j \leq \ell - 1,\ 1 \leq k \leq d')$ in the bifurcation graph Γ. (We recall that the elements $\gamma_{j,k}$ are the roots of the equation $g(x) = f(\delta_j)$, where δ_j is defined as in Sect. 1.) We choose $\alpha_{i_0}^+$ as base point, and we denote these generators by $\sigma(\alpha_i)$ and $\sigma(\gamma_{j,k})$. Then, $\sigma(\alpha_i)$ (respectively, $\sigma(\gamma_{j,k})$) is a loop in $\mathbb{C} \setminus \mathscr{S}$ surrounding the vertex α_i (respectively, $\gamma_{j,k}$). It is based at $\alpha_{i_0}^+$ and it runs along the edges of Γ avoiding the vertices corresponding to special lines (cf. Fig. 13). The monodromy relations around the special line L_{α_i} (respectively,

$L_{\gamma_{j,k}}$) are obtained by moving the generic fiber $L_{\alpha_{i_0}^+} \setminus C$ isotopically "above" the loop $\sigma(\alpha_i)$ (respectively, $\sigma(\gamma_{j,k})$) and by identifying each generator ξ_{j,r_j} ($1 \leq j \leq \ell$, $0 \leq r_j \leq \nu_j - 1$) of the group $\pi_1(L_{\alpha_{i_0}^+} \setminus C)$ with its image by the terminal homeomorphism of this isotopy (cf. [1,4,7]).

We start with the monodromy relations associated with the special line $L_{\alpha_{i_0}}$. These relations can be found using the local models $y^{\nu_j} = x^{\lambda_{i_0}}$ ($1 \leq j \leq \ell$). Precisely, if we write $\lambda_{i_0} = n_j \nu_j + r_j$, $n_j \in \mathbb{Z}$, $0 \leq r_j \leq \nu_j - 1$, they are given by

$$\begin{cases} \xi_{j,0} = \omega_j^{n_j} \cdot \xi_{j,r_j} \cdot \omega_j^{-n_j}, \\ \xi_{j,1} = \omega_j^{n_j} \cdot \xi_{j,r_j+1} \cdot \omega_j^{-n_j}, \\ \dots \\ \xi_{j,\nu_j-(r_j+1)} = \omega_j^{n_j} \cdot \xi_{j,\nu_j-1} \cdot \omega_j^{-n_j}, \\ \xi_{j,\nu_j-r_j} = \omega_j^{n_j+1} \cdot \xi_{j,0} \cdot \omega_j^{-(n_j+1)}, \\ \dots \\ \xi_{j,\nu_j-1} = \omega_j^{n_j+1} \cdot \xi_{j,r_j-1} \cdot \omega_j^{-(n_j+1)}. \end{cases}$$

By (8), these relations can be written more concisely as

$$\xi_{j,k_j} = \omega_j^{n_j} \cdot \xi_{j,k_j+r_j} \cdot \omega_j^{-n_j} = \xi_{j,k_j+\lambda_{i_0}} \quad \text{for} \quad 1 \leq j \leq \ell \text{ and } 0 \leq k_j \leq \nu_j - 1.$$

In fact, (8) shows that

$$\xi_{j,k} = \xi_{j,k+\lambda_{i_0}} \quad \text{for} \quad 1 \leq j \leq \ell \text{ and } k \in \mathbb{Z}. \tag{9}$$

Remark 3. If $\nu_j = 1$ for all $1 \leq j \leq \ell$, then $L_{\alpha_{i_0}}$ is not a special line, and hence, the corresponding monodromy relations are trivial. However, it is clear that the relations (9) remain valid. (Indeed, in this case, $\xi_{j,k} = \xi_{j,0}$ for all $k \in \mathbb{Z}$.)

Next, we look for the monodromy relations along the branches of $\Gamma(\alpha_{i_0})$. For $0 \leq q \leq 2\lambda_{i_0} - 1$, we denote by $B_{i_0,q}$ the q-th branch of $\Gamma(\alpha_{i_0})$. We suppose that the branches $B_{i_0,2q}$ (respectively, $B_{i_0,2q+1}$), $0 \leq q \leq \lambda_{i_0} - 1$, correspond to the positive part Σ_+ (respectively, the negative part Σ_-) of Σ through the correspondence $\Gamma(\alpha_{i_0}) \to \Sigma$ given by the restriction of g. We also suppose that the branch $B_{i_0,0}$ (respectively, $B_{i_0,1}$) contains the line segment $[\alpha_{i_0}, \gamma_{i_0}]$ if $g(\gamma_{i_0}) > 0$ (respectively, if $g(\gamma_{i_0}) < 0$). For instance, in the special case of the satellite $\Gamma(\alpha_2) = \Gamma(0)$ of Example 2, the branches $B_{2,q}$ are as in Fig. 14. For simplicity, hereafter we shall suppose $g(\gamma_{i_0}) > 0$. (The argument is similar in the case $g(\gamma_{i_0}) < 0$.)

Pick an element j_0 such that $1 \leq j_0 \leq \ell - 1$. If $f(\delta_{j_0}) > 0$, then, for each $0 \leq q \leq \lambda_{i_0} - 1$, there exists an unique vertex $\gamma_{i_0,j_0,2q} \in B_{i_0,2q}$ such that $g(\gamma_{i_0,j_0,2q}) = f(\delta_{j_0})$. For instance, in the special case of the satellite $\Gamma(\alpha_2) = \Gamma(0)$ of Example 2, $f(\delta_1) > 0$ and for $0 \leq q \leq 2$ there exists an unique vertex $\gamma_{2,1,2q} \in B_{2,2q}$ such that $g(\gamma_{2,1,2q}) = f(\delta_1)$ (cf. Fig. 14). As $\Gamma(\alpha_{i_0})$ is a regular satellite, $g(\gamma_{i_0}) \neq f(\delta_{j_0})$, and hence $\gamma_{i_0,j_0,0} \neq \gamma_{i_0}$. It follows that the monodromy relation associated with the

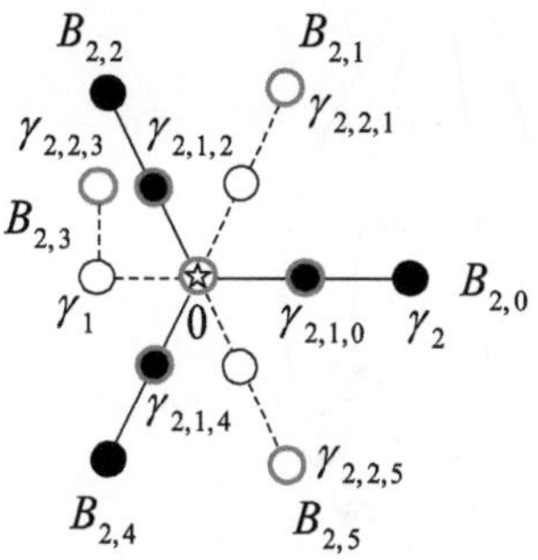

Fig. 14 Branches of the satellite $\Gamma(\alpha_2)$ of Example 2

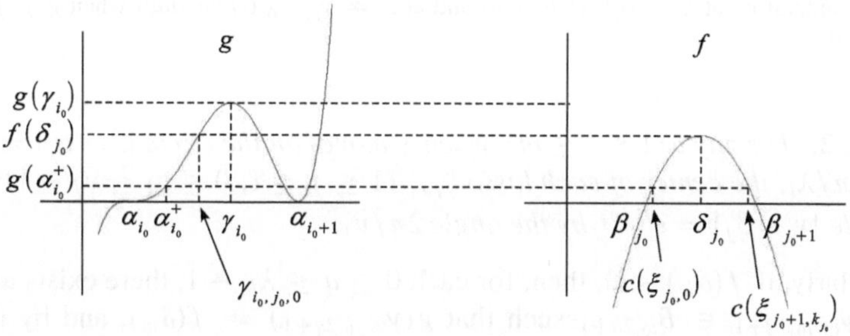

Fig. 15 Real graphs of g and f when $0 < f(\delta_{j0}) < g(\gamma_{i0})$

line $L_{\gamma_{i0,j0,0}}$ is a simple tangent relation (cf. (4)). Precisely, this relation is given by

$$\xi_{j0,0} = \xi_{j0+1,k_{j0}}, \tag{10}$$

where k_{j0} is some integer depending only on the first ordering of the elements $\xi_{j0+1,r_{j0+1}}$, $0 \leq r_{j0+1} \leq \nu_{j0+1} - 1$ (cf. Figs. 15 and 16). The graphs in Fig. 15 are the real graphs of g and f in neighborhoods of the intervals $[\alpha_{i0}, \alpha_{i0+1}]$ and $[\beta_{j0}, \beta_{j0+1}]$, respectively; $c(\xi_{j0,0})$ and $c(\xi_{j0+1,k_{j0}})$ are the centers of the lassos $\xi_{j0,0}$ and $\xi_{j0+1,k_{j0}}$, respectively. The picture on the left side (respectively, right side) of Fig. 16 represents the generators in a neighborhood of β_{j0} and β_{j0+1} (respectively, in a neighborhood of δ_{j0}) at $x = \alpha^+_{i0}$ (respectively, at $x = \gamma^-_{i0,j0,0}$).

Actually, as $\Gamma(\alpha_{i0})$ is regular, $\gamma_{i0,j0,2q} \notin \{\gamma_{i0-1}, \gamma_{i0}\}$ for any $0 \leq q \leq \lambda_{i0} - 1$, and the monodromy relation associated with the special line $L_{\gamma_{i0,j0,2q}}$ is a simple tangent relation given by

$$\xi_{j0,-q} = \xi_{j0+1,k_{j0}-q}. \tag{11}$$

This follows immediately from (10) and the following lemma.

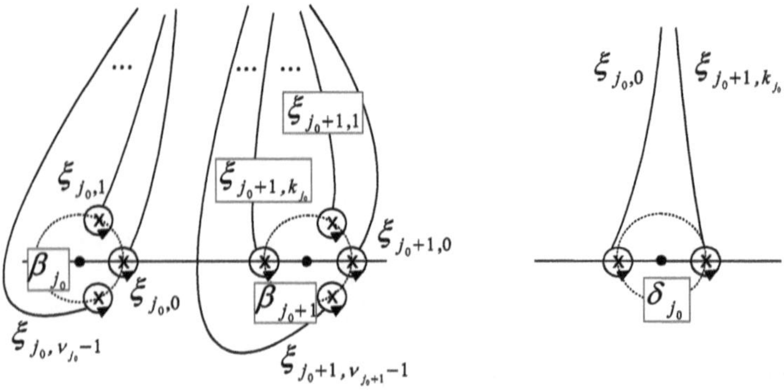

Fig. 16 Generators at $x = \alpha_{i_0}^{+}$ (*left side*) and at $x = \gamma_{i_0,j_0,0}^{-}$ (*right side*) when $g(\gamma_{i_0}) > 0$ and $f(\delta_{j_0}) > 0$

Lemma 2. *For any i $(1 \leq i \leq m)$, when x moves on the circle $|x - \alpha_i| = \varepsilon$ by the angle $2\pi/\lambda_i$, the center of each lasso ξ_{j,r_j} $(1 \leq j \leq \ell, 0 \leq r_j \leq \nu_j - 1)$ turns on the circle $|y - \beta_j| = \varepsilon^{\lambda_i/\nu_j}$ by the angle $2\pi/\nu_j$.*

Similarly, if $f(\delta_{j_0}) < 0$, then, for each $0 \leq q \leq \lambda_{i_0} - 1$, there exists an unique vertex $\gamma_{i_0,j_0,2q+1} \in B_{i_0,2q+1}$ such that $g(\gamma_{i_0,j_0,2q+1}) = f(\delta_{j_0})$, and by the same argument as above, the monodromy relation associated with the line $L_{\gamma_{i_0,j_0,2q+1}}$ is given by

$$\xi_{j_0,h_{j_0}-q} = \xi_{j_0+1,k_{j_0}-q}, \tag{12}$$

where h_{j_0} and k_{j_0} are integers depending only on the first ordering of the elements $\xi_{j_0,r_{j_0}}$ $(0 \leq r_{j_0} \leq \nu_{j_0} - 1)$ and $\xi_{j_0+1,r_{j_0+1}}$ $(0 \leq r_{j_0+1} \leq \nu_{j_0+1} - 1)$.

Remark 4. Note that the relations (11) can also be written under the form (12) by taking $h_{j_0} = 0$.

Combined with (9), the relations (12) imply

$$\xi_{j_0,h_{j_0}-k} = \xi_{j_0+1,k_{j_0}-k} \quad \text{for} \quad k \in \mathbb{Z},$$

and therefore,

$$\xi_{j_0,k} = \xi_{j_0+1,k+(k_{j_0}-h_{j_0})} \quad \text{for} \quad k \in \mathbb{Z}.$$

Then, by reordering the generators $\xi_{j,k}$ successively for $j = 2, \ldots, \ell$, we can assume that

$$\xi_{j_0,k} = \xi_{j_0+1,k} \quad \text{for} \quad k \in \mathbb{Z}, \tag{13}$$

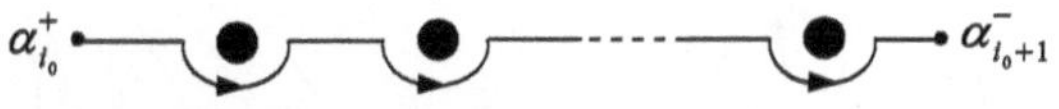

Fig. 17 The modified line segment $(\alpha^+_{i_0}, \alpha^-_{i_0+1})$

and hence, as j_0 is arbitrary, we can take, as generators, the elements

$$\xi_k := \xi_{j_0,k} \quad \text{for} \quad k \in \mathbb{Z}. \tag{14}$$

Then, the relations (9) are written as

$$\xi_k = \xi_{k+\lambda_{i_0}} \quad \text{for} \quad k \in \mathbb{Z}, \tag{15}$$

and, by applying Proposition 2, we have

$$\xi_{k+\nu_0} = \omega \xi_k \omega^{-1} \quad \text{for} \quad k \in \mathbb{Z}, \tag{16}$$

where $\omega := \xi_{\nu_0-1} \dots \xi_0$. Indeed, by Bezout's identity, there exist $k_1, \dots, k_\ell \in \mathbb{Z}$ such that $\nu_0 = k_1\nu_1 + \dots + k_\ell\nu_\ell$. Then, by (8),

$$\xi_{k+\nu_0} = (\omega_\ell^{k_\ell} \dots \omega_1^{k_1}) \cdot \xi_k \cdot (\omega_\ell^{k_\ell} \dots \omega_1^{k_1})^{-1},$$

while Proposition 2 shows that

$$\omega_\ell^{k_\ell} \dots \omega_1^{k_1} = \xi_{\nu_0-1} \dots \xi_0.$$

Remark 5. The relations (15) and (16) associated with the regular satellite $\Gamma(\alpha_{i_0})$ imply that the fundamental group $\pi_1(\mathbb{C}^2 \setminus C)$ is a quotient of the group $G(\nu_0; \lambda_{i_0})$.

Now, let us consider the monodromy relations associated with the other satellites. For simplicity, we still assume $g(\gamma_{i_0}) > 0$. We start with the satellite $\Gamma(\alpha_{i_0+1})$ and first look for the relations around the line $L_{\alpha_{i_0+1}}$. For this purpose, we need to know how the generators are deformed when x moves along the "modified" line segment $(\alpha^+_{i_0}, \alpha^-_{i_0+1})$. Here, "modified" means that x makes a half-turn counterclockwise around each vertex of $\Gamma \cap (\alpha^+_{i_0}, \alpha^-_{i_0+1})$ corresponding to a special line (cf. Fig. 17). Take an element j_0 such that $1 \leq j_0 \leq \ell - 1$. If $0 < f(\delta_{j_0}) < g(\gamma_{i_0})$, then there are exactly two vertices $\gamma_{i_0,j_0,0} \neq \gamma_{i_0}$ and $\gamma_{i_0+1,j_0,2q_0} \neq \gamma_{i_0}$ (for some $0 \leq q_0 \leq \lambda_{i_0+1} - 1$) on the line segment $(\alpha^+_{i_0}, \alpha^-_{i_0+1})$ that correspond to the special lines of the pencil associated with the critical value $f(\delta_{j_0})$ (i.e., $g(\gamma_{i_0,j_0,0}) = f(\delta_{j_0})$ and $g(\gamma_{i_0+1,j_0,2q_0}) = f(\delta_{j_0})$). The first one $\gamma_{i_0,j_0,0}$ is in $\Gamma(\alpha_{i_0})$ and the second one $\gamma_{i_0+1,j_0,2q_0}$ is in $\Gamma(\alpha_{i_0+1})$. Therefore, when x moves along the modified line segment $(\alpha^+_{i_0}, \alpha^-_{i_0+1})$, the generators are deformed as in Fig. 18. The picture on the left side of the figure represents the generators at $x = \alpha^+_{i_0}$ (i.e., before the deformation). The picture on the right side represents the generators at $x = \alpha^-_{i_0+1}$ (i.e., after the deformation). However, by (13), we can suppose that the generators in the fiber

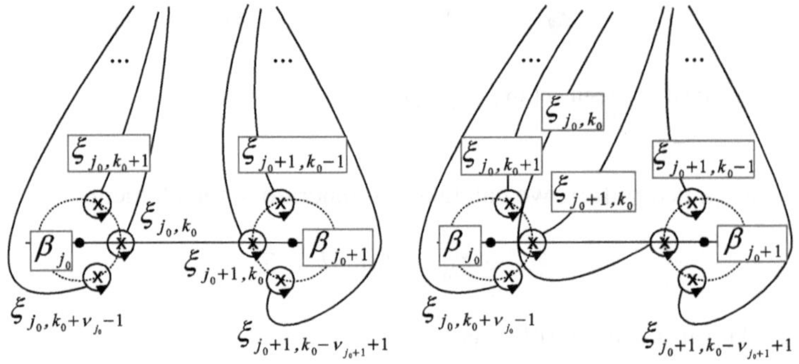

Fig. 18 Deformation of the generators when $0 < f(\delta_{j_0}) < g(\gamma_{i_0})$

$x = \alpha^-_{i_0+1}$ are still the same as in the fiber $x = \alpha^+_{i_0}$. In other words, the picture on the left side of Fig. 18 also represents the generators at $x = \alpha^-_{i_0+1}$. Hence, by the same argument as above, the monodromy relations associated with the special line $L_{\alpha_{i_0+1}}$ give the relations

$$\xi_k = \xi_{k+\lambda_{i_0+1}} \quad \text{for} \quad k \in \mathbb{Z}. \tag{17}$$

We get the same relations if $g(\gamma_{i_0}) < f(\delta_{j_0})$ or if $f(\delta_{j_0}) < 0$. Indeed, in these two cases, the set $g^{-1}(f(\delta_{j_0})) \cap (\alpha^+_{i_0}, \alpha^-_{i_0+1})$ is empty, and therefore the configuration of the generators is identical on the fibers $x = \alpha^-_{i_0+1}$ and $x = \alpha^+_{i_0}$.

The monodromy relations around the special lines corresponding to the vertices located on the branches of $\Gamma(\alpha_{i_0+1})$ do not give any new relation. This can be directly shown easily but it is not necessary. In fact, as we shall see below, it suffices to collect the monodromy relations associated with the special lines L_{α_i} for all i, $1 \le i \le m$. We already know that for $i = i_0$ and $i_0 + 1$, the monodromy relations around L_{α_i} are given by $\xi_k = \xi_{k+\lambda_i}$ for all $k \in \mathbb{Z}$. In fact, this is true for any i. For instance, let us show it for $i = i_0 + 2$. For this purpose, we need to know how the generators are deformed when x makes a half-turn on the circle $|x - \alpha_{i_0+1}| = \varepsilon$ from $\alpha^-_{i_0+1}$ to $\alpha^+_{i_0+1}$, and then moves along the modified line segment $(\alpha^+_{i_0+1}, \alpha^-_{i_0+2})$. Again, take j_0 such that $1 \le j_0 \le \ell - 1$, and for simplicity assume that $g(\gamma_{i_0})$, $g(\gamma_{i_0+1})$ and $f(\delta_{j_0})$ are positive. (The other cases are similar and left to the reader.) By Lemma 2, when x makes a half-turn on the circle $|x - \alpha_{i_0+1}| = \varepsilon$ from $\alpha^-_{i_0+1}$ to $\alpha^+_{i_0+1}$, the generators are deformed as in Fig. 19, where $k'_0 \in \mathbb{Z}$. That is, the configuration of the generators on the fiber $x = \alpha^+_{i_0+1}$ is just the parallel translation of that on the fiber $x = \alpha^-_{i_0+1}$. When x moves along the modified line segment $(\alpha^+_{i_0+1}, \alpha^-_{i_0+2})$, the generators are still as in Fig. 19. The proof of this fact is as above except that now $\gamma_{i_0+1} \in (\alpha^+_{i_0+1}, \alpha^-_{i_0+2})$ may correspond to an exceptional singularity (in particular, we may have $f(\delta_{j_0}) = g(\gamma_{i_0+1})$). But in this case, when x makes a half-turn on the circle $|x - \gamma_{i_0+1}| = \varepsilon$ from $\gamma^-_{i_0+1}$ to $\gamma^+_{i_0+1}$, the generators remains

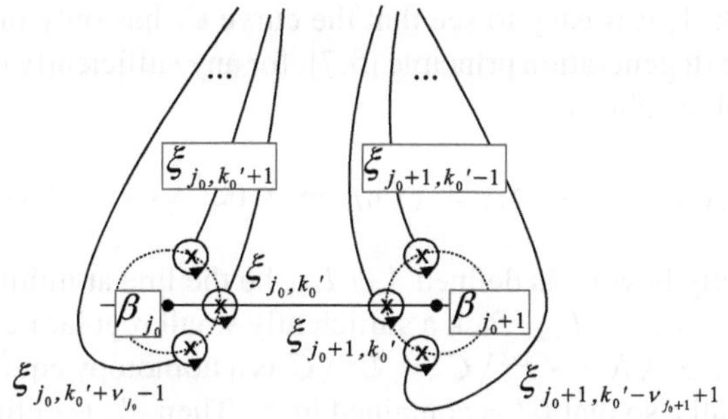

Fig. 19 Generators at $x = \alpha^+_{i_0+1}$ and $x = \alpha^-_{i_0+2}$ when $g(\gamma_{i_0})$, $g(\gamma_{i_0+1})$ and $f(\delta_{j_0})$ are > 0

unchanged, as, by (13), $\xi_{j_0,k_0'} = \xi_{j_0+1,k_0'}$. Finally, exactly as above, the monodromy relations associated with the special line $L_{\alpha_{i_0+2}}$ give the relations

$$\xi_k = \xi_{k+\lambda_{i_0+2}} \quad \text{for} \quad k \in \mathbb{Z}.$$

This argument can be repeated for all the other values of i, $1 \le i \le m$, so that the monodromy relations associated with the special line L_{α_i} for any i, $1 \le i \le m$, are written as

$$\xi_k = \xi_{k+\lambda_i} \quad \text{for} \quad k \in \mathbb{Z}. \tag{18}$$

By Proposition 3, the collection of relations (18), for $1 \le i \le m$, and the relation (16) are equivalent to

$$\begin{cases} \xi_k = \xi_{k+\lambda_0} \\ \xi_{k+\nu_0} = \omega \xi_k \omega^{-1} \end{cases} \quad \text{for} \quad k \in \mathbb{Z}.$$

In particular, this means that the fundamental group $\pi_1(\mathbb{C}^2 \setminus C)$ is presented by the generators ξ_k ($k \in \mathbb{Z}$) and ω and by a set of relations that includes the following relations:

$$\omega = \xi_{\nu_0 - 1} \cdots \xi_0, \tag{19}$$

$$\xi_{k+\lambda_0} = \xi_k \quad (k \in \mathbb{Z}), \tag{20}$$

$$\xi_{k+\nu_0} = \omega \xi_k \omega^{-1} \quad (k \in \mathbb{Z}). \tag{21}$$

In other words, $\pi_1(\mathbb{C}^2 \setminus C)$ is a quotient of the group $G(\nu_0; \lambda_0)$.

Now, consider the family $\{C_t\}_{0 \le t \ll 1}$ of ℝ-join type curves, where C_t is defined by the equation

$$f(y) = (1-t)g(x).$$

For any $0 < t \ll 1$, it is easy to see that the curve C_t has only inner singularities. Therefore, by the degeneration principle [5,7], for any sufficiently small $t > 0$, there is a canonical epimorphism

$$\psi_t \colon \pi_1(\mathbb{C}^2 \setminus C) = \pi_1(\mathbb{C}^2 \setminus C_0) \twoheadrightarrow \pi_1(\mathbb{C}^2 \setminus C_t) \simeq G(\nu_0; \lambda_0).$$

Let us recall briefly how ψ_t is defined. Let L_∞ be the line at infinity, and set $C' := C \cup L_\infty$ and $C'_t := C_t \cup L_\infty$. Pick a sufficiently small regular neighborhood N of C' in $\mathbb{P}^2$ so that $\iota \colon \mathbb{P}^2 \setminus N \hookrightarrow \mathbb{P}^2 \setminus C' = \mathbb{C}^2 \setminus C$ is a homotopy equivalence, and take a sufficiently small t so that C'_t is contained in N. Then, ψ_t is defined by taking the composition of $\iota_\sharp{}^{-1} \colon \pi_1(\mathbb{C}^2 \setminus C) \to \pi_1(\mathbb{P}^2 \setminus N)$ with the homomorphism induced by the inclusion $\mathbb{P}^2 \setminus N \hookrightarrow \mathbb{P}^2 \setminus C'_t = \mathbb{C}^2 \setminus C_t$. To distinguish the generators, we write $\xi_k(t)$ ($k \in \mathbb{Z}$) for the generators of $\pi_1(\mathbb{C}^2 \setminus C_t)$, which are represented by the same loops as ξ_k. Note that $\psi_t(\xi_k) = \xi_k(t)$. As C_t is generic, $\pi_1(\mathbb{C}^2 \setminus C_t)$ is presented by the generators $\xi_k(t)$ ($k \in \mathbb{Z}$) and $\omega(t) := \xi_{\nu_0-1}(t) \ldots \xi_0(t)$ and by the relations (19)–(21), replacing ξ_k by $\xi_k(t)$ and ω by $\omega(t)$. (We may assume that $N \cap L_{\alpha_{i_0}^+}$ is a copy of d disjoint (topologically) tiny 2-disks so that ξ_k ($0 \le k \le d-1$) are free generators of $\pi_1(L_{\alpha_{i_0}^+} \setminus N)$.) This implies that $\ker \psi_t$ is trivial, and hence

$$\pi_1(\mathbb{C}^2 \setminus C) \simeq G(\nu_0; \lambda_0).$$

(In particular, the branches of the satellites $\Gamma(\alpha_i)$, $i \neq i_0$, do not give any new relation.)

As for the fundamental group $\pi_1(\mathbb{P}^2 \setminus \tilde{C})$, we proceed as follows. If $d \ge d'$, then the base locus of the pencil $X = \gamma Z$ ($\gamma \in \mathbb{C}$) in $\mathbb{P}^2$ does not belong to the curve, and therefore the group $\pi_1(\mathbb{P}^2 \setminus \tilde{C})$ is obtained from the above presentation of $\pi_1(\mathbb{C}^2 \setminus C)$ by adding the vanishing relation at infinity $\omega_1 \ldots \omega_\ell = e$. By Proposition 2, the relations (19) and (21) imply $\omega_j = \omega^{\nu_j/\nu_0}$ ($1 \le j \le \ell$). Therefore, the relation $\omega_1 \ldots \omega_\ell = e$ can also be written as

$$\omega^{d/\nu_0} = e.$$

It follows that $\pi_1(\mathbb{P}^2 \setminus \tilde{C}) \simeq G(\nu_0; \lambda_0; d/\nu_0)$.

If $d' \ge d$, then we consider again the family $\{C_t\}_{0 \le t \ll 1}$, where C_t is defined by the equation $f(y) = (1-t)g(x)$. We use the same regular neighborhood N and the same isomorphism $\psi_t \colon \pi_1(\mathbb{C}^2 \setminus C) \to \pi_1(\mathbb{C}^2 \setminus C_t)$ for a sufficiently small $t > 0$. To compute $\pi_1(\mathbb{C}^2 \setminus C)$, this time we consider the pencil given by the horizontal lines $y = \delta$, where $\delta \in \mathbb{C}$. We fix a generic line $y = \delta_0$ and we choose geometric generators ρ_k ($0 \le k \le d' - 1$) as above so that the ρ_k's give generators of the fundamental group of the generic fiber of each complement $\mathbb{P}^2 \setminus N$, $\mathbb{C}^2 \setminus C$ and $\mathbb{C}^2 \setminus C_t$ simultaneously. Then we define elements τ and ρ_k, for $k \in \mathbb{Z}$, in the same way as we defined the elements ω and ξ_k ($k \in \mathbb{Z}$) above. Now, as ψ_t is an isomorphism and

C_t is generic, the generating relations for each group $\pi_1(\mathbb{C}^2 \setminus C_t)$, $\pi_1(\mathbb{C}^2 \setminus C)$ and $\pi_1(\mathbb{P}^2 \setminus N)$ are given by

$$\tau = \rho_{\lambda_0 - 1} \cdots \rho_0,$$
$$\rho_{k+\nu_0} = \rho_k \quad (k \in \mathbb{Z}),$$
$$\rho_{k+\lambda_0} = \tau \rho_k \tau^{-1} \quad (k \in \mathbb{Z}).$$

As $d' \geq d$, the base locus of the pencil $Y = \delta Z$ ($\delta \in \mathbb{C}$) in $\mathbb{P}^2$ does not belong to $\tilde{C}$, and hence the group $\pi_1(\mathbb{P}^2 \setminus \tilde{C})$ is obtained from the above presentation of $\pi_1(\mathbb{C}^2 \setminus C)$ by adding the vanishing relation at infinity $\tau^{d'/\lambda_0} = e$. Finally, we get $\pi_1(\mathbb{P}^2 \setminus \tilde{C}) \simeq G(\lambda_0; \nu_0; d'/\lambda_0)$.

6 Applications

In this section, we give two applications of Theorem 1.

6.1 *Maximal Irreducible Nodal Curves*

An irreducible curve is said to be *nodal* if it has only node singularities. By Plücker's formula, an irreducible nodal curve of degree d has at most $(d-1)(d-2)/2$ nodes. An irreducible nodal curve is called *maximal* if it has exactly $(d-1)(d-2)/2$ nodes (equivalently, if its genus is 0). A method to construct such curves is given in [3]. There, the irreducibility is obtained using the braid group action. Hereafter, we repeat the construction of [3] but apply Theorem 1 to show the irreducibility.

For simplicity, let us suppose that $d = 2n + 1$, $n \in \mathbb{Z}$. (The construction is similar when d is even.) Consider the *Chebyshev* polynomial of degree d, which is defined by $T_d(z) := \cos(d \arccos(z))$. This polynomial has $2n + 1$ simple real roots $\beta_1 < \ldots < \beta_{2n+1}$ and $2n$ critical points $\delta_1, \ldots, \delta_{2n}$, with $\delta_j \in (\beta_j, \beta_{j+1})$, such that $T_d(\delta_1) = T_d(\delta_3) = \ldots = T_d(\delta_{2n-1}) = 1$ and $T_d(\delta_2) = T_d(\delta_4) = \ldots = T_d(\delta_{2n}) = -1$. Now, take a small deformation $\tilde{T}_d(z)$ of $T_d(z)$ so that:

1. $\tilde{T}_d(z)$ has n critical points $\gamma_1, \gamma_3, \ldots, \gamma_{2n-1}$ such that $\tilde{T}_d(\gamma_1) = \tilde{T}_d(\gamma_3) = \ldots = \tilde{T}_d(\gamma_{2n-1}) = 1$;
2. $\tilde{T}_d(z)$ has $n-1$ critical points $\gamma_2, \gamma_4, \ldots, \gamma_{2n-2}$ such that $\tilde{T}_d(\gamma_2) = \tilde{T}_d(\gamma_4) = \ldots = \tilde{T}_d(\gamma_{2n-2}) = -1$;
3. $\tilde{T}_d(z)$ has a critical point γ_{2n} with $\tilde{T}_d(\gamma_{2n}) < -1$.

The existence of such a polynomial $\tilde{T}_d(z)$ is due to R. Thom [6]. It has $2n + 1$ simple real roots $\alpha_1 < \ldots < \alpha_{2n+1}$, and $\gamma_i \in (\alpha_i, \alpha_{i+1})$. Then, consider the $\mathbb{R}$-join-type curve C defined by $T_d(y) = \tilde{T}_d(x)$. Clearly, the satellite $\Gamma(\alpha_{2n+1})$ (of the bifurcation graph of C with respect to $\tilde{T}_d$) is regular and the numbers ν_0, λ_0 which appear in Theorem 1 are equal to 1. Hence, $\pi_1(\mathbb{C}^2 \setminus C) \simeq \mathbb{Z}$ and $\pi_1(\mathbb{P}^2 \setminus \tilde{C}) \simeq \mathbb{Z}_d$. In particular, the curve C is irreducible. The number of nodes is given by the cardinality of the set

$$\{(\gamma_{2i-1}, \delta_{2j-1}) \mid 1 \leq i, j \leq n\} \cup \{(\gamma_{2i}, \delta_{2j}) \mid 1 \leq i \leq n-1,\ 1 \leq j \leq n\},$$

that is,

$$n^2 + (n-1)n = \frac{(d-1)(d-2)}{2}.$$

In other words, C is a maximal irreducible nodal curve.

6.2 Curves with Node and Cusp Singularities

Let $C\colon f(y) = g(x)$ be an $\mathbb{R}$-join-type curve with only nodes and cusps as singularities. For simplicity, we suppose that C has degree $d = 6n$, $n \in \mathbb{Z}$. The maximal number of cusps on such a curve is obtained when f and g have the form $f(y) = a(y-\beta_1)^3(y-\beta_2)^3\dots(y-\beta_{2n})^3$ and $g(x) = b(x-\alpha_1)^2(x-\alpha_2)^2\dots(x-\alpha_{3n})^2$, in which case we have $6n^2$ cusps. (As usual, we suppose $\beta_j < \beta_{j+1}$ and $\alpha_i < \alpha_{i+1}$ for all i, j.) By the result of R. Thom [6], f can be chosen so that its $2n-1$ critical points $\delta_1 < \dots < \delta_{2n-1}$ satisfy $f(\delta_1) = f(\delta_3) = \dots = f(\delta_{2n-1}) = -1$ and $f(\delta_2) = f(\delta_4) = \dots = f(\delta_{2n-2}) = 1$. Similarly, g can be chosen so that its $3n-1$ critical points $\gamma_1 < \dots < \gamma_{3n-1}$ satisfy $g(\gamma_1) = g(\gamma_2) = \dots = g(\gamma_{3n-2}) = -1 > g(\gamma_{3n-1})$. In this case, the curve also has $n(3n-2)$ nodes. Clearly, the satellite $\Gamma(\alpha_{3n})$ (of the bifurcation graph of C with respect to g) is regular, and hence, by Theorem 1, $\pi_1(\mathbb{C}^2 \setminus C) \simeq G(3;2) \simeq B(3)$ (the braid group on 3 strings), while $\pi_1(\mathbb{P}^2 \setminus \tilde{C}) \simeq G(3;2;2n)$. Note that by Theorem (2.12) of [2], we have a central extension

$$0 \to \mathbb{Z}_n \to G(3;2;2n) \to \mathbb{Z}_3 * \mathbb{Z}_2 \to 0,$$

where $\mathbb{Z}_n$ is generated by ω^2.

Acknowledgements This research was partially supported by the *Institut des Hautes Études Scientifiques*, Bures-sur-Yvette, France.

References

1. van Kampen, E.R.: On the fundamental group of an algebraic curve. Am. J. Math. **55**, 255–260 (1933)
2. Oka, M.: On the fundamental group of the complement of certain plane curves. J. Math. Soc. Jpn. **30**, 579–597 (1978)

3. Oka, M.: Symmetric plane curves with nodes and cups. J. Math. Soc. Jpn. **44**, 375–414 (1992)
4. Oka, M.: Two transforms of plane curves and their fundamental groups. J. Math. Sci. Univ. Tokyo **3**, 399–443 (1996)
5. Oka, M.: A survey on Alexander polynomials of plane curves. Séminaires Congrès **10**, 209–232 (2005)
6. Thom, R.: L'équivalence d'une fonction différentiable et d'un polynôme. Topology **3**, 297–307 (1965)
7. Zariski, O.: On the problem of existence of algebraic functions of two variables possessing a given branch curve. Am. J. Math. **51**, 305–328 (1929)

Some Remarks on the Realizability Spaces of (3,4)-Nets

Denis Ibadula and Daniela Anca Măcinic

Dedicated to Professors Alexandru Dimca and Ştefan Papadima on the occasion of their 60th birthday

Abstract We prove that in the class of $(3,4)$-nets with double and triple points, lattice isomorphism actually translates into lattice isotopy. We disprove the existence of Zariski pairs involving an example of Yoshinaga of a $(3,6)$-net with 48 triple points.

Keywords (p,q)-nets • Zariski pairs • Realization spaces

Symmetric structures of (multi)net type appear in the theory of complex hyperplane arrangements in relation to the nontriviality of the resonance varieties (Falk–Yuzvinsky [7]), or, for instance, related to the nontriviality and combinatoriality of the algebraic monodromy of the Milnor fiber associated with an arrangement [5, 10].

Definition 1 (Yuzvinsky [18]). Let $p \geq 3 \in \mathbb{Z}$. A p-net in a projective plane $\mathbb{P}^2$ is a pair $(\mathcal{A}, \chi)$, where $\mathcal{A}$ is an arrangement of lines in $\mathbb{P}^2$ partitioned into p blocks $\mathcal{A} = \cup_{i=1}^{p} \mathcal{A}_i$ and χ is the set of intersection points of all couples of lines from distinct blocks such that for every $X \in \chi$ and every $i \in \overline{1,p}$ there is a unique $L \in \mathcal{A}_i$ passing through X.

It is immediate that all the subarrangements $\mathcal{A}_i$ have the same cardinality q. It can be proved easily that $q = |\chi \cap L|$, for $L \in \mathcal{A}$ arbitrary line, and $|\chi| = q^2$. We will call $\mathcal{A}$ a *(p,q)-net*.

It can be seen also as the union of k completely reducible fibers of a pencil of degree q complex plane curves; in which case the base point set of the pencil is χ.

D. Ibadula (✉)
Faculty of Mathematics and Informatics, Ovidius University, BD. Mamaia 124, 900527 Constanta, Romania
e-mail: denis@univ-ovidius.ro

D.A. Măcinic
Simion Stoilow Institute of Mathematics, P.O. Box 1-764, RO-014700 Bucharest, Romania
e-mail: Anca.Macinic@imar.ro

D. Ibadula, W. Veys (eds.), *Bridging Algebra, Geometry, and Topology*, Springer Proceedings in Mathematics & Statistics 96,
DOI 10.1007/978-3-319-09186-0_10

The notion of net is also assimilated to the abstract configuration of lines with pre-set intersection points described by χ. The problem of realizability of (p,q)-nets, i.e., the existence of an arrangement of lines that realizes the combinatorial intersection pattern described by χ, was studied in [8, 9, 15, 16, 18]. Yuzvinsky showed that an arbitrary (p,q)-net may only be realizable in $\mathbb{P}^2\mathbb{C}$ for $(p = 3, q \geq 2)$, $(p = 4, q \geq 3)$, $(p = 5, q \geq 6)$, and Stipins later showed that $(5,q)$-nets are not realizable by lines in $\mathbb{P}^2\mathbb{C}$.

Another way to define (p,q)-nets is by relation to Latin squares. *Latin squares* are basically multiplication tables for finite quasigroups, although they can be directly defined as $q \times q$ matrices filled with q different symbols (for instance, integers from 1 to q), each occurring exactly once in each row and exactly once in each column. A (p,q)-net is associated with an orthogonal set of $(p-2)$ Latin squares of type $q \times q$ (see [4]).

Let us describe in detail the way the q^2 triple points in χ of a $(3,q)$-net $\mathcal{A} = \mathcal{A}_1 \cup \mathcal{A}_2 \cup \mathcal{A}_3$ are encoded by a $q \times q$ Latin square. Choose an arbitrary order for the lines on each of the subarrangements $\mathcal{A}_1, \mathcal{A}_2, \mathcal{A}_3$, say, $\mathcal{A}_j = \{L_1^j, L_2^j, \ldots, L_q^j\}$, $1 \leq j \leq 3$. Then, the element at the intersection of the line k and column l in the corresponding Latin square is the label $\alpha(k,l) \in \{1,2,\ldots,q\}$ of the unique line in $\mathcal{A}_3$ that passes through the intersection $L_k^1 \cap L_l^2$, i.e. such that $L_k^1 \cap L_l^2 \cap L_{\alpha(k,l)}^3 \in \chi$ (see, for instance, [15]).

However, since the particular choice of labels $\{1,2,3\}$ for the subarrangements, or for the order of the lines inside each subarrangement is not relevant for the realization of $\mathcal{A}$ as a curve, one works with an equivalence relation on the set of Latin squares of a given dimension q. The equivalence class of a given Latin square M contains all the Latin squares obtained by rearranging the lines, columns or symbols of M (this corresponds to reordering the lines inside the subarrangements $\mathcal{A}_i$), respectively by permuting the sets of lines, columns and symbols among them (this corresponds to reordering the labels $\{1,2,3\}$ of the subarrangements $\mathcal{A}_1$, $\mathcal{A}_2$, $\mathcal{A}_3$).

For instance, to restrict our attention to $(3,4)$-nets, it is known that there are two equivalence classes of 4×4 Latin squares:

$$M_1 := \begin{matrix} 1\,2\,3\,4 \\ 2\,3\,4\,1 \\ 3\,4\,1\,2 \\ 4\,1\,2\,3 \end{matrix} \qquad M_2 := \begin{matrix} 1\,2\,3\,4 \\ 2\,1\,4\,3 \\ 3\,4\,1\,2 \\ 4\,3\,2\,1 \end{matrix} \tag{1}$$

where M_1 is the multiplication table of the abelian group $\mathbb{Z}_4$ and M_2 is the multiplication table of the abelian group $\mathbb{Z}_2 \oplus \mathbb{Z}_2$.

We proved in [6] the following:

Theorem 1 ([6, Thm 2.5]). *Let $\mathcal{A} = \mathcal{A}_1 \cup \mathcal{A}_2 \cup \mathcal{A}_3$ be an arrangement of* 12 *lines in $\mathbb{P}^2\mathbb{C}$ defined by a* $(3,4)$*-net via a Latin square M_j, $j \in \{1,2\}$, and having only double and triple points. Then, for each j, $\mathcal{A}$ can have either* 16, 17 *or* 19 *triple points, and the number of triple points decides the lattice isomorphism type of $\mathcal{A}$.*

Our main result in Sect. 1, Theorem 2, strengthens the claim of the previous result from lattice isomorphism to lattice isotopy.

To give a motivation for our result, let us recall that a *Zariski pair*, in the sense of [1], is a pair of lattice isomorphic arrangements that have non-isomorphic fundamental groups of the complement. The well-known result of Rybnikov [14] showing that the fundamental group of the complement of complex hyperplane arrangements is not a combinatorial invariant involves a Zariski pair of arrangements of 13 lines. It is not known if there are Zariski pairs with fewer lines. Nazeer–Yoshinaga show in [11] that there are none such pairs among arrangements of up to 9 lines (see also [17]), and in [1] and [2] the authors consider theoretic configurations of 10 lines (with some restrictions on the multiplicity of points) and make a classification of their realization spaces, again with the purpose of detecting potential Zariski pairs.

As consequence of Theorem 2, we show that there are no Zariski pairs among arrangements of 12 lines with only double and triple points supporting a structure of $(3,4)$-net.

1 (3,4)-Nets

In [16] Urzua gives a one-to-one correspondence between arrangements of d lines in $\mathbb{P}^2\mathbb{C}$ and lines in $\mathbb{P}^{d-2}\mathbb{C}$, which turns out to be very useful to describe the moduli space of an abstract combinatorial type (the realization space for all arrangements having as intersection lattice a given abstract lattice), and applies this to abstract $(3,q)$-net structures.

The starting point for our computations are [6, Thm 2.5] and the before mentioned correspondence from [16, Sect. 4].

For ease of exposition, we call an arrangement with a $(3,4)$-net structure associated with the Latin square M_i to be *of type* M_i, $i = 1,2$. As in [6], we partition the multiple points of an arrangement into *mixed* points, if they are points in χ (see Definition 1) and *non-mixed* points, otherwise.

By [16, Sect. 4], in the same notations used there, the equations of the 12 lines of a $(3,4)$-net $\mathcal{A} = \mathcal{A}_1 \cup \mathcal{A}_2 \cup \mathcal{A}_3$ corresponding to the Latin square M_i are $L_1 = (y)$, $L_2 = (a_1x + y + z)$, $L_3 = (a_2x + b_1y + z)$, $L_4 = (a_3x + z)$, $L_5 = (x)$, $L_6 = (x + b_2y + z)$, $L_7 = (a_4x + b_3y + z)$, $L_8 = (b_4y + z)$, $L_9 = (ax - bc^{2-i}y)$, $L_{10} = (x + y + z)$, $L_{11} = (a_4x + b_1y + z)$, $L_{12} = (z)$, where $a_1 = \frac{a(b-1)}{bc}$, $a_2 = \frac{ab}{abc+ab-a-bc}$, $a_3 = a$, $a_4 = \frac{a^2(b-1)}{abc+ab-a-bc}$, $b_1 = \frac{b^2(a-1)c}{abc+ab-a-bc}$, $b_2 = \frac{bc(a-1)}{a}$, $b_3 = \frac{abc}{abc+ab-a-bc}$, $b_4 = b$ for M_1 type arrangements, respectively $a_1 = \frac{a(b-c)}{bc}$, $a_2 = \frac{abc}{abc+ab-bc-ac}$, $a_3 = a$, $a_4 = \frac{a^2(b-c)}{abc+ab-bc-ac}$, $b_1 = \frac{b^2(a-c)}{abc+ab-bc-ac}$, $b_2 = \frac{b(a-c)}{ac}$, $b_3 = \frac{abc}{abc+ab-bc-ac}$, $b_4 = b$ for M_2 type arrangements and $\mathcal{A}_1 = \{L_1, \dots, L_4\}$, $\mathcal{A}_2 = \{L_5, \dots, L_8\}$, $\mathcal{A}_3 = \{L_9, \dots, L_{12}\}$, $(a,b,c) \in \mathbb{C}^3$.

Obviously, there are at most three quadruple points in $\mathcal{A}$, and all of them must be non-mixed points.

We denote by $\mathsf{m}_i(\mathcal{A}_j), i = \overline{1,4}$, the 3×3 minors that encode the intersections of triplets of lines inside the subarrangement $\mathcal{A}_j$. $\mathsf{m}_1(\mathcal{A}_1)$ is given by the coefficients of the equations of the lines (L_1, L_2, L_3), and $\mathsf{m}_2(\mathcal{A}_1), \mathsf{m}_3(\mathcal{A}_1), \mathsf{m}_4(\mathcal{A}_1)$ are associated in the same manner to the triplets of lines $(L_1, L_2, L_4), (L_1, L_3, L_4)$, respectively (L_2, L_3, L_4). The 3×3 minors associated with the other two subarrangements are defined accordingly.

Remark 1. An important observation used in the proof of [6, Thm 2.5] was that the cancellation of any two minors associated with distinct subarrangements $\mathcal{A}_i, \mathcal{A}_{j\neq i}$ implies the cancellation of a minor in the third subarrangement $\mathcal{A}_{k\neq i\neq j}$.

For instance, if $\mathsf{m}_1(\mathcal{A}_1) = 0$, then $c = -\frac{a(b-1)^2}{b(ab-a-2b+1)}$ and one gets $\mathsf{m}_1(\mathcal{A}_2) = -\frac{(b-1)(a+b-1)(ab-a-b)}{b(ab-a-2b+1)}$, $\mathsf{m}_2(\mathcal{A}_2) = -\frac{2ab^2-3ab+a-3b^2+3b-1}{ab-a-2b+1}$, $\mathsf{m}_3(\mathcal{A}_2) = \frac{-b^2+ab-a}{b}$, $\mathsf{m}_4(\mathcal{A}_2) = -\frac{(2ab-a-b)(ab-a-b)}{b}$, $\mathsf{m}_1(\mathcal{A}_3) = -\frac{(ab-a-b)(2ab^2-3ab+a-3b^2+3b-1)a}{b(ab-a-2b+1)}$, $\mathsf{m}_2(\mathcal{A}_3) = \frac{a(-b^2+ab-a)}{ab-a-2b+1}$, $\mathsf{m}_3(\mathcal{A}_3) = \frac{a(b-1)(2ab-a-b)}{b}$ and $\mathsf{m}_4(\mathcal{A}_3) = \frac{(a+b-1)(ab-a-b)}{b}$.

Lemma 1. *Let $\mathcal{A}$ be of type M_1. Then the existence of two quadruple points in $\mathcal{A}$ implies the existence of a third one.*

Proof. The existence of a quadruple point inside the subarrangement $\mathcal{A}_1$ amounts to the restrictions $a = \frac{1+c}{2c}, b = \frac{1}{1-c}$, while the existence of a quadruple point inside $\mathcal{A}_2$, respectively $\mathcal{A}_3$ translate into $a = \frac{c}{c-1}, b = \frac{c+1}{2}$, respectively $a = \frac{1-c}{2}, b = \frac{c-1}{2c}$. The simultaneous existence of any two quadruple points implies $c = \pm i$, and the conclusion of the lemma follows.

Remark 2. The two arrangements satisfying the hypothesis of the above lemma, corresponding to $c = \pm i$, are lattice isomorphic to the Ceva type arrangement of equation $(x^4 - y^4)(y^4 - z^4)(z^4 - x^4)$ [18, Proposition 3.3].

Lemma 2. *Let $\mathcal{A}$ be of type M_1. If $\mathcal{A}$ contains exactly one quadruple point, then all the other non-mixed points are double points.*

Proof. Assume the quadruple point sits inside a subarrangement $\mathcal{A}_i$, and assume moreover that another subarrangement $\mathcal{A}_{j\neq i}$ contains a triple non-mixed point. Then the minors $\mathsf{m}_*(\mathcal{A}_i) = 0$, $* = \overline{1,4}$ and $\mathsf{m}_k(\mathcal{A}_j) = 0$, for some $k \in \overline{1,4}$. By the computations from the proof of Theorem [6, Thm 2.5] (see also Remark 1) this implies the cancellations of all minors corresponding to the third subarrangement $\mathsf{m}_*(\mathcal{A}_l)$, $l \neq i \neq j$, hence the existence of another quadruple point, contradiction.

Lemma 3. *An arrangement of type M_2 cannot have quadruple points.*

Proof. In the notations from the beginning of $\mathcal{A}_1$ contains a quadruple point iff $a = 1, b = \frac{c}{1-c}$. But this implies $a_1 = 1$, hence $L_2 \equiv L_{10}$, contradiction!

The case when $\mathcal{A}_2$ contains a quadruple point is symmetric to this one, and translates into $b = 1, a = \frac{c}{1-c}$. But this implies $b_2 = 1$, hence $L_6 \equiv L_{10}$, again contradiction.

Finally, the system of equations that amount to the existence of a quadruple point inside the subarrangement $\mathcal{A}_3$ has no solution.

Theorem 2. *Let $\mathcal{A}$ be a $(3,4)$-net with only double and triple points. Then, if $\mathcal{A}$ is of a given type $M_i, i = 1, 2$, the isotopy type of $\mathcal{A}$ is classified by the number of triple points of the arrangement.*

Proof. We make a discussion by the (class of the) Latin square describing the net structure of $\mathcal{A}$.

I. Assume $\mathcal{A}$ is associated with the Latin square M_1, and has exactly 19 triple points. We know from [6, Thm 2.5] that there is a unique lattice isomorphism class of $(3,4)$-nets with only double and triple points satisfying these conditions. According to the proof of [6, Thm 2.5], we may assume that the non-mixed triple points are the intersections of (L_1, L_2, L_3), (L_5, L_6, L_7), respectively (L_{10}, L_{11}, L_{12}). Now we may use Urzua's procedure to describe the realizability space of this theoretic lattice. Since the existence of two non-mixed triple points implies the existence of a third one, this essentially adds to the context described in two more equations, $a_1 = a_2$ and $b_2 = b_3$, corresponding to the cancellation of the minors $\mathsf{m}_1(\mathcal{A}_1)$ and $\mathsf{m}_1(\mathcal{A}_2)$.

Recall that if $\mathsf{m}_1(\mathcal{A}_1) = 0$, then $\mathsf{m}_1(\mathcal{A}_2) = -\frac{(b-1)(a+b-1)(ab-a-b)}{b(ab-a-2b+1)}$ (Remark 1). Then the cancellation of $\mathsf{m}_1(\mathcal{A}_2)$ implies $b + a - 1 = 0$ or $ab - a - b = 0$, and the second equality, by 1 leads to the existence of quadruple points in $\mathcal{A}_2$ and $\mathcal{A}_3$, hence by Lemma 1, is one of the two arrangements from Remark 2.

We cut out the points $b - 1 = 0$, since this equality implies $a_1 = a_4 = 0, b_1 = 1$, hence $L_2 \equiv L_{11}$, contradiction.

The remaining equality, $b + a - 1 = 0$, describes a connected 1-dimensional (over $\mathbb{C}$) open set: $a = 1 - b, c = -(\frac{b-1}{b})^3$. This completes the description of the realization space for $\mathcal{A}$ of type M_1, having 19 triple points.

According to [6, Thm 2.5], arrangements $\mathcal{A}$ of type M_1 with only double and triple points, having 17 triple points, are lattice isomorphic. To apply Urzua's procedure to a theoretic lattice that contains one triple non-mixed point apart from the 16 mixed ones encoded by the Latin square M_1, means to describe the realization space of all arrangements of this type having either at least 17 triple points, or at least one quadruple point. The only additional equation one needs to add to is $a_1 = a_2$ (corresponds to the cancellation of the minor $\mathsf{m}_1(\mathcal{A}_1)$). This describes a connected 2-dimensional open set (over $\mathbb{C}$), so the subsets one may need to subtract from this one (the realization space for arrangements with quadruple points, the realization space for arrangements with 19 triple points, other isolated points), which are at most 1-dimensional, do not disconnect the space.

Finally, consider the realization space for arrangements $\mathcal{A}$ of type M_1 with only double and triple points, having exactly 16 triple points, which are lattice isomorphic by [6, Thm 2.5].

We know that the realization space for the part of the lattice corresponding to the 16 mixed triple points, the net structure, is a connected, open 3-dimensional (over $\mathbb{C}$) set [16, Sect. 4]. Then the realization space for arrangements $\mathcal{A}$ of type M_1 with only double and triple points, having exactly 16 triple points is connected, since it can be obtained by subtracting from this 3-dimensional set subsets of dimension at most 2, the ones described in the previous cases.

II. Let $\mathcal{A}$ be of type M_2. As we know from Lemma 3, there are no quadruple points in a $(3,4)$-net of type M_2.

We also know that all arrangements $\mathcal{A}$ of type M_2 with 19 triple points are lattice isomorphic and the realization space for the partial intersection lattice corresponding to the 16 mixed triple points, the net structure, is a connected, open, 3-dimensional (over $\mathbb{C}$) set [16, Sect. 4].

To obtain the realization space for an abstract lattice as before, with 19 triple points, it is enough to add to results referred to in conditions of cancellation for two minors, $\mathsf{m}_i(\mathcal{A}_k), \mathsf{m}_j(\mathcal{A}_l), k \neq l$, take for instance for ease of computation $\mathsf{m}_3(\mathcal{A}_1)$ and $\mathsf{m}_3(\mathcal{A}_2)$. We obtain $b = \frac{1}{1-c}$ and $a = \frac{c}{c-1}$, so the realization space is a complex line with eventually a finite number of points removed.

Adding to the condition of cancellation for just one of the minors, and taking into account [6, Thm 2.5] and Lemma 3, gives the 2-dimensional connected realization space for M_2 type arrangements with at least 17 triple points. The connected realization space for arrangements of type M_2 with exactly 17 triple points is then obtained by subtracting the 1-dimensional parameter space that realizes arrangements of type M_2 with 19 triple points.

Finally, subtracting from the 3-dimensional realization space of the net structure the above 2-dimensional space one obtains precisely the (connected) realization space for arrangements with precisely 16 triple points.

Corollary 1. *If two $(3,4)$-nets with only double and triple points, of the same given type $M_i, i = 1,2$, have the same number of triple points, then their complements are diffeomorphic.*

Proof. The claim follows by Randell's isotopy theorem [13].

An immediate consequence is the following:

Corollary 2. *There are no Zariski pairs among arrangements of type $(3,4)$-net with only double and triple points.*

Remark 3. Let us consider the $(3,6)$-net type arrangement $\mathcal{A}$ from [5, Example 2.2] of equation $f_1 f_2 f_3 = 0$,

$$f_1(x,y,z) = x^6 - y^6 + 3cx^4yz - 3cxy^4z - c^3x^3z^3 + c^3y^3z^3,$$

$f_2(x,y,z) = f_1(y,z,x)$, $f_3(x,y,z) = f_1(z,x,y)$, where $a = \exp(2\pi i/6)$ is a primitive 6-th root of unity, $c \in \mathbb{R}$ is a large real number and $\epsilon = \exp(2\pi i/3)$ a primitive 3-th root of unity. Explicitly, $f_1(x,y,z) = (x-y)(x-\epsilon y)(x-\epsilon^2 y)(x+y-cz)(ax+a^5y+cz)(a^5x+ay+cz)$, $f_2(x,y,z) = (y-z)(y-\epsilon z)(y-\epsilon^2 z)(y+$

$z - cx)(ay + a^5z + cx)(a^5y + az + cx)$, $f_3(x, y, z) = -(x - z)(\epsilon^2 x - z)(\epsilon x - z)(x - cy + z)(ax + cy + a^5 z)(a^5 x + cy + az)$.

Each of the subarrangements $\mathcal{A}_i$ (described by $f_i = 0$, $i = \overline{1,3}$) is lattice equivalent to the braid A_3 arrangement.

By studying the mixed points in the intersection lattice of the arrangement $\mathcal{A}$, it follows that $\mathcal{A}$ is a $(3, 6)$-net of type M_9 (see for instance the classification of the main classes of Latin squares from [16]).

This example realizes the theoretic configuration described by [5, Thm 1.1 (ii), 1.3]. Taking into account [5, Remark 1.2], it makes sense to investigate the lattice isomorphism class of this arrangement, to search for possible counterexamples of combinatorial determination of the cohomology (in degree 1) of the Milnor fiber. We show that there are no such counterexamples in this class.

The existence of a maximal number of triple points (48 triple points; in particular this means there are no quadruple points) allows us to make use of the correspondence briefly recalled in the beginning of Sect. 1, to obtain the moduli space for the entire intersection lattice of the arrangement.

A Singular assisted computation shows that the realization space for this intersection lattice consists of two zero-dimensional conjugated connected components. More precisely, we compute a reduced Grobner basis of the ideal I defined by the equations of the system describing the realization space. It consists of equations as described in [16], that appear from the constraints of the net structure, and additional equations setting the non-mixed triple points. We compute the solution of the system by replacing its equations with the equations that define the Grobner basis of the ideal I with respect to a suitable choice of LEX ordering.

To conclude, in this lattice isomorphism class we have two conjugated arrangements. Their Milnor fibers are linked through the conjugation diffeomorphism, so the cohomology of the Milnor fiber is combinatorial. Another way to see that this could not be a counterexample for combinatorial determination of the cohomology of the Milnor fiber is from the recent result [12, Thm 1.2].

The absence of Zariski pairs involving these arrangements follows from [3, Thm 3.9], since the complements of these arrangements are again diffeomorphic.

Acknowledgements The authors would like to thank Professor Viviana Ene for her kind assistance in dealing with Singular and Grobner basis arguments in the last example.

References

1. Amram, M., Cohen, M., Teicher, M., Ye, F.: Moduli spaces of ten-line arrangements with double and triple points. arXiv:1306.6105 (2013)
2. Amram, M., Teicher, M., Ye, F.: Moduli spaces of arrangements of 10 projective lines with quadruple points. Adv. Appl. Math. **51**, 392–418 (2013)
3. Cohen, D., Suciu, A.: The braid monodromy of plane algebraic curves and hyperplane arrangements. Comment. Math. Helv. **72**(2), 285–315 (1997)
4. Denes, J., Keedwell, A.D.: Latin Squares and Their applications. Academic, New York (1974)
5. Dimca, A.: Monodromy of triple point line arrangements. arXiv:1107.2214 (2012)

6. Dimca, A., Ibadula, D., Macinic, D.A.: Pencil type line arrangements of low degree: classification and monodromy. Ann. Sc. Norm. Super. Pisa CI. Sci. (2014) (to appear). arXiv: 1305.5092
7. Falk, M., Yuzvinsky, S.: Multinets, resonance varieties, and pencils of plane curves. Compositio Math. **143**(4), 1069–1088 (2007)
8. Korchmaros, G., Nagy, G., Pace, N.: 3-nets realizing a group in a projective plain, Journal of Algebraic Combinatorics, **39**(4), 939–966 (2014)
9. Korchmaros, G., Nagy, G., Pace, N.: k-nets embedded in a projective plane over a field. Available at arXiv:1306.5779 (2013)
10. Libgober, A.: On combinatorial invariance of the cohomology of Milnor fiber of arrangements and Catalan equation over function field, Arrangements of Hyperplanes - Sapporo 2009, 175–187, Adv. Stud. Pure Math., 62, Math. Soc. Japan, Tokyo (2012)
11. Nazir, S., Yoshinaga, M.: On the connectivity of the realization spaces of line arrangements. Ann. Scuola Norm. Sup. Pisa **XI**(4), 921–937 (2012)
12. Papadima, S., Suciu, A.: The Milnor fibration of a hyperplane arrangement: from modular resonance to algebraic monodromy. arXiv:1401.0868 (2014)
13. Randell, R.: Lattice-isotopic arrangements are topologically isomorphic. Proc. Am. Math. Soc. **107**(2), 555–559 (1989)
14. Rybnikov, G.: On the fundamental group of the complement of a complex hyperplane arrangement. DIMACS Technical Report 94-13, pp. 33–50 (1994). Available at arXiv: math.AG/9805056.
15. Stipins, J.: Old and new examples of k-nets in P^2. arXiv:math/0701046v1 (2007)
16. Urzúa, G.: On line arrangements with applications to 3-nets. Adv. Geom. **2**, 287–310 (2012)
17. Ye, F.: Classification of moduli spaces of arrangements of 9 projective lines. arXiv:1112.4306v2. Pac. J. Math. (2013, to appear)
18. Yuzvinsky, S.: Realization of finite abelian groups by nets in P^2. Compos. Math. **140**(6), 1614–1624 (2004)

Critical Points of Master Functions and the mKdV Hierarchy of Type $A_2^{(2)}$

A. Varchenko, T. Woodruff, and D. Wright

Abstract We consider the population of critical points generated from the critical point of the master function with no variables, which is associated with the trivial representation of the affine Lie algebra $A_2^{(2)}$. The population consists of a sequence of m-parameter families of critical points, where $m = 0, 1, \ldots$. We embed such a family into the space $\mathscr{M}(A_2^{(2)})$ of Miura opers of type $A_2^{(2)}$. We show that the embedding defines a variety which is invariant with respect to all mKdV flows on $\mathscr{M}(A_2^{(2)})$, and that variety is point-wise fixed by all flows of the index big enough.

Keywords Affine Lie algebras • mKdV integrable hierarchies • Miura opers • Master functions • Critical points

MSC2010 : 82B23 (17B80, 81R10)

1 Introduction

Let $\mathfrak{g}$ be a Kac–Moody algebra with invariant scalar product $(\,,\,)$, $\mathfrak{h} \subset \mathfrak{g}$ Cartan subalgebra, $\alpha_0, \ldots, \alpha_N$ simple roots. Let $\Lambda_1, \ldots, \Lambda_n$ be dominant integral weights, $k_0, \ldots, k_N$ nonnegative integers, $k = k_0 + \cdots + k_N$.

Consider $\mathbb{C}^n$ with coordinates $z = (z_1, \ldots, z_n)$. Consider $\mathbb{C}^k$ with coordinates u collected into $N + 1$ groups, the jth group consisting of k_j variables,

$$u = (u^{(0)}, \ldots, u^{(N)}), \qquad u^{(j)} = (u_1^{(j)}, \ldots, u_{k_j}^{(j)}).$$

A. Varchenko (✉) • T. Woodruff • D. Wright
Department of Mathematics, University of North Carolina at Chapel Hill, Chapel Hill, NC 27599-3250, USA
e-mail: anv@email.unc.edu; twoodruf@live.unc.edu; wrightd@live.unc.edu

D. Ibadula, W. Veys (eds.), *Bridging Algebra, Geometry, and Topology*, Springer Proceedings in Mathematics & Statistics 96,
DOI 10.1007/978-3-319-09186-0_11

The *master function* is the multivalued function on $\mathbb{C}^k \times \mathbb{C}^n$ defined by the formula

$$\Phi(u,z) = \sum_{a<b} (\Lambda_a, \Lambda_b) \ln(z_a - z_b) - \sum_{a,i,j} (\alpha_j, \Lambda_a) \ln(u_i^{(j)} - z_a) +$$

$$+ \sum_{j<j'} \sum_{i,i'} (\alpha_j, \alpha_{j'}) \ln(u_i^{(j)} - u_{i'}^{(j')}) + \sum_{j} \sum_{i<i'} (\alpha_j, \alpha_j) \ln(u_i^{(j)} - u_{i'}^{(j)}),$$

with singularities at the places where the arguments of the logarithms are equal to zero.

Examples of master functions associated with $\mathfrak{g} = \mathfrak{sl}_2$ were considered by Stieltjes and Heine in the nineteenth century, see [Sz]. Master functions were introduced in [SV] to construct integral representations for solutions of the KZ equations, see also [V1, V2]. The critical points of master functions with respect to u-variables were used to find eigenvectors in the associated Gaudin models by the Bethe ansatz method, see [BF, RV, V3]. In important cases the algebra of functions on the critical set of master functions is closely related to Schubert calculus, see [MTV].

In [ScV, MV1] it was observed that the critical points of master functions with respect to the u-variables can be deformed and form families. Having one critical point, one can construct a family of new critical points. The family is called a population of critical points. A point of the population is a critical point of the same master function or of another master function associated with the same $\mathfrak{g}, \Lambda_1, \dots, \Lambda_n$ but with a different integer parameters $k_0, \dots, k_N$. The population is a variety isomorphic to the flag variety of the Kac–Moody algebra $\mathfrak{g}^t$ Langlands dual to $\mathfrak{g}$, see [MV1, MV2, F].

In [VW], it was discovered that the population originated from the critical point of the master function associated with the affine Lie algebra $\widehat{\mathfrak{sl}}_{N+1}$ and the parameters $n = 0,\ k_0 = \dots = k_N = 0$ is connected with the mKdV integrable hierarchy associated with $\widehat{\mathfrak{sl}}_{N+1}$. Namely, the population consists of a sequence of m-parameter families of critical points, where $m = 0, 1, \dots$. One can naturally embed such a family into the space $\mathscr{M}(\widehat{\mathfrak{sl}}_{N+1})$ of Miura opers of type $\widehat{\mathfrak{sl}}_{N+1}$ and show that the embedding defines a variety which is invariant with respect to all mKdV flows on $\mathscr{M}(\widehat{\mathfrak{sl}}_{N+1})$. Moreover, that variety is point-wise fixed by all flows of the index big enough. For $N = 1$, that result follows from the classical paper by Adler and Moser [AM]. In [VW], two proofs of the results for $\widehat{\mathfrak{sl}}_{N+1}$ are given. The first proof is a development of the ideas from [AM]. The second proof is based on studying the corresponding tau-functions and, in particular, Schur polynomials.

The problem addressed in this paper is if a similar connection between the critical points of master functions and integrable hierarchies holds for other affine Lie algebras? In this paper we give the positive answer for the twisted affine Lie algebra $A_2^{(2)}$. The proofs in this paper follow the ideas from [AM] and the first proof in [VW].

In Sects. 2 and 3, we follow the paper [DS] by Drinfled and Sokolov and review the Lie algebras of types $A_2^{(2)}$, $A_2^{(1)}$ and the associated mKdV hierarchies.

In Sect. 4 formula (10), we introduce our master functions associated with $A_2^{(2)}$,

$$\Phi(u;k_0,k_1)=2\sum_{i<i'}\ln(u_i^{(0)}-u_{i'}^{(0)})+8\sum_{i<i'}\ln(u_i^{(1)}-u_{i'}^{(1)})-4\sum_{i,i'}\ln(u_i^{(0)}-u_{i'}^{(1)}).$$

Following [MV1, MV2, VW], we describe the generation procedure of new critical points starting from a given one. We define the population of critical points generated from the critical point of the function with no variables, namely, the function corresponding to the parameters $k_0 = k_1 = 0$. That population is partitioned into complex cells $\mathbb{C}^m$ labeled by finite sequences $J = (j_1, \ldots, i_m)$, $m \geqslant 0$, of the form $(0,1,0,1,\ldots)$ or $(1,0,1,0,\ldots)$. Such sequences are called basic.

In Sects. 5, to every basic sequence J we assign a map $\mu^J : \mathbb{C}^m \to \mathcal{M}(A_2^{(2)})$ of that cell to the space $\mathcal{M}(A_2^{(2)})$ of Miura opers of type $A_2^{(2)}$. We describe some properties of that map. In Sect. 6, we formulate and prove our main result. Theorem 6.1 says that for any basic sequence, the variety $\mu^J(\mathbb{C}^m)$ is invariant with respect to all mKdV flows on $\mathcal{M}(A_2^{(2)})$ and that variety is point-wise fixed by all flows $\frac{\partial}{\partial t_r}$ with index r greater than $3m+1$.

In the next papers we plan to extend this result to arbitrary affine Lie algebras.

We thank the referee for helpful remarks and indicating misprints.

2 Kac–Moody Algebras of Types $A_2^{(2)}$ and $A_2^{(1)}$

In this section we introduce the Lie algebras $\mathfrak{g}(A_2^{(2)})$ and $\mathfrak{g}(A_2^{(1)})$, their realizations and connections. We introduce elements $\Lambda^{(2)} \in \mathfrak{g}(A_2^{(2)})$ and $\Lambda^{(1)} \in \mathfrak{g}(A_2^{(1)})$, which are used later to define the corresponding mKdV hierarchies. We prove necessary properties of elements $\Lambda^{(2)}$ and $\Lambda^{(1)}$ in Sects. 2.1.3 and 2.2.3.

In this section we follow the exposition on Lie algebras in [DS, Section 5] and adjust the exposition to our case of the Lie algebras $\mathfrak{g}(A_2^{(2)})$ and $\mathfrak{g}(A_2^{(1)})$.

2.1 *Kac–Moody Algebra of Type $A_2^{(2)}$*

2.1.1 Definition

Consider the Cartan matrix of type $A_2^{(2)}$,

$$A_2^{(2)}=\begin{pmatrix} a_{0,0} & a_{0,1}\\ a_{1,0} & a_{1,1}\end{pmatrix}=\begin{pmatrix} 2 & -1\\ -4 & 2\end{pmatrix}.$$

The diagonal matrix $D = \mathrm{diag}(d_0, d_2) = \mathrm{diag}(4, 1)$ is such that $B = DA_2^{(2)} = \begin{pmatrix} 8 & -4 \\ -4 & 2 \end{pmatrix}$ is symmetric. The Kac–Moody algebra $\mathfrak{g} = \mathfrak{g}(A_2^{(2)})$ *of type* $A_2^{(2)}$ is the Lie algebra with *canonical generators* $e_i, h_i, f_i \in \mathfrak{g}$, $i = 0, 1$, subject to the relations

$$[e_i, f_j] = \delta_{i,j} h_i, \qquad [h_i, e_j] = a_{i,j} e_j, \qquad [h_i, f_j] = -a_{i,j} f_j\,,$$
$$(\mathrm{ad}\, e_i)^{1-a_{i,j}} e_j = 0, \qquad (\mathrm{ad}\, f_i)^{1-a_{i,j}} f_j = 0, \qquad 2h_0 + h_1 = 0,$$

see this definition in [DS, Section 5]. More precisely, we have

$$[h_0, e_0] = 2e_0, \qquad [h_0, e_1] = -e_1, \qquad [h_1, e_0] = -4e_0, \qquad [h_1, e_1] = 2e_1,$$
$$[h_0, f_0] = -2f_0, \qquad [h_0, f_1] = f_1, \qquad [h_1, f_0] = 4f_0, \qquad [h_1, f_1] = -2f_1,$$
$$(\mathrm{ad}\, e_0)^2 e_1 = 0, \qquad (\mathrm{ad}\, e_1)^5 e_0 = 0, \qquad (\mathrm{ad}\, f_0)^2 f_1 = 0, \qquad (\mathrm{ad}\, f_1)^5 f_0 = 0.$$

The Lie algebra $\mathfrak{g}(A_2^{(2)})$ is graded with respect to the *standard grading*, $\deg e_i = 1, \deg f_i = -1, i = 0, 1$. Let $\mathfrak{g}(A_2^{(2)})^j = \{x \in \mathfrak{g}(A_2^{(2)}) \mid \deg x = j\}$, then $\mathfrak{g}(A_2^{(2)}) = \oplus_{j \in \mathbb{Z}} \mathfrak{g}(A_2^{(2)})^j$.

Notice that $\mathfrak{g}(A_2^{(2)})^0$ is the one-dimensional space generated by h_0, h_1. Denote $\mathfrak{h} = \mathfrak{g}(A_2^{(2)})^0$. Introduce elements α_0, α_1 of the dual space $\mathfrak{h}^*$ by the conditions $\langle \alpha_j, h_i \rangle = a_{i,j}$ for $i, j = 0, 1$. More precisely,

$$\langle \alpha_0, h_0 \rangle = 2, \qquad \langle \alpha_1, h_1 \rangle = 2, \qquad \langle \alpha_0, h_1 \rangle = -4, \qquad \langle \alpha_1, h_0 \rangle = -1.$$

2.1.2 Realizations of $\mathfrak{g}(A_2^{(2)})$

Consider the complex Lie algebra $\mathfrak{gl}_3$ with standard basis $e_{i,j}$, $i, j = 1, 2, 3$.

Let $w = e^{\pi i/3}$. Define the *Coxeter automorphism* $C : \mathfrak{sl}_3 \to \mathfrak{sl}_3$ of order 6 by the formula

$$e_{1,2} \mapsto w^{-1} e_{2,3}, \qquad e_{2,3} \mapsto w^{-1} e_{1,2}, \qquad e_{2,1} \mapsto w e_{3,2}, \qquad e_{3,2} \mapsto w e_{2,1}.$$

Denote $(\mathfrak{sl}_3)_j = \{x \in \mathfrak{sl}_3 \mid Cx = w^j x\}$. Then

$$(\mathfrak{sl}_3)_0 = \langle e_{1,1} - e_{3,3} \rangle, \quad (\mathfrak{sl}_3)_1 = \langle e_{2,1} + e_{3,2}, e_{1,3} \rangle, \quad (\mathfrak{sl}_3)_2 = \langle e_{1,2} - e_{2,3} \rangle,$$
$$(\mathfrak{sl}_3)_3 = \langle e_{1,1} - 2e_{2,2} + e_{3,3} \rangle, \quad (\mathfrak{sl}_3)_4 = \langle e_{2,1} - e_{3,2} \rangle, \quad (\mathfrak{sl}_3)_5 = \langle e_{1,2} + e_{2,3}, e_{3,1} \rangle.$$

The twisted Lie subalgebra $L(\mathfrak{sl}_3, C) \subset \mathfrak{sl}_3[\xi, \xi^{-1}]$ is the subalgebra

$$L(\mathfrak{sl}_3, C) = \oplus_{j \in \mathbb{Z}} \xi^j \otimes (\mathfrak{sl}_3)_{j \bmod 6}.$$

The isomorphism $\tau_C : \mathfrak{g}(A_2^{(2)}) \to L(\mathfrak{sl}_3, C)$ is defined by the formula

$$e_0 \mapsto \xi \otimes e_{1,3}, \qquad e_1 \mapsto \xi \otimes (e_{2,1} + e_{3,2}),$$
$$f_0 \mapsto \xi^{-1} \otimes e_{3,1}, \qquad f_1 \mapsto \xi^{-1} \otimes (2e_{1,2} + 2e_{2,3}),$$
$$h_0 \mapsto 1 \otimes (e_{1,1} - e_{3,3}), \qquad h_1 \mapsto 1 \otimes (-2e_{1,1} + 2e_{3,3}).$$

Under this isomorphism we have $\mathfrak{g}(A_2^{(2)})^j = \xi^j \otimes (\mathfrak{sl}_3)_j$.

Define the *standard automorphism* $\sigma_0 : \mathfrak{sl}_3 \to \mathfrak{sl}_3$ of order 2 by the formula

$$e_{1,2} \mapsto e_{2,3}, \qquad e_{2,3} \mapsto e_{1,2}, \qquad e_{2,1} \mapsto e_{3,2}, \qquad e_{3,2} \mapsto e_{2,1}.$$

Let $(\mathfrak{sl}_3)_{0,j} = \{x \in \mathfrak{sl}_3 \mid \sigma_0 x = (-1)^j x\}$. Then

$$(\mathfrak{sl}_3)_{0,0} = \langle e_{1,1} - e_{3,3}, e_{1,2} + e_{2,3}, e_{2,1} + e_{3,2} \rangle,$$
$$(\mathfrak{sl}_3)_{0,1} = \langle e_{1,2} - e_{2,3}, e_{2,1} - e_{3,2}, e_{1,3}, e_{3,1} \rangle.$$

The twisted Lie subalgebra $L(\mathfrak{sl}_3, \sigma_0) \subset \mathfrak{sl}_3[\lambda, \lambda^{-1}]$ is the subalgebra

$$L(\mathfrak{sl}_3, \sigma_0) = \oplus_{j \in \mathbb{Z}} \lambda^j \otimes (\mathfrak{sl}_3)_{0,\, j \bmod 2}.$$

The isomorphism $\tau_0 : \mathfrak{g}(A_2^{(2)}) \to L(\mathfrak{sl}_3, \sigma_0)$ is defined by the formula

$$e_0 \mapsto \lambda \otimes e_{1,3}, \qquad e_1 \mapsto 1 \otimes (e_{2,1} + e_{3,2}),$$
$$f_0 \mapsto \lambda^{-1} \otimes e_{3,1}, \qquad f_1 \mapsto 1 \otimes (2e_{1,2} + 2e_{2,3}),$$
$$h_0 \mapsto 1 \otimes (e_{1,1} - e_{3,3}), \qquad h_1 \mapsto 1 \otimes (-2e_{1,1} + 2e_{3,3}).$$

Define the *standard automorphism* $\sigma_1 : \mathfrak{sl}_3 \to \mathfrak{sl}_3$ of order 4,

$$e_{1,2} \mapsto i e_{2,3}, \qquad e_{2,3} \mapsto i e_{1,2}, \qquad e_{2,1} \mapsto i^{-1} e_{3,2}, \qquad e_{3,2} \mapsto i^{-1} e_{2,1},$$

where $i = \sqrt{-1}$. Let $(\mathfrak{sl}_3)_{1,j} = \{x \in \mathfrak{sl}_3 \mid \sigma_0 x = i^j x\}$. Then

$$(\mathfrak{sl}_3)_{1,0} = \langle e_{1,3}, e_{3,1}, e_{1,1} - e_{3,3} \rangle, \qquad (\mathfrak{sl}_3)_{1,1} = \langle e_{1,2} - e_{2,3}, e_{2,1} + e_{3,2} \rangle,$$
$$(\mathfrak{sl}_3)_{1,2} = \langle e_{1,1} - 2e_{2,2} + e_{3,3} \rangle, \qquad (\mathfrak{sl}_3)_{1,3} = \langle e_{2,1} - e_{3,2}, e_{1,2} + e_{2,3} \rangle.$$

The twisted Lie subalgebra $L(\mathfrak{sl}_3, \sigma_1) \subset \mathfrak{sl}_3[\mu, \mu^{-1}]$ is the subalgebra

$$L(\mathfrak{sl}_3, \sigma_1) = \oplus_{j \in \mathbb{Z}} \mu^j \otimes (\mathfrak{sl}_3)_{1,\, j \bmod 4}.$$

The isomorphism $\tau_1 : \mathfrak{g}(A_2^{(2)}) \to L(\mathfrak{sl}_3, \sigma_1)$ is defined by the formula

$$e_0 \mapsto 1 \otimes e_{1,3}, \qquad e_1 \mapsto \mu \otimes (e_{2,1} + e_{3,2}),$$
$$f_0 \mapsto 1 \otimes e_{3,1}, \qquad f_1 \mapsto \mu^{-1} \otimes (2e_{1,2} + 2e_{2,3}),$$
$$h_0 \mapsto 1 \otimes (e_{1,1} - e_{3,3}), \qquad h_1 \mapsto 1 \otimes (-2e_{1,1} + 2e_{3,3}).$$

The composition isomorphism $L(\mathfrak{sl}_3, \sigma_0) \to L(\mathfrak{sl}_3, C)$ is given by the formula $\lambda^m \otimes e_{k,l} \mapsto \xi^{3m+k-l} \otimes e_{k,l}$. The composition isomorphism $L(\mathfrak{sl}_3, \sigma_0) \to L(\mathfrak{sl}_3, \sigma_1)$ is given by the formula $\lambda^m \otimes e_{k,l} \mapsto \mu^{2m+l-k} \otimes e_{k,l}$.

Remark 1. The standard automorphisms σ_0, σ_1 correspond to the two vertices of the Dynkin diagram of type $A_2^{(2)}$, see [DS, Section 5].

2.1.3 Element $\Lambda^{(2)}$

Denote by $\Lambda = \Lambda^{(2)}$ the element $e_0 + e_1 \in \mathfrak{g}(A_2^{(2)})$. Then $\mathfrak{z}(A_2^{(2)}) = \{x \in \mathfrak{g}(A_2^{(2)}) \mid [\Lambda, x] = 0\}$ is an Abelian Lie subalgebra of $\mathfrak{g}(A_2^{(2)})$. Denote $\mathfrak{z}(A_2^{(1)})^j = \mathfrak{z}(A_2^{(2)}) \cap \mathfrak{g}(A_2^{(2)})^j$, then $\mathfrak{z}(A_2^{(2)}) = \oplus_{j\in\mathbb{Z}}\, \mathfrak{z}(A_2^{(2)})^j$. We have $\dim \mathfrak{z}(A_2^{(2)})^j = 1$ if $j = 1, 5 \bmod 6$ and $\dim \mathfrak{z}(A_2^{(2)})^j = 0$ otherwise.

If $\mathfrak{g}(A_2^{(2)})$ is realized as $L(\mathfrak{sl}_3, C)$, then the element $\xi^{6m+1} \otimes (e_{2,1} + e_{3,2} + e_{1,3})$ generates $\mathfrak{z}^{6m+1}$ and the element $\xi^{6m-1} \otimes (e_{1,2} + e_{2,3} + e_{3,1})$ generates $\mathfrak{z}^{6m-1}$.

If $\mathfrak{g}(A_2^{(2)})$ is realized as $L(\mathfrak{sl}_3, \sigma_0)$, then the element $\lambda^{2m} \otimes (e_{2,1} + e_{3,2}) + \xi^{2m+1} \otimes e_{1,3}$ generates $\mathfrak{z}(A_2^{(2)})^{6m+1}$ and the element $\lambda^{2m} \otimes (e_{2,1} + e_{3,2}) + \lambda^{2m-1} \otimes e_{3,1})$ generates $\mathfrak{z}(A_2^{(2)})^{6m-1}$.

Lemma 2.1. *For any $m \in \mathbb{Z}$, the elements*

$$(\tau_C)^{-1}(\xi^{6m+1} \otimes (e_{2,1} + e_{3,2} + e_{1,3})), \quad (\tau_0)^{-1}(\lambda^{2m} \otimes (e_{2,1} + e_{3,2}) + \xi^{2m+1} \otimes e_{1,3})$$

of $\mathfrak{z}(A_2^{(2)})^{6m+1}$ are equal. Similarly, the elements

$$(\tau_C)^{-1}(\xi^{6m-1} \otimes (e_{1,2} + e_{2,3} + e_{3,1})), \quad (\tau_0)^{-1}(\lambda^{2m} \otimes (e_{2,1} + e_{3,2}) + \lambda^{2m-1} \otimes e_{3,1})$$

of $\mathfrak{z}(A_2^{(2)})^{6m-1}$ are equal. □

Denote the elements $(\tau_C)^{-1}(\xi^{6m+1} \otimes (e_{2,1} + e_{3,2} + e_{1,3}))$, $(\tau_C)^{-1}(\xi^{6m-1} \otimes (e_{1,2} + e_{2,3} + e_{3,1}))$ of $\mathfrak{g}(A_2^{(2)})$ by Λ_{6m+1} and Λ_{6m-1}, respectively. Notice that $\Lambda_1 = e_0 + e_1 = \Lambda$.

We set $\Lambda_j = 0$ if $j \neq 1, 5 \bmod 6$.

Lemma 2.2. *Let us consider the elements* $\xi^{6m+1} \otimes (e_{2,1} + e_{3,2} + e_{1,3})$, $\xi^{6m-1} \otimes (e_{1,2} + e_{2,3} + e_{3,1})$ *as* 3×3 *matrices,*

$$A_{6m+1} = \begin{pmatrix} 0 & 0 & \xi^{6m+1} \\ \xi^{6m+1} & 0 & 0 \\ 0 & \xi^{6m+1} & 0 \end{pmatrix}, \qquad A_{6m-1} = \begin{pmatrix} 0 & \xi^{6m-1} & 0 \\ 0 & 0 & \xi^{6m-1} \\ \xi^{6m-1} & 0 & 0 \end{pmatrix},$$

respectively. Then $A_{6m+1} = (A_1)^{6m+1}$ *and* $A_{6m-1} = (A_1)^{6m-1}$.

Similarly, let us consider the elements $\lambda^{2m} \otimes (e_{2,1} + e_{3,2}) + \xi^{2m+1} \otimes e_{1,3}$, $\lambda^{2m} \otimes (e_{2,1} + e_{3,2}) + \lambda^{2m-1} \otimes e_{3,1}$ *as* 3×3 *matrices,*

$$B_{6m+1} = \begin{pmatrix} 0 & 0 & \lambda^{2m+1} \\ \lambda^{2m} & 0 & 0 \\ 0 & \lambda^{2m} & 0 \end{pmatrix}, \qquad B_{6m-1} = \begin{pmatrix} 0 & \lambda^{2m} & 0 \\ 0 & 0 & \lambda^{2m} \\ \lambda^{2m-1} & 0 & 0 \end{pmatrix},$$

respectively. Then $B_{6m+1} = (B_1)^{6m+1}$ *and* $B_{6m-1} = (B_1)^{6m-1}$. □

2.2 *Kac–Moody Algebra of Type* $A_2^{(1)}$

2.2.1 Definition

Consider the Cartan matrix of type $A_2^{(1)}$,

$$A_2^{(1)} = \begin{pmatrix} a_{0,0} & a_{0,1} & a_{0,2} \\ a_{1,0} & a_{1,1} & a_{1,2} \\ a_{2,0} & a_{2,1} & a_{2,2} \end{pmatrix} = \begin{pmatrix} 2 & -1 & -1 \\ -1 & 2 & -1 \\ -1 & -1 & 2 \end{pmatrix}.$$

The Kac–Moody algebra $\mathfrak{g}(A_2^{(1)})$ *of type* $A_2^{(1)}$ is the Lie algebra with *canonical generators* $E_i, H_i, F_i \in \mathfrak{g}(A_2^{(1)})$, $i = 0, 1, 2$, subject to the relations

$$[E_i, F_j] = \delta_{i,j} H_i,$$

$$[H_i, E_j] = a_{i,j} E_j, \qquad [H_i, F_j] = -a_{i,j} F_j,$$

$$(\operatorname{ad} E_i)^{1-a_{i,j}} E_j = 0, \qquad (\operatorname{ad} F_i)^{1-a_{i,j}} F_j = 0,$$

$$H_0 + H_1 + H_2 = 0,$$

see this definition in [DS, Section 5]. The Lie algebra $\mathfrak{g}(A_2^{(1)})$ is graded with respect to the *standard grading*, $\deg E_i = 1, \deg F_i = -1, i = 0, 1, 2$. Let $\mathfrak{g}(A_2^{(1)})^j = \{x \in \mathfrak{g}(A_2^{(1)}) \mid \deg x = j\}$, then $\mathfrak{g}(A_2^{(1)}) = \oplus_{j\in\mathbb{Z}}\, \mathfrak{g}(A_2^{(1)})^j$.

For $j = 0, 1, 2$, we denote by $\mathfrak{n}_j^- \subset \mathfrak{g}(A_2^{(1)})$ the Lie subalgebra generated by F_i, $i \in \{0, 1, 2, \}$, $i \neq j$. For example, $\mathfrak{n}_0^-$ is generated by F_1, F_2.

2.2.2 Realizations of $\mathfrak{g}(A_2^{(1)})$

Consider the Lie algebra $\mathfrak{sl}_3[\lambda, \lambda^{-1}]$. The isomorphism $\tau_0^{(0)} : \mathfrak{g}(A_2^{(1)}) \to \mathfrak{sl}_3[\lambda, \lambda^{-1}]$ is defined by the formula

$$E_0 \mapsto \lambda \otimes e_{1,3}, \qquad E_1 \mapsto 1 \otimes e_{2,1}, \qquad E_2 \mapsto 1 \otimes e_{3,2},$$
$$F_0 \mapsto \lambda^{-1} \otimes e_{3,1}, \qquad F_1 \mapsto 1 \otimes e_{1,2}, \qquad F_2 \mapsto 1 \otimes e_{2,3},$$
$$H_0 \mapsto 1 \otimes (e_{1,1} - e_{3,3}), \quad H_1 \mapsto 1 \otimes (e_{2,2} - e_{1,1}), \quad H_2 \mapsto 1 \otimes (e_{3,3} - e_{2,2}).$$

Let $\epsilon = e^{2\pi i/3}$. Define the *Coxeter automorphism of type* $A_2^{(1)}$, $C^{(1)} : \mathfrak{sl}_3 \to \mathfrak{sl}_3$ of order 3 by the formula

$$e_{1,2} \mapsto \epsilon^{-1} e_{1,2}, \qquad e_{2,3} \mapsto \epsilon^{-1} e_{2,3}, \qquad e_{2,1} \mapsto \epsilon e_{2,1}, \qquad e_{3,2} \mapsto \epsilon e_{3,2}.$$

Denote $(\mathfrak{sl}_3)_j^{(1)} = \{x \in \mathfrak{sl}_3 \mid C^{(1)} x = \epsilon^j x\}$. Then

$$(\mathfrak{sl}_3)_0^{(1)} = \langle e_{1,1} - e_{2,2}, e_{2,2} - e_{3,3}\rangle, \quad (\mathfrak{sl}_3)_1^{(1)} = \langle e_{2,1}, e_{3,2}, e_{1,3}\rangle,$$
$$(\mathfrak{sl}_3)_2^{(1)} = \langle e_{1,2}, e_{2,3}, e_{3,1}\rangle.$$

The twisted Lie subalgebra $L(\mathfrak{sl}_3, C^{(1)}) \subset \mathfrak{sl}_3[\xi, \xi^{-1}]$ is the subalgebra

$$L(\mathfrak{sl}_3, C^{(1)}) = \oplus_{j\in\mathbb{Z}}\, \xi^j \otimes (\mathfrak{sl}_3)^{(1)}_{j \bmod 3}.$$

The isomorphism $\tau_{C^{(1)}} : \mathfrak{g}(A_2^{(1)}) \to L(\mathfrak{sl}_3, C^{(1)})$ is defined by the formula

$$E_0 \mapsto \xi \otimes e_{1,3}, \qquad E_1 \mapsto \xi \otimes e_{2,1}, \qquad E_2 \mapsto \xi \otimes e_{3,2},$$
$$F_0 \mapsto \xi^{-1} \otimes e_{3,1}, \qquad F_1 \mapsto \xi^{-1} \otimes e_{1,2}, \qquad F_2 \mapsto \xi^{-1} \otimes e_{2,3},$$
$$H_0 \mapsto 1 \otimes (e_{1,1} - e_{3,3}), \quad H_1 \mapsto 1 \otimes (e_{2,2} - e_{1,1}), \quad H_2 \mapsto 1 \otimes (e_{3,3} - e_{2,2}).$$

Under this isomorphism we have $\mathfrak{g}(A_2^{(1)})^j = \xi^j \otimes (\mathfrak{sl}_3)_j^{(1)}$.

The composition isomorphism $\mathfrak{sl}_3[\lambda, \lambda^{-1}] \to L(\mathfrak{sl}_3, C^{(1)})$ is given by the formula $\lambda^m \otimes e_{k,l} \mapsto \xi^{3m+k-l} \otimes e_{k,l}$.

2.2.3 Element $\Lambda^{(1)}$

Denote by $\Lambda^{(1)}$ the element $E_0 + E_1 + E_2 \in \mathfrak{g}(A_2^{(1)})$. Then $\mathfrak{z}(A_2^{(1)}) = \{x \in \mathfrak{g}(A_2^{(1)}) \,|\, [\Lambda^{(1)}, x] = 0\}$ is an Abelian Lie subalgebra of $\mathfrak{g}(A_2^{(1)})$. Denote $\mathfrak{z}(A_2^{(1)})^j = \mathfrak{z}(A_2^{(1)}) \cap \mathfrak{g}(A_2^{(1)})^j$, then $\mathfrak{z}(A_2^{(1)}) = \oplus_{j\in\mathbb{Z}}\, \mathfrak{z}(A_2^{(1)})^j$. We have $\dim \mathfrak{z}(A_2^{(1)})^j = 1$ if $j \neq 0 \bmod 3$ and $\dim \mathfrak{z}(A_2^{(1)})^j = 0$ otherwise. If $\mathfrak{g}(A_2^{(1)})$ is realized as $\mathfrak{sl}_3[\lambda, \lambda^{-1}]$ or $L(\mathfrak{sl}_3, C^{(1)})$, then a basis of $\mathfrak{z}(A_2^{(1)})$ is formed by the matrices $(\Lambda^{(1)})^j$ with $j \neq 0 \bmod 3$.

Let $\mathfrak{g}(A_2^{(1)})$ be realized as $\mathfrak{sl}_3[\lambda, \lambda^{-1}]$. Consider $\Lambda^{(1)} = e_{2,1} + e_{3,2} + \lambda \otimes e_{1,3}$ as a 3×3 matrix depending on the parameter λ. Let $T = \sum_{j=-\infty}^{n} T_j$ be a formal series with $T_j \in \mathfrak{g}(A_2^{(1)})^j$. Denote $T^+ = \sum_{j=0}^{n} T_j,\ T^- = \sum_{j<0} T_j$.

By [DS, Lemma 3.4], we may represent T uniquely in the form $T = \sum_{j=-\infty}^{m} b_j\, (\Lambda^{(1)})^j$, $b_j \in$ Diag, where Diag $\subset \mathfrak{gl}_3$ is the space of diagonal 3×3 matrices. Denote $(T)^+_\Lambda = \sum_{j=0}^{n} b_j\, (\Lambda^{(1)})^j,\ (T)^-_\Lambda = \sum_{j<0} b_j\, (\Lambda^{(1)})^j$.

Lemma 2.3. *We have* $(T)^+_\Lambda = T^+$, $(T)^-_\Lambda = T^-$, $b_0 = T^0$.

Proof. The isomorphism $\iota : \mathfrak{sl}_3[\lambda, \lambda^{-1}] \to L(\mathfrak{sl}_3, C^{(1)})$ is given by the formula $\lambda^m \otimes e_{k,l} \mapsto \xi^{3m+l-k}$. We have $\iota(b_0) = \iota(b_0^1 e_{1,1} + b_0^2 e_{2,2} + b_0^3 e_{3,3}) = 1 \otimes (b_0^1 e_{1,1} + b_0^2 e_{2,2} + b_0^3 e_{3,3}) \in \mathfrak{g}(A_2^{(1)})^0$, $\iota(b_1 \Lambda^{(1)}) = \iota((b_1^1 e_{1,1} + b_1^2 e_{2,2} + b_1^3 e_{3,3})(e_{2,1} + e_{3,2} + \lambda e_{1,3})) = \iota(b_1^1 \lambda e_{1,3} + b_1^2 e_{2,1} + b_1^3 e_{3,2}) = \xi \otimes (b_1^1 e_{1,3} + b_1^2 e_{2,1} + b_1^3 e_{3,2}) \in \mathfrak{g}(A_2^{(1)})^1$, $\iota(b_{-1}(\Lambda^{(1)})^{-1}) = \iota((b_{-1}^1 e_{1,1} + b_{-1}^2 e_{2,2} + b_{-1}^3 e_{3,3})(e_{1,2} + e_{2,3} + \lambda^{-1} e_{3,1})) = \iota(b_{-1}^1 e_{1,2} + b_{-1}^2 e_{2,3} + b_{-1}^3 \lambda^{-1} e_{3,1}) = \xi \otimes (b_{-1}^1 e_{1,2} + b_{-1}^2 e_{2,3} + b_{-1}^3 e_{3,1}) \in \mathfrak{g}(A_2^{(1)})^{-1}$. Similarly one checks that $\iota(b_j\, (\Lambda^{(1)})^j) \in \mathfrak{g}(A_2^{(1)})^j$ for any j.

Let us consider the elements $F_0, 2F_1 + 2F_2$ as the 3×3 matrices $\lambda^{-1} e_{3,1}, 2e_{1,2} + 2e_{2,3}$, respectively.

Lemma 2.4. *Let* $g \in \mathbb{C}$. *Then*

$$e^{gF_0} = 1 + g e_{3,3} (\Lambda^{(1)})^{-1}, \tag{1}$$
$$e^{g(2F_1+2F_2)} = 1 + 2g(e_{1,1} + e_{2,2})(\Lambda^{(1)})^{-1} + 2g^2 e_{1,1} (\Lambda^{(1)})^{-2}.$$

□

Lemma 2.5. *We have* $(\Lambda^{(1)})^{-1} = e_{1,2} + e_{2,3} + \lambda^{-1} e_{3,1}$, *and*

$$e_{i+1,i+1} \Lambda^{(1)} = \Lambda^{(1)} e_{i,i}, \qquad e_{i,i} (\Lambda^{(1)})^{-1} = (\Lambda^{(1)})^{-1} e_{i+1,i+1} \tag{2}$$

for all i, *where we set* $e_{4,4} = e_{1,1}$. □

2.2.4 Lie Algebra $\mathfrak{g}(A_2^{(2)})$ as a Subalgebra of $\mathfrak{g}(A_2^{(1)})$

The map $\varrho : \mathfrak{g}(A_2^{(2)}) \to \mathfrak{g}(A_2^{(1)})$,

$$e_0 \mapsto E_0, \qquad e_1 \mapsto E_1 + E_2, \qquad f_0 \mapsto F_0, \qquad f_1 \mapsto 2F_1 + 2F_2,$$

realizes the Lie algebra $\mathfrak{g}(A_2^{(2)})$ as a subalgebra of $\mathfrak{g}(A_2^{(1)})$. This embedding preserves the standard grading and $\varrho(\Lambda^{(2)}) = \Lambda^{(1)}$. We have $\varrho(\mathfrak{z}(A_2^{(2)})) \subset \mathfrak{z}(A_2^{(1)})$.

If there is no confusion, we will consider $\mathfrak{g}(A_2^{(2)})$ as a subalgebra of $\mathfrak{g}(A_2^{(1)})$.

3 mKdV Equations

In this section we introduce the mKdV equations of type $A_2^{(2)}$, mKdV and KdV equations of type $A_2^{(1)}$ and describe relations between the equations. We follow the exposition in [DS].

3.1 *mKdV Equations of Type* $A_2^{(2)}$

Denote by $\mathscr{B}$ the space of complex-valued functions of one variable x. Given a finite dimensional vector space W, denote by $\mathscr{B}(W)$ the space of W-valued functions of x. Denote by ∂ the differential operator $\frac{d}{dx}$.

Consider the Lie algebra $\tilde{\mathfrak{g}}(A_2^{(2)})$ of formal differential operators of the form $c\partial + \sum_{i=-\infty}^{n} p_i$, $c \in \mathbb{C}$, $p_i \in \mathscr{B}(\mathfrak{g}(A_2^{(2)})^i)$. Let $U = \sum_{i<0} U_i$, $U_i \in \mathscr{B}(\mathfrak{g}(A_2^{(2)})^i)$. If $\mathscr{L} \in \tilde{\mathfrak{g}}(A_2^{(2)})$, define

$$e^{\mathrm{ad}U}(\mathscr{L}) = \mathscr{L} + [U, \mathscr{L}] + \frac{1}{2!}[U, [U, \mathscr{L}]] + \dots .$$

The operator $e^{\mathrm{ad}U}(\mathscr{L})$ belongs to $\tilde{\mathfrak{g}}(A_2^{(2)})$. The map $e^{\mathrm{ad}U}$ is an automorphism of the Lie algebra $\tilde{\mathfrak{g}}(A_2^{(2)})$. The automorphisms of this type form a group.

If elements of $\mathfrak{g}(A_2^{(2)})$ are realized as matrices depending on a parameter as in Sect. 2.1.2, then $e^{\mathrm{ad}U}(\mathscr{L}) = e^U \mathscr{L} e^{-U}$.

A *Miura oper* of type $A_2^{(2)}$ is a differential operator of the form

$$\mathscr{L} = \partial + \Lambda + V \tag{3}$$

where $\Lambda = \Lambda^{(2)} = e_0 + e_1 \in \mathfrak{g}$ and $V \in \mathscr{B}(\mathfrak{g}^0)$. Any Miura oper of type $A_2^{(2)}$ is an element of $\tilde{\mathfrak{g}}(A_2^{(2)})$. Denote by $\mathscr{M}(A_2^{(2)})$ the space of all Miura opers of type $A_2^{(2)}$.

Proposition 3.1 ([DS, Proposition 6.2]). *For any Miura oper $\mathcal{L}$ of type $A_2^{(2)}$ there exists an element $U = \sum_{i<0} U_i$, $U_i \in \mathcal{B}(\mathfrak{g}(A_2^{(2)})^i)$, such that the operator $\mathcal{L}_0 = e^{\mathrm{ad}U}(\mathcal{L})$ has the form*

$$\mathcal{L}_0 = \partial + \Lambda + H,$$

where $H = \sum_{j<0} H_j$, $H_j \in \mathcal{B}(\mathfrak{z}(A_2^{(2)})^j)$. If $U, \tilde{U}$ are two such elements, then $e^{\mathrm{ad}U} e^{-\mathrm{ad}\tilde{U}} = e^{\mathrm{ad}T}$, where $T = \sum_{j<0} T_j$, $T_j \in \mathfrak{z}(A_2^{(2)})^j$.

Let $\mathcal{L}, U$ be as in Proposition 3.1. Let $r = 1, 5 \bmod 6$. The element $\phi(\Lambda_r) = e^{-\mathrm{ad}U}(\Lambda_r)$ does not depend on the choice of U in Proposition 3.1.

The element $\phi(\Lambda_r)$ is of the form $\sum_{i=-\infty}^{n} \phi(\Lambda_r)^i$, $\phi(\Lambda_r)^i \in \mathfrak{g}(A_2^{(2)})^i$. We set $\phi(\Lambda_r)^+ = \sum_{i=0}^{n} \phi(\Lambda_r)^i$, $\phi(\Lambda_r)^- = \sum_{i<0} \phi(\Lambda_r)^i$.

Let $r \in \mathbb{Z}_{>0}$ and $r = 1, 5 \bmod 6$. The differential equation

$$\frac{\partial \mathcal{L}}{\partial t_r} = [\phi(\Lambda_r)^+, \mathcal{L}] \tag{4}$$

is called the *rth mKdV equation* of type $A_2^{(2)}$.

Equation (4) defines a vector field $\frac{\partial}{\partial t_r}$ on the space $\mathcal{M}(A_2^{(2)})$ of Miura opers of type $A_2^{(2)}$. For all r, s, the vector fields $\frac{\partial}{\partial t_r}, \frac{\partial}{\partial t_s}$ commute, see [DS, Section 6].

Lemma 3.2 ([DS]). *We have*

$$\frac{\partial \mathcal{L}}{\partial t_r} = \frac{d}{dx}\phi(\Lambda_r)^0. \tag{5}$$

See the proof of Proposition 6.8 in [DS].

3.2 mKdV Equations of Type $A_2^{(1)}$

A *Miura oper* of type $A_2^{(1)}$ is a differential operator of the form

$$\mathcal{L} = \partial + \Lambda^{(1)} + V \tag{6}$$

where $\Lambda^{(1)} = E_0 + E_1 + E_2 \in \mathfrak{g}(A_2^{(1)})$ and $V \in \mathcal{B}(\mathfrak{g}(A_2^{(1)})^0)$. Denote by $\mathcal{M}(A_2^{(1)})$ the space of all Miura opers of type $A_2^{(1)}$.

Proposition 3.3 ([DS, Proposition 6.2]). *For any Miura oper $\mathscr{L}$ of type $A_2^{(1)}$ there exists an element $U = \sum_{i<0} U_i$, $U_i \in \mathscr{B}(\mathfrak{g}(A_2^{(1)})^i)$, such that the operator $\mathscr{L}_0 = e^{\operatorname{ad} U}(\mathscr{L})$ has the form*

$$\mathscr{L}_0 = \partial + \Lambda^{(1)} + H,$$

where $H = \sum_{j<0} H_j, H_j \in \mathscr{B}(\mathfrak{z}(A_2^{(1)})^j)$. If $U, \tilde{U}$ are two such elements, then $e^{\operatorname{ad} U} e^{-\operatorname{ad} \tilde{U}} = e^{\operatorname{ad} T}$, where $T = \sum_{j<0} T_j$, $T_j \in \mathfrak{z}(A_2^{(1)})^j$.

Let $\mathscr{L}, U$ be as in Proposition 3.3. Let $r \not\equiv 0 \bmod 3$. The element $\phi((\Lambda^{(1)})^r) = e^{-\operatorname{ad} U}((\Lambda^{(1)})^r)$ does not depend on the choice of U in Proposition 3.3.

The element $\phi((\Lambda^{(1)})^r)$ is of the form $\sum_{i=-\infty}^{n} \phi((\Lambda^{(1)})^r)^i$, $\phi((\Lambda^{(1)})^r)^i \in \mathfrak{g}(A_2^{(1)})^i$. We set $\phi((\Lambda^{(1)})^r)^+ = \sum_{i=0}^{n} \phi((\Lambda^{(1)})^r)^i$, $\phi((\Lambda^{(1)})^r)^- = \sum_{i<0} \phi((\Lambda^{(1)})^r)^i$.

Let $r \in \mathbb{Z}_{>0}$ and $r \not\equiv 0 \bmod 3$. The differential equation

$$\frac{\partial \mathscr{L}}{\partial t_r} = [\phi((\Lambda^{(1)})^r)^+, \mathscr{L}] \tag{7}$$

is called the *rth mKdV equation* of type $A_2^{(1)}$.

Equation (7) defines a vector field $\frac{\partial}{\partial t_r}$ on the space $\mathscr{M}(A_2^{(1)})$ of Miura opers of type $A_2^{(1)}$. For all r, s, the vector fields $\frac{\partial}{\partial t_r}, \frac{\partial}{\partial t_s}$ commute, see [DS, Section 6].

3.3 Comparison of mKdV Equations of Types $A_2^{(2)}$ and $A_2^{(1)}$

Consider $\mathfrak{g}(A_2^{(2)})$ as a Lie subalgebra of $\mathfrak{g}(A_2^{(1)})$, see Sect. 2.2.4. Let $\mathscr{L}$ be a Miura oper of type $A_2^{(2)}$. Then $\mathscr{L}$ is a Miura oper of type $A_2^{(1)}$.

Lemma 3.4. *Let $r = 1, 5 \bmod 6$, $r > 0$. Let $\mathscr{L}^{A_2^{(2)}}(t_r)$ be the solution of the rth mKdV equation of type $A_2^{(2)}$ with initial condition $\mathscr{L}^{A_2^{(2)}}(0) = \mathscr{L}$. Let $\mathscr{L}^{A_2^{(1)}}(t_r)$ be the solution of the rth mKdV equation of type $A_2^{(1)}$ with initial condition $\mathscr{L}^{A_2^{(1)}}(0) = \mathscr{L}$. Then $\mathscr{L}^{A_2^{(2)}}(t_r) = \mathscr{L}^{A_2^{(1)}}(t_r)$ for all values of t_r.* □

Proof. The element U in Proposition 3.1 which is used to construct the mKdV equation of type $A_2^{(2)}$ can be used also to construct the mKdV equation of type $A_2^{(1)}$.

3.4 KdV Equations of Type $A_2^{(1)}$

Let $\mathscr{B}((\partial^{-1}))$ be the algebra of formal pseudodifferential operators of the form $a = \sum_{i\in\mathbb{Z}} a_i\partial^i$, with $a_i \in \mathscr{B}$ and finitely many terms with $i > 0$. The relations in this algebra are

$$\partial^k u - u\partial^k = \sum_{i=1}^{\infty} k(k-1)\dots(k-i+1)\frac{d^i u}{dx^i}\partial^{k-i}$$

for any $k \in \mathbb{Z}$ and $u \in \mathscr{B}$. For $a = \sum_{i\in\mathbb{Z}} a_i\partial^i \in \mathscr{B}((\partial^{-1}))$, define $a^+ = \sum_{i\geqslant 0} a_i\partial^i$.

Denote $\mathscr{B}[\partial] \subset \mathscr{B}((\partial^{-1}))$ the subalgebra of differential operators $a = \sum_{i=0}^{m} a_i\partial^i$ with $m \in \mathbb{Z}_{\geqslant 0}$. Denote $\mathfrak{d} \subset \mathscr{B}[\partial]$ the affine subspace of the differential operators of the form $L = \partial^3 + u_1\partial + u_0$.

For $L \in \mathfrak{d}$, there exists a unique $L^{\frac{1}{3}} = \partial + \sum_{i\leqslant 0} a_i\partial^i \in \mathscr{B}((\partial^{-1}))$ such that $(L^{\frac{1}{3}})^3 = L$. For $r \in \mathbb{N}$, we have $L^{\frac{r}{3}} = \partial^r + \sum_{i=-\infty}^{r-1} b_i\partial^i$, $b_i \in \mathscr{B}$. We set $(L^{\frac{r}{3}})^+ = \partial^r + \sum_{i=0}^{r-1} b_i\partial^i$.

For $r \in \mathbb{Z}_{>0}$, the differential equation

$$\frac{\partial L}{\partial t_r} = [L, (L^{\frac{r}{3}})^+] \tag{8}$$

is called the *rth KdV equation* of type $A_2^{(1)}$.

Equation (8) defines flows $\frac{\partial}{\partial t_r}$ on the space $\mathfrak{d}$. For all r, s the flows $\frac{\partial}{\partial t_r}$ and $\frac{\partial}{\partial t_s}$ commute, see [DS].

3.5 Miura Maps

Let $\mathscr{L} = \partial + \Lambda^{(1)} + V$ be a Miura oper of type $A_2^{(1)}$ with $V = \sum_{k=1}^{3} v_k e_{k,k}$, $\sum_{k=1}^{3} v_k = 0$. For $i = 0, 1, 2$, define the scalar differential operator $L_i = \partial^3 + u_{1,i}\partial + u_{0,i} \in \mathfrak{d}$ by the formula

$$L_0 = (\partial - v_3)(\partial - v_2)(\partial - v_1), \quad L_1 = (\partial - v_1)(\partial - v_3)(\partial - v_2), \tag{9}$$
$$L_2 = (\partial - v_2)(\partial - v_1)(\partial - v_3).$$

Theorem 3.5 ([DS, Proposition 3.18]). *Let a Miura oper $\mathscr{L}$ satisfy the mKdV equation* (7) *for some j. Then for every $i = 0, 1, 2$, the differential operator L_i satisfies the KdV equation* (8).

We define the *ith Miura map* by the formula

$$\mathfrak{m}_i : \mathscr{M}(A_2^{(1)}) \to \mathfrak{d}, \quad \mathscr{L} \mapsto L_i,$$

see (9).

For $i = 0, 1, 2$, an *i-oper* is a differential operator of the form

$$\mathscr{L} = \partial + \Lambda^{(1)} + V + W$$

with $V \in \mathscr{B}(\mathfrak{g}(A_2^{(1)})^0)$ and $W \in \mathscr{B}(\mathfrak{n}_i^-)$. For $w \in \mathscr{B}(\mathfrak{n}_i^-)$ and an i-oper $\mathscr{L}$, the differential operator $e^{\operatorname{ad} w}(\mathscr{L})$ is an i-oper. The i-opers $\mathscr{L}$ and $e^{\operatorname{ad} w}(\mathscr{L})$ are called *i-gauge equivalent.* A Miura oper is an i-oper for any i.

Theorem 3.6 ([DS, Proposition 3.10]). *If Miura opers $\mathscr{L}$ and $\tilde{\mathscr{L}}$ are i-gauge equivalent, then $\mathfrak{m}_i(\mathscr{L}) = \mathfrak{m}_i(\tilde{\mathscr{L}})$.*

4 Critical Points of Master Functions and Generation of Pairs of Polynomials

In this section we collect necessary constructions and facts from [MV1, MV2, VW] and apply them to our situation. For functions $f(x), g(x)$, we denote

$$\operatorname{Wr}(f, g) = f(x)g'(x) - f'(x)g(x),$$

the Wronskian determinant.

4.1 Master Function

Choose a pair of nonnegative integers $\boldsymbol{k} = (k_0, k_1)$. Consider variables $u = (u_i^{(j)})$, where $j = 0, 1$, and $i = 1, \ldots, k_j$. The *master function* $\Phi(u; \boldsymbol{k})$ is defined by the formula

$$\Phi(u, \boldsymbol{k}) = 2\sum_{i<i'} \ln(u_i^{(0)} - u_{i'}^{(0)}) + 8\sum_{i<i'} \ln(u_i^{(1)} - u_{i'}^{(1)}) - 4\sum_{i,i'} \ln(u_i^{(0)} - u_{i'}^{(1)}). \quad (10)$$

The product of symmetric groups $\Sigma_{\boldsymbol{k}} = \Sigma_{k_0} \times \Sigma_{k_1}$ acts on the set of variables by permuting the coordinates with the same upper index. The function Φ is symmetric with respect to the $\Sigma_{\boldsymbol{k}}$-action.

A point u is a *critical point* if $d\Phi = 0$ at u. In other words, u is a critical point if

$$\sum_{i' \neq i} \frac{2}{u_i^{(0)} - u_{i'}^{(0)}} - \sum_{i'=1}^{k_1} \frac{4}{u_i^{(0)} - u_{i'}^{(1)}} = 0, \qquad i = 1, \ldots, k_0,$$

$$\sum_{i' \neq i} \frac{8}{u_i^{(1)} - u_{i'}^{(1)}} - \sum_{i'=1}^{k_0} \frac{4}{u_i^{(1)} - u_{i'}^{(0)}} = 0, \qquad i = 1, \ldots, k_1. \quad (11)$$

The critical set is $\Sigma_{\boldsymbol{k}}$-invariant. All orbits have the same cardinality $k_0!k_1!$. We do not make distinction between critical points in the same orbit.

Remark 2. The master functions $\Phi(u, \boldsymbol{k})$ for all vectors $\boldsymbol{k}$ are associated with the Kac–Moody algebra with Cartan matrix $\begin{pmatrix} 2 & -4 \\ -1 & 2 \end{pmatrix}$, which is Langlands dual to the Cartan matrix $A_2^{(2)}$, see [SV, MV1, MV2].

4.2 Polynomials Representing Critical Points

Let $u = (u_i^{(j)})$ be a critical point of the master function Φ. Introduce the pair of polynomials $y = (y_0(x), y_1(x))$,

$$y_j(x) \;=\; \prod_{i=1}^{k_j}(x - u_i^{(j)}). \tag{12}$$

Each polynomial is considered up to multiplication by a nonzero number. The pair defines a point in the direct product $(\mathbb{C}[x])^2$. We say that the pair $y = (y_0(x), y_1(x))$ *represents the critical point* $u = (u_i^{(j)})$.

It is convenient to think that the pair $y^{\emptyset} = (1, 1)$ of constant polynomials represents in $(\mathbb{C}[x])^2$ the critical point of the master function with no variables. This corresponds to the case $\boldsymbol{k} = (0, 0)$.

We say that a given pair $y \in (\mathbb{C}[x])^2$ is *generic* if each polynomial $y_j(x)$ of the pair has no multiple roots and the polynomials $y_0(x)$ and $y_1(x)$ have no common roots. If a pair represents a critical point, then it is generic, see (11). For example, the pair $y^{\emptyset}$ is generic.

4.3 Elementary Generation

The pair is called *fertile* if there exist polynomials $\tilde{y}_0, \tilde{y}_1 \in \mathbb{C}[x]$ such that the following two equations are satisfied,

$$\mathrm{Wr}(y_0, \tilde{y}_0) = y_1^4, \qquad \mathrm{Wr}(y_1, \tilde{y}_1) = y_0. \tag{13}$$

These equations can be written as

$$\mathrm{Wr}(y_j, \tilde{y}_j) = \prod_{i \neq j} y_i^{-a_{i,j}}, \qquad j = 0, 1. \tag{14}$$

For example, the pair $y^{\emptyset}$ is fertile and $\tilde{y}_0 = x + c_1, \tilde{y}_1 = x + c_2$, where c_1, c_2 are arbitrary numbers.

Assume that a pair of polynomials $y = (y_0, y_1)$ is fertile. Equation $\mathrm{Wr}(y_0, \tilde{y}_0) = y_1^4$ is a first order inhomogeneous differential equation with respect to $\tilde{y}_0$. Its solutions are

$$\tilde{y}_0 = y_0 \int \frac{y_1^4}{y_0^2} dx + c y_0, \tag{15}$$

where c is any number. The pairs

$$y^{(0)}(x,c) = (\tilde{y}_0(x,c), y_1(x)) \in (\mathbb{C}[x])^2 \tag{16}$$

form a one-parameter family. This family is called the *generation of pairs from y in the 0th direction*. A pair of this family is called an *immediate descendant* of y in the 0th direction.

Similarly, equation $\mathrm{Wr}(y_1, \tilde{y}_1) = y_0$ is a first order inhomogeneous differential equation with respect to $\tilde{y}_1$. Its solutions are

$$\tilde{y}_1 = y_1 \int \frac{y_0}{y_1^2} dx + c y_1, \tag{17}$$

where c is any number. The pairs

$$y^{(1)}(x,c) = (y_0(x), \tilde{y}_1(x,c)) \quad \in \quad (\mathbb{C}[x])^2 \tag{18}$$

form a one-parameter family. This family is called the *generation of pairs from y in the first direction*. A pair of this family is called an *immediate descendant* of y in the first direction.

Theorem 4.1 ([MV1]). *(i) A generic pair $y = (y_0, y_1)$ represents a critical point of a master function if and only if y is fertile.*

(ii) If y represents a critical point, then for any $c \in \mathbb{C}$ the pairs $y^{(0)}(x,c)$ and $y^{(1)}(x,c)$ are fertile.

(iii) If y is generic and fertile, then for almost all values of the parameter $c \in \mathbb{C}$ both pairs $y^{(0)}(x,c)$ and $y^{(1)}(x,c)$ are generic. The exceptions form a finite set in $\mathbb{C}$.

(iv) Assume that a sequence $f_i = (f_{0,i}(x), f_{1,i}(x)), i = 1, 2, \ldots,$ of fertile pairs has a limit $f_\infty = (f_{0,\infty}(x), f_{1,\infty}(x))$ in $(\mathbb{C}[x])^2$ as i tends to infinity.

(a) Then the limiting pair f_∞ is fertile.

(b) For $j = 0, 1$, let $f_\infty^{(j)}$ be an immediate descendant of f_∞. Then for $j = 0, 1$, there exist immediate descendants $f_i^{(j)}$ of f_i such that $f_\infty^{(j)}$ is the limit of $f_i^{(j)}$ as i tends to infinity.

4.4 Degree Increasing Generation

Let $y = (y_0, y_1)$ be a generic fertile pair of polynomials. For $j = 0, 1$, define $k_j = \deg y_j$.

The polynomial $\tilde{y}_0$ in (13) is of degree k_0 or $\tilde{k}_0 = 4k_1 + 1 - k_0$. We say that the generation $(y_0, y_1) \to (\tilde{y}_0, y_1)$ is *degree increasing* in the 0th direction if $\tilde{k}_0 > k_0$. In that case $\deg \tilde{y}_0 = \tilde{k}_0$ for all c.

If the generation is degree increasing in the 0th direction we normalize family (16) and construct a map $Y_{y,0} : \mathbb{C} \to (\mathbb{C}[x])^2$ as follows. First we multiply the polynomials y_0, y_1 by numbers to make them monic. Then we choose the monic polynomial $y_{0,0}$ satisfying the equation $\mathrm{Wr}(y_0, y_{0,0}) = a\, y_1^4$, for some $a \in \mathbb{C}^\times$, and such that the coefficient of x^{k_0} in $y_{0,0}$ equals zero. Such $y_{0,0}$ exists and is unique. Set

$$\tilde{y}_0(x, c) = y_{0,0}(x) + c y_0(x) \tag{19}$$

and define

$$Y_{y,0} : \mathbb{C} \to (\mathbb{C}[x])^2, \qquad c \mapsto y^{(0)}(x, c) = (\tilde{y}_0(x, c), y_1(x)). \tag{20}$$

Both polynomials of the pair $y^{(0)}(x, c)$ are monic.

The polynomial $\tilde{y}_1$ in (13) is of degree k_1 or $\tilde{k}_1 = k_0 + 1 - k_1$. We say that the generation $(y_0, y_1) \to (y_0, \tilde{y}_1)$ is *degree increasing* in the first direction if $\tilde{k}_1 > k_1$. In that case $\deg \tilde{y}_1 = \tilde{k}_1$ for all c.

If the generation is degree increasing in the first direction we normalize family (18) and construct a map $Y_{y,1} : \mathbb{C} \to (\mathbb{C}[x])^2$ as follows. First we multiply the polynomials y_0, y_1 by numbers to make them monic. Then we choose the monic polynomial $y_{1,0}$ satisfying the equation $\mathrm{Wr}(y_1, y_{1,0}) = a\, y_0$, for some $a \in \mathbb{C}^\times$, and such that the coefficient of x^{k_1} in $y_{1,0}$ equals zero. Such $y_{1,0}$ exists and is unique. Set

$$\tilde{y}_1(x, c) = y_{1,0}(x) + c y_1(x) \tag{21}$$

and define

$$Y_{y,1} : \mathbb{C} \to (\mathbb{C}[x])^2, \qquad c \mapsto y^{(1)}(x, c) = (y_0(x), \tilde{y}_1(x, c)). \tag{22}$$

Both polynomials of the pair $y^{(1)}(x, c)$ are monic.

4.5 Degree-Transformations and Generation of Vectors of Integers

The degree-transformations

$$\boldsymbol{k} = (k_0, k_1) \mapsto \boldsymbol{k}^{(0)} = (4k_1 + 1 - k_0, k_1), \tag{23}$$
$$\boldsymbol{k} = (k_0, k_1) \mapsto \boldsymbol{k}^{(1)} = (k_0, k_0 + 1 - k_1)$$

correspond to the shifted action of reflections $w_0, w_1 \in W_{A_2^{(2)}}$, where $W_{A_2^{(2)}}$ is the Weyl group of type $A_2^{(2)}$ and w_0, w_1 are its standard generators, see Lemma 3.11 in [MV1] for more detail.

We take formula (23) as the definition of *degree-transformations*:

$$w_0 : \boldsymbol{k} = (k_0, k_1) \mapsto \boldsymbol{k}^{(0)} = (4k_1 + 1 - k_0, k_1), \tag{24}$$
$$w_1 : \boldsymbol{k} = (k_0, k_1) \mapsto \boldsymbol{k}^{(1)} = (k_0, k_0 + 1 - k_1)$$

acting on arbitrary vectors $\boldsymbol{k} = (k_0, k_1)$.

We start with the vector $\boldsymbol{k}^{\emptyset} = (0, 0)$ and a sequence $J = (j_1, j_2, \dots, j_m)$ of integers, where $J = (0, 1, 0, 1, 0, 1, \dots)$ or $J = (1, 0, 1, 0, 1, 0, \dots)$. We apply the corresponding degree transformations to $\boldsymbol{k}^{\emptyset}$ and obtain the sequence of vectors $\boldsymbol{k}^{\emptyset}$, $\boldsymbol{k}^{(j_1)} = w_{j_1}\boldsymbol{k}^{\emptyset}$, $\boldsymbol{k}^{(j_1,j_2)} = w_{j_2}w_{j_1}\boldsymbol{k}^{\emptyset}, \dots,$

$$\boldsymbol{k}^{J} = w_{j_m} \dots w_{j_2} w_{j_1} \boldsymbol{k}^{\emptyset}. \tag{25}$$

We say that the *vector* $\boldsymbol{k}^J$ *is generated from* $(0, \dots, 0)$ *in the direction of* J.

For example, for $J = (0, 1, 0, 1, 0, 1)$ we get the sequence $(0, 0)$, $(1, 0)$, $(1, 2)$, $(8, 2)$, $(8, 7)$, $(21, 7)$, $(21, 15)$. If $J = (1, 0, 1, 0, 1, 0)$, then the sequence is $(0, 0)$, $(0, 1)$, $(5, 1)$, $(5, 5)$, $(16, 5)$, $(16, 12)$, $(33, 12)$.

We call a sequence J *degree increasing* if for every i the transformation w_{j_i} applied to $w_{j_{i-1}} \dots w_{j_1}\boldsymbol{k}^{\emptyset}$ increases the j_ith coordinate.

Lemma 4.2. *If* $J = (0, 1, 0, 1, 0, 1, \dots)$, *then after* $2n$ *steps of this procedure the degree vector is* $(3n^2 - 2n, (3n^2 + n)/2)$. *If* $J = (1, 0, 1, 0, 1, 0, \dots)$, *then after* $2n + 1$ *steps of this procedure the degree vector is* $(3n^2 + 2n, (3n^2 + 5n + 2)/2)$.

□

Corollary 4.3. *Both sequences* $(0, 1, 0, 1, 0, 1, \dots)$ *and* $(1, 0, 1, 0, 1, 0, \dots)$ *are degree increasing.* □

The sequences $(0, 1, 0, 1, 0, 1, \dots)$ and $(1, 0, 1, 0, 1, 0, \dots)$ will be called the *basic sequences*.

4.6 Multistep Generation

Let $J = (j_1, \dots, j_m)$ be a basic sequence. Starting from $y^{\emptyset} = (1,1)$ and J, we construct by induction on m a map

$$Y^J : \mathbb{C}^m \to (\mathbb{C}[x])^2.$$

If $J = \emptyset$, the map $Y^{\emptyset}$ is the map $\mathbb{C}^0 = (pt) \mapsto y^{\emptyset}$. If $m = 1$ and $J = (j_1)$, the map $Y^{(j_1)} : \mathbb{C} \to (\mathbb{C}[x])^N$ is given by one of the formulas (20) or (22) for $y = y^{\emptyset}$ and $j = j_1$. More precisely, equation $\mathrm{Wr}(y_0, \tilde{y}_0) = y_1^4$ takes the form $\mathrm{Wr}(1, \tilde{y}_0) = 1$. Then $\tilde{y}_{0,0} = x$ and

$$Y^{(0)} : \mathbb{C} \mapsto (\mathbb{C}[x])^2, \qquad c \mapsto (x + c, 1).$$

By Theorem 4.1 all pairs in the image are fertile and almost all pairs are generic (in this example all pairs are generic). Similarly, equation $\mathrm{Wr}(y_1, \tilde{y}_1) = y_0$ takes the form $\mathrm{Wr}(1, \tilde{y}_1) = 1$. Then $\tilde{y}_{1,0} = x$ and

$$Y^{(1)} : \mathbb{C} \mapsto (\mathbb{C}[x])^2, \qquad c \mapsto (1, x + c).$$

Assume that for $\tilde{J} = (j_1, \dots, j_{m-1})$, the map $Y^{\tilde{J}}$ is constructed. To obtain Y^J we apply the generation procedure in the j_mth direction to every pair of the image of $Y^{\tilde{J}}$. More precisely, if

$$Y^{\tilde{J}} : \tilde{c} = (c_1, \dots, c_{m-1}) \mapsto (y_0(x, \tilde{c}), y_1(x, \tilde{c})), \tag{26}$$

then

$$Y^J : \mathbb{C}^m \mapsto (\mathbb{C}[x])^2, \quad (\tilde{c}, c_m) \mapsto (y_{0,0}(x, \tilde{c}) + c_m y_0(x, \tilde{c}), y_1(x, \tilde{c})), \qquad \text{if } j_m = 0, \tag{27}$$
$$Y^J : \mathbb{C}^m \mapsto (\mathbb{C}[x])^2, \quad (\tilde{c}, c_m) \mapsto (y_0(x, \tilde{c}), y_{1,0}(x, \tilde{c}) + c_m y_1(x, \tilde{c})), \qquad \text{if } j_m = 1,$$

see formulas (19), (21). The map Y^J is called the *generation of pairs from $y^{\emptyset}$ in the Jth direction.*

Lemma 4.4. *All pairs in the image of Y^J are fertile and almost all pairs are generic. For any $c \in \mathbb{C}^m$ the pair $Y^J(c)$ consists of monic polynomials. The degree vector of this pair equals $\mathbf{k}^J$, see (25).* □

Lemma 4.5. *The map Y^J sends distinct points of $\mathbb{C}^m$ to distinct points of $(\mathbb{C}[x])^2$.*

Proof. The lemma is easily proved by induction on m.

Example 1. If $J = (0, 1)$, then

$$Y^{(0)}(c_1) = (x + c_1, 1), \qquad Y^{(0,1)}(c_1, c_2) = (x + c_1, (x + c_1)^2 + c_2 - c_1^2).$$

If $J = (1, 0)$, then

$$Y^{(1)}(c_1) = (1, x + c_1), \qquad Y^{(1,0)}(c_1, c_2) = ((x + c_1)^4 + c_2 - c_1^4, x + c_1).$$

The set of all pairs $(y_0, y_1) \in (\mathbb{C}[x])^2$ obtained from $y^{\emptyset} = (1, 1)$ by generations in all degree increasing directions is called the *population of pairs* generated from $y^{\emptyset}$, c.f. [MV1].

5 Critical Points of Master Functions and Miura Opers

In this section we assign a Miura oper of type $A_2^{(2)}$ to a pair of polynomials $y = (y_0, y_1)$. In Theorem 5.3 we remind the connection between the generation procedure of critical points of master functions and deformability of Miura opers. The main results of this section are Lemmas 5.7 and 5.8.

5.1 Miura Oper Associated with a Pair of Polynomials, [MV2]

Define a map

$$\mu : (\mathbb{C}[x])^2 \to \mathscr{M}(A_2^{(2)}),$$

which sends a pair $y = (y_0, y_1)$ to the Miura oper $\mathscr{L} = \partial + \Lambda^{(2)} + V$ with

$$V = \ln'\Big(\frac{y_1^2}{y_0}\Big)\, h_0\,,$$

where for a function $f(x)$ we denote $\ln'(f(x)) = f'(x)/f(x)$. We say that the Miura oper $\mu(y)$ is *associated with the pair of polynomials* y. For example,

$$\mathscr{L}^{\emptyset} = \partial + \Lambda^{(2)}$$

is associated with the pair $y^{\emptyset} = (1, 1)$.

We have

$$\langle \alpha_0, V\rangle = \ln'\Big(\frac{y_1^4}{y_0^2}\Big), \qquad \langle \alpha_1, V\rangle = \ln'\Big(\frac{y_0}{y_1^2}\Big). \tag{28}$$

Equations (28) can be written as

$$\langle \alpha_j, V\rangle = \ln'\Big(\prod_{i=0}^{1} y_i^{-a_{i,j}}\Big), \tag{29}$$

see [MV2].

5.2 Deformations of Miura Opers of Type $A_2^{(2)}$, [MV2]

Lemma 5.1 ([MV2]). *Let* $\mathcal{L} = \partial + \Lambda^{(2)} + V$ *be a Miura oper of type* $A_2^{(2)}$*. Let* $g \in \mathcal{B}$*. Let* f_j, $j \in \{0, 1\}$*, be one of canonical generators of* $\mathfrak{g}(A_2^{(2)})$*, see Sect. 2.1.1. Then*

$$e^{\text{ad}\, gf_j}(\mathcal{L}) = \partial + \Lambda^{(2)} + V - gh_j - (g' - \langle \alpha_j, V \rangle g + g^2) f_j. \tag{30}$$

Corollary 5.2 ([MV2]). *Let* $\mathcal{L} = \partial + \Lambda^{(2)} + V$ *be a Miura oper. Then* $e^{\text{ad}\, gf_j}(\mathcal{L})$ *is a Miura oper if and only if the scalar function* g *satisfies the Ricatti equation*

$$g' - \langle \alpha_j, V \rangle g + g^2 = 0\,. \tag{31}$$

Let $\mathcal{L} = \partial + \Lambda^{(2)} + V$ be a Miura oper with $V = vh_0$. Assume that the functions v is a rational functions of x. For $j \in \{0, 1\}$, we say that $\mathcal{L}$ is *deformable in the* j*th direction* if Eq. (31) has a nonzero solution g, which is a rational function.

Theorem 5.3 ([MV2]). *Let the Miura oper* $\mathcal{L} = \partial + \Lambda^{(2)} + V$ *be associated with a pair of polynomials* $y = (y_0, y_1)$*. Let* $j \in \{0, 1\}$*. Then* $\mathcal{L}$ *is deformable in the* j*th direction if and only if there exists a polynomial* $\tilde{y}_j$ *satisfying Eq.* (14)*. Moreover, in that case any nonzero rational solution* g *of the Ricatti equation* (31) *has the form* $g = \ln'(\tilde{y}_j / y_j)$ *where* $\tilde{y}_j$ *is a solution of Eq.* (14)*. If* $g = \ln'(\tilde{y}_i / y_i)$*, then the Miura oper*

$$e^{\text{ad}\, gf_i}(\mathcal{L}) = \partial + \Lambda^{(2)} + V - gh_j \tag{32}$$

is associated the pair $y^{(j)}$*, which is obtained from the pair* y *by replacing* y_j *with* $\tilde{y}_j$*.*

Proof. Write (31) as

$$g'/g + g = \ln'\Big(\prod_{j=0}^{1} y_i^{-a_{i,j}}\Big)\,. \tag{33}$$

If g is a rational function, then $g \to 0$ as $x \to \infty$ and all poles of g are simple. Moreover, the residue of g at any point is an integer. Hence $g = b'/b$ for a suitable rational function $b(x)$. Then

$$b(x) = \int \prod_{j=0}^{1} y_j(x)^{-a_{i,j}}\, dx \tag{34}$$

and Eq. (14) has a polynomial solution $\tilde{y}_j = -b(x) y_j$. Conversely if Eq. (14) has a polynomial solution $\tilde{y}_i$, then the function $b(x)$ in (34) is rational. Then $g = b'/b$ is a rational solution of Eq. (31).

Let $g = \ln' b = \ln'(\tilde{y}_i/y_i)$, where $\tilde{y}_i$ is a solution of (14). Then

$$e^{\mathrm{ad}\, g f_j}(\mathscr{L}) = \partial + \Lambda^{(2)} + V - \ln'(\tilde{y}_j/y_j)h_j$$

and

$$\begin{aligned}\langle \alpha_k, V\rangle - \langle \alpha_k, h_j\rangle \ln'(\tilde{y}_j/y_j) &= \ln'\Big(\prod_{i=0}^{1} y_i^{-a_{i,k}}\Big) - a_{j,k}\ln'(\tilde{y}_j/y_j)\\ &= \ln'\Big(\tilde{y}_j^{-a_{j,k}} \prod_{i=0,\, i\neq j}^{1} y_i^{-a_{i,k}}\Big).\end{aligned}$$

Note that if Eq. (31) has one nonzero rational solution $g = b'/b$ with rational $b(x)$, then other nonzero (rational) solutions have the form $g = b'/(b + \mathrm{const})$.

5.3 Miura Opers Associated with the Generation Procedure

Let $J = (j_1, \dots, j_m)$ be a basic sequence, see Sect. 4.5. Let $Y^J : \mathbb{C}^m \to (\mathbb{C}[x])^2$ be the generation of pairs from $y^{\emptyset}$ in the Jth direction. We define the associated family of Miura opers by the formula:

$$\mu^J : \mathbb{C}^m \to \mathscr{M}(A_2^{(2)}), \qquad c \mapsto \mu(Y^J(c)).$$

The map μ^J is called the *generation of Miura opers from $\mathscr{L}^{\emptyset}$ in the Jth direction.*

For $\ell = 1, \dots, m$, denote $J_\ell = (j_1, \dots, j_\ell)$ the beginning ℓ-interval of the sequence J. Consider the associated map $Y^{J_\ell} : \mathbb{C}^\ell \to (\mathbb{C}[x])^2$. Denote

$$Y^{J_\ell}(c_1, \dots, c_\ell) = (y_0(x, c_1, \dots, c_\ell, \ell), y_1(x, c_1, \dots, c_\ell, \ell)).$$

Introduce

$$\begin{aligned} g_1(x, c_1, \dots, c_m) &= \ln'(y_{j_1}(x, c_1, 1)), \\ g_\ell(x, c_1, \dots, c_m) &= \ln'(y_{j_\ell}(x, c_1 \dots, c_\ell, \ell)) - \ln'(y_{j_\ell}(x, c_1, \dots, c_{\ell-1}, \ell - 1)), \end{aligned} \tag{35}$$

for $\ell = 2, \dots, m$. For $c \in \mathbb{C}^m$, define the element $U^J(c) = \sum_{i<0}(U^J(c))_i$, $(U^J(c))_i \in \mathscr{B}(\mathfrak{g}^i)$, depending on $c \in \mathbb{C}^m$, by the formula

$$e^{-\mathrm{ad} U^J(c)} = e^{\mathrm{ad} g_m(x,c) f_{j_m}} \cdots e^{\mathrm{ad} g_1(x,c) f_{j_1}}. \tag{36}$$

Lemma 5.4. *For $c \in \mathbb{C}^m$, we have*

$$\mu^J(c) = e^{-\operatorname{ad} U^J(c)}(\mathscr{L}^\emptyset) \tag{37}$$

and

$$\mu^J(c) = \partial + \Lambda^{(2)} - \sum_{\ell=1}^{m} g_\ell(x,c) h_{j_\ell}. \tag{38}$$

Proof. The lemma follows from Theorem 5.3.

Corollary 5.5. *Let $r > 0$ and $r = 1, 5 \bmod 6$. Let $c \in \mathbb{C}^m$. Let $\frac{\partial}{\partial t_r}\big|_{\mu^J(c)}$ be the value at $\mu^J(c)$ of the vector field of the rth mKdV flow on the space $\mathscr{M}(A_2^{(2)})$, see (4). Then*

$$\frac{\partial}{\partial t_r}\Big|_{\mu^J(c)} = -\frac{\partial}{\partial x}(e^{-\operatorname{ad} U^J(c)}(\Lambda_r))^0. \tag{39}$$

Proof. The corollary follows from (4) and (37).

We have a natural embedding $\mathscr{M}(A_2^{(2)}) \hookrightarrow \mathscr{M}(A_2^{(1)})$. Let $\mathfrak{m}_i : \mathscr{M}(A_2^{(1)}) \to \mathrm{d}$, $\mathscr{L} \mapsto L_i$, be the Miura maps defined in Sect. 3.5 for $i = 0, 1$. Below we consider the composition of the embedding and a Miura map.

Denote $\tilde{J} = (j_1, \dots, j_{m-1})$. Consider the associated family $\mu^{\tilde{J}} : \mathbb{C}^{m-1} \to \mathscr{M}(A_2^{(2)})$. Denote $\tilde{c} = (c_1, \dots, c_{m-1})$.

Lemma 5.6. *For all $(\tilde{c}, c_m) \in \mathbb{C}^m$, we have $\mathfrak{m}_1 \circ \mu^J(\tilde{c}, c_m) = \mathfrak{m}_1 \circ \mu^{\tilde{J}}(\tilde{c})$ if $j_m = 0$ and we have $\mathfrak{m}_0 \circ \mu^J(\tilde{c}, c_m) = \mathfrak{m}_0 \circ \mu^{\tilde{J}}(\tilde{c})$ if $j_m = 1$.*

Proof. The lemma follows from formula (37) and Theorem 3.6.

Lemma 5.7. *If $j_m = 0$, then*

$$\frac{\partial \mu^J}{\partial c_m}(\tilde{c}, c_m) = -a\, \frac{y_1(x, \tilde{c}, m-1)^4}{y_0(x, \tilde{c}, c_m, m)^2}\, h_0 \tag{40}$$

for some $a \in \mathbb{C}^\times$. If $j_m = 1$, then

$$\frac{\partial \mu^J}{\partial c_m}(\tilde{c}, c_m) = -a\, \frac{y_0(x, \tilde{c}, m-1)}{y_1(x, \tilde{c}, c_m, m)^2}\, h_1 \tag{41}$$

for some $a \in \mathbb{C}^\times$.

Proof. Let $j_m = 0$. Then $y_0(x, \tilde{c}, c_m, m) = y_{0,0}(x, \tilde{c}) + c_m y_0(x, \tilde{c}, m-1)$, where $y_{0,0}(x, \tilde{c})$ is such that

$$\mathrm{Wr}(y_0(x, \tilde{c}, m-1), y_{0,0}(x, \tilde{c})) = a\, y_1(x, \tilde{c}, m-1)^4,$$

for some $a \in \mathbb{C}^\times$, see (19). We have $g_m = \ln'(y_0(x, \tilde{c}, c_m, m)) - \ln'(y_0(x, \tilde{c}, m-1))$.

By formula (38), we have

$$\frac{\partial \mu^J}{\partial c_m}(\tilde{c}, c_m) = -\frac{\partial g_m}{\partial c_m}(\tilde{c}, c_m)\, h_0$$

and

$$\frac{\partial g_m}{\partial c_m}(\tilde{c}, c_m) = \frac{\partial}{\partial c_m}\left(\frac{y_{0,0}'(x,\tilde{c}) + c_m y_0'(x,\tilde{c}, m-1)}{y_{0,0}(x,\tilde{c}) + c_m y_0(x,\tilde{c}, m-1)}\right) =$$
$$= \frac{\mathrm{Wr}(y_{0,0}(x,\tilde{c}), y_0(x,\tilde{c}, m-1))}{(y_{0,0}(x,\tilde{c}) + c_m y_0(x,\tilde{c}, m-1))^2} = a\,\frac{y_1(x,\tilde{c}, m-1)^4}{y_0(x,\tilde{c}, c_m, m)^2}\,.$$

This proves formula (40). Formula (41) is proved similarly.

Let us describe the kernels of the differentials of the Miura maps $\mathfrak{m}_i, i = 0, 1$, restricted to Miura opers of type $A_2^{(2)}$. A Miura oper $\mathcal{L} = \partial + \Lambda + v h_0$ of type $A_2^{(2)}$ is mapped to the differential operator $(\partial + v)\partial(\partial - v) = \partial^3 - (2v' + v^2)\partial - (v'' + vv')$ by the Miura map $\mathfrak{m}_0$ and to the differential operator $(\partial - v)(\partial + v)\partial = \partial^3 + (v' - v^2)\partial$ by the Miura map $\mathfrak{m}_1$. The derivative maps are

$$d\mathfrak{m}_0 : X h_0 \mapsto -(2X' + 2vX)\partial - (X'' + vX' + v'X),$$
$$d\mathfrak{m}_1 : X h_0 \mapsto (X' - 2vX)\partial,$$

where $x \in \mathcal{B}$.

Lemma 5.8. *Assume that $\mathcal{L}$ is associated with a pair (y_0, y_1), that is, $v = \ln'\left(\frac{y_1^2}{y_0}\right)$. Then the kernel of $d\mathfrak{m}_0$ at $\mathcal{L}$ is one-dimensional and is generated by the function $\frac{y_0}{y_1^2} h_0$. Also the kernel of $d\mathfrak{m}_1$ at $\mathcal{L}$ is one-dimensional and is generated by the function $\frac{y_1^4}{y_0^2} h_0$.*

Proof. We have $X \in \ker d\mathfrak{m}_0$ if and only if $X' + vX = 0$. This implies the first statement. Similarly, $X \in \ker d\mathfrak{m}_1$ if and only if $X' - 2vX = 0$. This implies the second statement.

6 Vector Fields

In this section we formulate and prove the main result of this paper, Theorem 6.1.

6.1 Statement

Let $r > 0$ and $r = 1, 5 \mod 6$. Recall that we denote by $\frac{\partial}{\partial t_r}$ the rth mKdV vector field on the space $\mathcal{M}(A_2^{(2)})$ of Miura opers of type $A_2^{(2)}$. We also denote by $\frac{\partial}{\partial t_r}$ the rth mKdV vector field of type $A_2^{(1)}$ on the space $\mathscr{M}(A_2^{(1)})$ of Miura opers of type $A_2^{(1)}$. We have a natural embedding $\mathscr{M}(A_2^{(2)}) \hookrightarrow \mathscr{M}(A_2^{(1)})$. Under this embedding the vector $\frac{\partial}{\partial t_r}$ on $\mathscr{M}(A_2^{(2)})$ equals the vector field $\frac{\partial}{\partial t_r}$ on $\mathscr{M}(A_2^{(1)})$ restricted to $\mathscr{M}(A_2^{(2)})$, see Sect. 3.3. We also denote by $\frac{\partial}{\partial t_r}$ the rth KdV vector field on the space d of the differential operators of the form $L = \partial^3 + u_1\partial + u_0$, see Sect. 3.4.

For a Miura map $\mathfrak{m}_i : \mathscr{M} \to \mathrm{d}$, $\mathscr{L} \mapsto L_i$, denote by $d\mathfrak{m}_i$ the associated derivative map $T\mathscr{M}(A_2^{(1)}) \to T\mathrm{d}$. By Theorem 3.5 we have $d\mathfrak{m}_i : \frac{\partial}{\partial t_r}\big|_{\mathscr{L}} \mapsto \frac{\partial}{\partial t_r}\big|_{L_i}$.

Fix a basic sequence $J = (j_1, \dots, j_m)$. Consider the associated family $\mu^J : \mathbb{C}^m \to \mathscr{M}(A_2^{(2)})$ of Miura opers. For a vector field Γ on $\mathbb{C}^m$, we denote by $\frac{\partial \mu^J}{\partial \Gamma}$ the derivative of μ^J along the vector field. The derivative is well defined since $\mathscr{M}(A_2^{(2)})$ is an affine space.

Theorem 6.1. *Let $r > 0$ and $r = 1, 5$ mod 6. Then there exists a polynomial vector field Γ_r on $\mathbb{C}^m$ such that*

$$\frac{\partial}{\partial t_r}\Big|_{\mu^J(c)} = \frac{\partial \mu^J}{\partial \Gamma_r}(c) \tag{42}$$

for all $c \in \mathbb{C}^m$. If m is even and $r > 3m$, then $\frac{\partial}{\partial t_r}\big|_{\mu^J(c)} = 0$ for all $c \in \mathbb{C}^m$ and, hence, $\Gamma_r = 0$. If m is odd and $j_1 = j_m = 0$, then for $r > 3m - 2$ we have $\frac{\partial}{\partial t_r}\big|_{\mu^J(c)} = 0$ for all $c \in \mathbb{C}^m$ and, hence, $\Gamma_r = 0$. If m is odd and $j_1 = j_m = 1$, then for $r > 3m + 1$ we have $\frac{\partial}{\partial t_r}\big|_{\mu^J(c)} = 0$ for all $c \in \mathbb{C}^m$ and, hence, $\Gamma_r = 0$.

Corollary 6.2. *The family μ^J of Miura opers is invariant with respect to all mKdV flows of type $A_2^{(2)}$ and is point-wise fixed by flows with $r > 3m + 1$.*

6.2 Proof of Theorem 6.1 for $m = 1$

If $m = 1$, then $J = (0)$ or $J = (1)$.

Let $J = (0)$. Then

$$\begin{aligned}\mu^J(c_1) &= e^{g_1 F_0}\mathscr{L}^{\emptyset} e^{-g_1 F_0} = (1 + g_1 e_{3,3}(\Lambda^{(2)})^{-1})\mathscr{L}^{\emptyset}(1 - g_1 e_{3,3}(\Lambda^{(2)})^{-1}) = \\ &= \partial + \Lambda^{(2)} + g_1(e_{3,3} - e_{1,1}) = \partial + \Lambda^{(2)} - g_1 h_0,\end{aligned}$$

where $g_1 = \frac{1}{x + c_1}$. By formula (39),

$$\frac{\partial}{\partial t_r}\Big|_{\mu^J(c_1)} = -\frac{\partial}{\partial x}((1 + g_1 e_{3,3}(\Lambda^{(2)})^{-1})(\Lambda^{(2)})^r(1 - g_1 e_{3,3}(\Lambda^{(2)})^{-1}))^0.$$

It follows from Lemma 2.5 that $\frac{\partial}{\partial t_r}\big|_{\mu^J(c_1)} = 0$ for $r > 1$ and hence $\Gamma_r = 0$. For $r = 1$, we have $\frac{\partial}{\partial t_1}\big|_{\mu^J(c_1)} = -\frac{1}{(x+c_1)^2}(e_{1,1} - e_{3,3})$. On the other hand, $\frac{\partial}{\partial c_1}\mu^J(c_1) = -\frac{\partial g_1}{\partial c_1}h_0 = \frac{1}{(x+c_1)^2}h_0$. Hence $\Gamma_1 = -\frac{\partial}{\partial c_1}$. Theorem 6.1 is proved for $m = 1$, $J = (0)$.

Let $J = (1)$. Then

$$\mu^J(c_1) = e^{\mathrm{ad} g_1(2F_1+2F_2)}(\mathcal{L}^{\emptyset}) = \partial + \Lambda^{(2)} - g_1 h_1,$$

where $g_1 = \frac{1}{x+c_1}$. By formula (39),

$$\frac{\partial}{\partial t_r}\Big|_{\mu^J(c_1)} = -\frac{\partial}{\partial x}\Big((1 + g_1(e_{1,1} + e_{2,2})(\Lambda^{(2)})^{-1} + 2g_1^2 e_{1,1}(\Lambda^{(2)})^{-2}) \times$$
$$\times (\Lambda^{(2)})^r (1 - g_1(e_{1,1} + e_{2,2})(\Lambda^{(2)})^{-1} + 2g_1^2 e_{1,1}(\Lambda^{(2)})^{-2})\Big)^0.$$

It follows from Lemma 2.5 that $\frac{\partial}{\partial t_r}\big|_{\mu^J(c_1)} = 0$ for $r > 4$ and hence $\Gamma_r = 0$. For $r = 1$, we have $\frac{\partial}{\partial t_1}\big|_{\mu^J(c_1)} = \frac{1}{(x+c_1)^2}(e_{1,1} - e_{3,3})$. On the other hand, $\frac{\partial}{\partial c_1}\mu^J(c_1) = -\frac{\partial g_1}{\partial c_1}h_1 = \frac{1}{(x+c_1)^2}h_1$. Hence $\Gamma_1 = -\frac{1}{2}\frac{\partial}{\partial c_1}$. Theorem 6.1 is proved for $m = 1$, $J = (1)$.

6.3 *Proof of Theorem 6.1 for m > 1*

Lemma 6.3. *If m is even and $r > 3m$, then $\frac{\partial}{\partial t_r}\big|_{\mu^J(c)} = 0$ for all $c \in \mathbb{C}^m$ and, hence, $\Gamma_r = 0$. If m is odd and $j_1 = j_m = 0$, then for $r > 3m-2$ we have $\frac{\partial}{\partial t_r}\big|_{\mu^J(c)} = 0$ for all $c \in \mathbb{C}^m$ and, hence, $\Gamma_r = 0$. If m is odd and $j_1 = j_m = 1$, then for $r > 3m+1$ we have $\frac{\partial}{\partial t_r}\big|_{\mu^J(c)} = 0$ for all $c \in \mathbb{C}^m$ and, hence, $\Gamma_r = 0$.*

Proof. The vector $\frac{\partial}{\partial t_r}\big|_{\mu^J(c)}$ equals the right-hand side of formula (39). By Lemmas 2.4 and 2.5 the right-hand side of (39) is zero if r is as described in the lemma.

We prove the first statement of Theorem 6.1 by induction on m. Assume that the statement is proved for $\tilde{J} = (j_1, \dots, j_{m-1})$. Let

$$Y^{\tilde{J}} : \tilde{c} = (c_1, \dots, c_{m-1}) \mapsto (y_0(x, \tilde{c}), y_1(x, \tilde{c}))$$

be the generation of pairs in the $\tilde{J}$th direction. Then the generation of pairs in the Jth direction is

$$Y^J : \mathbb{C}^m \mapsto (\mathbb{C}[x])^2, \quad (\tilde{c}, c_m) \mapsto (\dots, y_{j_m,0}(x, \tilde{c}) + c_m y_{j_m}(x, \tilde{c}), \dots),$$

see (26) and (27). We have $g_m = \ln'(y_{j_m,0}(x, \tilde{c}) + c_m y_{j_m}(x, \tilde{c})) - \ln'(y_{j_m}(x, \tilde{c}))$, see (35).

By the induction assumption, there exists a polynomial vector field $\Gamma_{r,\tilde{J}} = \sum_{i=1}^{m-1} \gamma_i(\tilde{c}) \frac{\partial}{\partial c_i}$ on $\mathbb{C}^{m-1}$ such that

$$\frac{\partial}{\partial t_r}\Big|_{\mu^{\tilde{J}}(\tilde{c})} = \frac{\partial \mu^{\tilde{J}}}{\partial \Gamma_{r,\tilde{J}}}(\tilde{c}) \tag{43}$$

for all $\tilde{c} \in \mathbb{C}^{m-1}$.

Theorem 6.4. *There exists a scalar polynomial $\gamma_m(\tilde{c}, c_m)$ on $\mathbb{C}^m$ such that the vector field $\Gamma_r = \Gamma_{r,\tilde{J}} + \gamma_m(\tilde{c}, c_m)\frac{\partial}{\partial c_m}$ satisfies (42) for all $(\tilde{c}, c_m) \in \mathbb{C}^m$.*

The first statement of Theorem 6.1 follows from Theorem 6.4.

6.4 Proof of Theorem 6.4

Lemma 6.5. *If $j_m = 1$, then for all $(\tilde{c}, c_m) \in \mathbb{C}^m$, we have*

$$d\mathfrak{m}_0\big|_{\mu^J(\tilde{c},c_m)} \left(\frac{\partial}{\partial t_r}\Big|_{\mu^J(\tilde{c},c_m)} - \frac{\partial \mu^J}{\partial \Gamma_{r,\tilde{J}}}(\tilde{c}, c_m) \right) = 0\,, \tag{44}$$

If $j_m = 0$, then for all $(\tilde{c}, c_m) \in \mathbb{C}^m$, we have

$$d\mathfrak{m}_1\big|_{\mu^J(\tilde{c},c_m)} \left(\frac{\partial}{\partial t_r}\Big|_{\mu^J(\tilde{c},c_m)} - \frac{\partial \mu^J}{\partial \Gamma_{r,\tilde{J}}}(\tilde{c}, c_m) \right) = 0\,. \tag{45}$$

Proof. The proof of this lemma is the same as the proof of Lemma 5.5 in [VW].

Let $j_m = 1$. By Lemma 6.5, the vector $\frac{\partial}{\partial t_r}\big|_{\mu^J(\tilde{c},c_m)} - \frac{\partial \mu^J}{\partial \Gamma_{r,\tilde{J}}}(\tilde{c}, c_m)$ lies in the kernel of the map $d\mathfrak{m}_0\big|_{\mu^J(\tilde{c},c_m)}$. By Lemma 5.8, this kernel is generated by $\frac{y_0(x,\tilde{c},m-1)}{y_1(x,\tilde{c},c_m)^2} h_0$. By Lemma 5.7, we have $\frac{\partial \mu^J}{\partial c_m}(\tilde{c}, c_m) = a\frac{y_0(x,\tilde{c},m-1)}{y_1(x,\tilde{c},c_m,m)^2} h_0$ for some $a \in \mathbb{C}^\times$. Hence there exists a number $\gamma_m(\tilde{c}, c_m)$ such that $\frac{\partial}{\partial t_r}\big|_{\mu^J(\tilde{c},c_m)} = \Gamma_{r,\tilde{J}}\big|_{\tilde{c}} + \gamma_m(\tilde{c}, c_m)\frac{\partial}{\partial c_m}$.

Let $j_m = 0$. By Lemma 6.5, the vector $\frac{\partial}{\partial t_r}\big|_{\mu^J(\tilde{c},c_m)} - \frac{\partial \mu^J}{\partial \Gamma_{r,\tilde{J}}}(\tilde{c}, c_m)$ lies in the kernel of the map $d\mathfrak{m}_1\big|_{\mu^J(\tilde{c},c_m)}$. By Lemma 5.8, this kernel is generated by the polynomial $\frac{y_1(x,\tilde{c},m-1)^4}{y_0(x,\tilde{c},c_m)^2} h_0$. By Lemma 5.7, we have $\frac{\partial \mu^J}{\partial c_m}(\tilde{c}, c_m) = a\frac{y_1(x,\tilde{c},m-1)^4}{y_0(x,\tilde{c},c_m,m)^2} h_0$ for some $a \in \mathbb{C}^\times$. Hence there exists a number $\gamma_m(\tilde{c}, c_m)$ such that $\frac{\partial}{\partial t_r}\big|_{\mu^J(\tilde{c},c_m)} = \Gamma_{r,\tilde{J}}\big|_{\tilde{c}} + \gamma_m(\tilde{c}, c_m)\frac{\partial}{\partial c_m}$.

Proposition 6.6. *The function $\gamma_m(\tilde{c}, c_m)$ is a polynomial on $\mathbb{C}^m$.*

Proof. The proof is similar to the proof of Proposition 5.9 in [VW].

More precisely, let $g = x^d + \sum_{i=0}^{d-1} A_i(c_1, \dots, c_m)x^i$ be a polynomial in $x, c_1, \dots, c_m$. Denote $h = \ln' g$ the logarithmic derivative of g with respect to x. Consider the Laurent expansion of h at $x = \infty$, $h = \sum_{i=1}^{\infty} B_i(c_1, \dots, c_m)x^{-i}$.

Lemma 6.7. *All coefficients B_i are polynomials in $c_1, \dots, c_m$.* □

The vector $Y = \frac{\partial}{\partial t_r}\Big|_{\mu^J(\tilde{c}, c_m)}$ is a 3×3 diagonal matrix depending on $x, c_1, \dots, c_m$, $Y = Y_1(e_{1,1} - e_{3,3})$ where Y_1 is a scalar function.

Lemma 6.8. *The function Y_1 is a rational function in $x, c_1, \dots, c_m$ which has a Laurent expansion of the form $Y_1 = \sum_{i=1}^{\infty} B_i(c_1, \dots, c_m)x^{-i}$ where all coefficients B_i are polynomials in $c_1, \dots, c_m$.* □

The vector $Y = \frac{\partial \mu^J}{\partial \Gamma_{r,\bar{J}}}(\tilde{c}, c_m)$ is a 3×3 diagonal matrix depending on $x, c_1, \dots, c_m$, $Y = Y_1(e_{1,1} - e_{3,3})$ where Y_1 is a scalar function.

Lemma 6.9. *The function Y_1 is a rational function of $x, c_1, \dots, c_m$ which has a Laurent expansion of the form $Y_1 = \sum_{i=1}^{\infty} B_i(c_1, \dots, c_m)x^{-i}$ where all coefficients B_i are polynomials in $c_1, \dots, c_m$.* □

Let us finish the proof of Proposition 6.6. The function $\gamma_m(\tilde{c}, c_m)$ is determined from the equation

$$\frac{\partial}{\partial t_r}\Big|_{\mu^J(\tilde{c}, c_m)} - \frac{\partial \mu^J}{\partial \Gamma_{r,\bar{J}}}(\tilde{c}, c_m) = a_1 \gamma_m(\tilde{c}, c_m) \frac{y_0(x, \tilde{c}, m-1)}{y_1(x, \tilde{c}, c_m, m)^2} h_0$$

if $j_m = 1$ and from the equation

$$\frac{\partial}{\partial t_r}\Big|_{\mu^J(\tilde{c}, c_m)} - \frac{\partial \mu^J}{\partial \Gamma_{r,\bar{J}}}(\tilde{c}, c_m) = a_2 \gamma_m(\tilde{c}, c_m) \frac{y_1(x, \tilde{c}, m-1)^4}{y_0(x, \tilde{c}, c_m, m)^2} h_0$$

if $j_m = 0$. Here a_1, a_2 are nonzero complex numbers independent of $\tilde{c}, c_m$.

The function $\frac{y_0(x, \tilde{c}, m-1)}{y_1(x, \tilde{c}, c_m, m)^2}$ has the Laurent expansion of the form $\sum_{i=1}^{\infty} B_i (c_1, \dots, c_m)x^{-i}$ and the first nonzero coefficient B_i of this expansion is 1 since the polynomials y_0, y_1 are monic polynomials. Hence γ_m is a polynomial if $j_m = 1$. Similarly, the function $\frac{y_1(x, \tilde{c}, m-1)^4}{y_0(x, \tilde{c}, c_m, m)^2}$ has the Laurent expansion of the form $\sum_{i=1}^{\infty} B_i(c_1, \dots, c_m)x^{-i}$ and the first nonzero coefficient B_i of this expansion is 1 since the polynomials y_0, y_1 are monic polynomials. Hence γ_m is a polynomial if $j_m = 0$.

Theorem 6.1 is proved.

6.5 *Critical Points and the Population Generated from* $y^{\emptyset}$

Theorem 6.10 ([MV3]). *If a pair of polynomials* (y_0, y_1) *represents a critical point of the master function* $\Phi(u, \boldsymbol{k})$ *in* (10) *for some parameters* $\boldsymbol{k} = (k_0, k_1)$, *then* (y_0, y_1) *is a point of the population of pairs generated from* $y^{\emptyset}$.

The theorem says that the critical points discussed in this paper and, in particular, in Theorem 6.1 exhaust all critical points of the considered master functions.

References

[AM] Adler, M., Moser, J.: On a class of polynomials connected with the Korteweg-de Vries equation. Commun. Math. Phys. **61** (1978), 1–30

[BF] Babujian, H., Flume, R.: Off-shell Bethe ansatz equation for Gaudin magnets and solutions of Knizhnik-Zamolodchikov equations. Mod. Phys. Lett. A **9**, 2029–2039 (1994)

[DS] Drinfel′d, V.G., Sokolov, V.V.: Lie algebras and equations of Korteweg-de Vries type. In: Current problems in mathematics. Itogi Nauki i Tekhniki, vol. 24, pp. 81–180. Akad. Nauk SSSR Vsesoyuz. Inst. Nauchn. i Tekhn. Inform., Moscow (1984)

[F] Frenkel, E.: Opers on the projective line, flag manifolds and Bethe ansatz. Mosc. Math. J. **4**(3), 655–705, 783 (2004)

[MTV] Mukhin, E., Tarasov, V., Varchenko, A.: Schubert calculus and representations of general linear group. J. Am. Math. Soc. **22**(4), 909–940 (2009)

[MV1] Mukhin, E., Varchenko, A.: Critical points of master functions and flag varieties. Commun. Contemp. Math. **6**, 111–163 (2004)

[MV2] Mukhin, E., Varchenko, A.: Miura opers and critical points of master functions. Cent. Eur. J. Math. **3**, 155–182 (2005) (electronic)

[MV3] Mukhin, E., Varchenko, A.: On critical points of master functions associated with affine Lie algebras. *J. Singul.* **8**, 31–38 (2014)

[RV] Reshetikhin, N., Varchenko, A.: Quasiclassical asymptotics of solutions to the KZ equations. In: Geometry, Topology and Physics for R. Bott, pp. 293–322. International Press, Cambridge (1995)

[SV] Schechtman, V., Varchenko, A.: Arrangements of Hyperplanes and Lie Algebra Homology. Invent. Math. **106**, 139–194 (1991)

[ScV] Scherbak, I., Varchenko, A.: Critical points of functions, sl_2 representations, and Fuchsian differential equations with only univalued solutions. Mosc. Math. J. **3**(2), 621–645 (2003)

[Sz] Szego, G.: Orthogonal Polynomials. AMS, New York (1939)

[V1] Varchenko, A.: Multidimensional Hypergeometric Functions and Representation Theory of Lie Algebras and Quantum Groups. Advanced Series in Mathematical Physics, vol. 21. World Scientific, Singapore (1995)

[V2] Varchenko, A.: Special functions, KZ type equations, and Representation theory. In: CBMS, Regional Conference Series in Math, vol. 98. AMS, Providence (2003)

[V3] Varchenko, A.: Quantum integrable model of an arrangement of hyperplanes. SIGMA Symmetry Integrability Geom. Methods Appl. **7**, Paper 032, 55 pp. (2011)

[VW] Varchenko, A.,Wright, D.: Critical points of master functions and integrable hierarchies. Adv. Math. **263**, 178–229 (2014)

Gauss–Lucas and Kuo–Lu Theorems

Laurenţiu Păunescu

Abstract An elementary algebraic calculation over the Newton–Puiseux field, only employing its contact order structure, shows that the Kuo–Lu theorem is in fact a Gauss–Lucas type theorem, via a new notion of convexity over the Newton–Puiseux field.

Keywords Contact order • Newton–Puiseux field

1 Introduction

Recall that the classical Newton–Puiseux Theorem asserts that the field $\mathbb{F}$ of Puiseux series in an indeterminate y is algebraically closed [4]. A nonzero element of $\mathbb{F}$ is a convergent series $\alpha(y) = a_0 y^{n_0/N} + \cdots + a_i y^{n_i/N} + \cdots, n_0 < n_1 < \cdots,\ n_i \in \mathbb{Z}$, where $0 \neq a_i \in \mathbb{C}$.

We define the *conjugate* of α by $\overline{\alpha}(y) := \sum \overline{a_i} y^{n_i/N}$, where the coefficients are conjugated in $\mathbb{C}$, as usual. We can easily see that $\alpha \cdot \overline{\alpha}$ has all coefficients real, with the initial one, $a_0\overline{a_0}$ strictly positive (if $\alpha \not\equiv 0$).

The *order* of α is $O_y(\alpha) := n_0/N$, $O_y(0) := +\infty$. By the Newton–Puiseux Theorem, any polynomial of degree m, say ϕ, has m roots in $\mathbb{F}$ (counting multiplicities), and ϕ has $m-1$ critical points in $\mathbb{F}$, $\phi(\zeta) = unit \cdot \prod_{i=1}^{m}(\zeta - \zeta_i(y))$.

2 Convexity

In the following we will introduce a notion of convexity, suitable for our Newton–Puiseux field.

Let us consider $\mathcal{R} := \{\zeta \in \mathbb{F} | \zeta = \overline{\zeta}\}$, a subfield of $\mathbb{F} = \mathcal{R} + i\mathcal{R}$. As usual we may consider $\alpha : \mathbb{F} \to \mathcal{R}$, $\mathcal{R}$-linear mappings; they are either surjective

L. Păunescu (✉)
School of Mathematics and Statistics, University of Sydney, Sydney, NSW, 2006, Australia
e-mail: laurent@maths.usyd.edu.au

D. Ibadula, W. Veys (eds.), *Bridging Algebra, Geometry, and Topology*,
Springer Proceedings in Mathematics & Statistics 96,
DOI 10.1007/978-3-319-09186-0_12

or identical zero. (A basic example is taking the real part of the coefficients, $\alpha(\sum_{i\geq 0} a_i y^{n_i}) = \sum_{i\geq 0} \mathcal{R}e(a_i) y^{n_i}$.) Accordingly, we can define the associated half-spaces $\alpha^{-1}(\mathcal{R}_+); \mathcal{R}_+ := \{\zeta \in \mathcal{R}, In(\zeta) \geq 0\}$, where $In(\sum_{i\geq 0} a_i y^{n_i}) = a_0, In(\zeta) = 0$ only for $\zeta \equiv 0$. Their boundaries are $\mathcal{R}$–lines given by $\alpha^{-1}(0)$, subspaces isomorphic to $\mathcal{R}$. Accordingly we can introduce the open half-spaces as $\alpha^{-1}(\mathcal{R}_+) \setminus \alpha^{-1}(0) = \alpha^{-1}(\mathcal{R}_+^*), \mathcal{R}_+^* = \mathcal{R}_+ \setminus \{0\}$. We will also call half-spaces (open half-spaces) their translates like $\eta + \alpha^{-1}(\mathcal{R}_+)$, (respectively $\eta + \alpha^{-1}(\mathcal{R}_+^*)$), $\eta \in \mathbb{F}$. The half-spaces, their boundaries and the open half-spaces, are convex in the following sense. A subset $A \subseteq \mathbb{F}$ is convex iff it is closed under taking arbitrary convex combinations. More specifically in our case, given any $c_i \in \mathcal{R}_+, \sum c_i = 1, \eta_i \in A$, then $\sum c_i \eta_i$ is still in A.

3 Results

The following identity is well known,

$$\frac{\phi' := \phi_\zeta(\zeta)}{\phi(\zeta)} = \sum \frac{1}{\zeta - \zeta_i} = \sum \frac{\overline{\zeta} - \overline{\zeta_i}}{(\zeta - \zeta_i)(\overline{\zeta} - \overline{\zeta_i})}.$$

If w is a root of ϕ' which is not a root of ϕ, then the above becomes,

$$\sum \frac{1}{w - \zeta_i} = 0, \text{ so} \sum \frac{\overline{w} - \overline{\zeta_i}}{(w - \zeta_i)(\overline{w} - \overline{\zeta_i})} = 0,$$

i.e.

$$\sum \frac{\overline{w}}{(w - \zeta_i)(\overline{w} - \overline{\zeta_i})} = \sum \frac{\overline{\zeta_i}}{(w - \zeta_i)(\overline{w} - \overline{\zeta_i})},$$

and conjugating again we get

$$\sum \frac{\zeta_i}{(w - \zeta_i)(\overline{w} - \overline{\zeta_i})} = \sum \frac{w}{(w - \zeta_i)(\overline{w} - \overline{\zeta_i})}.$$

We put $A_i = \frac{1}{(w-\zeta_i)(\overline{w}-\overline{\zeta_i})}$ and $A = \sum A_i$. By construction, $A_i, A \in \mathcal{R}_+^*$, so all $\neq 0$, thus we can finally write a Gauss–Lucas type result, namely:

Theorem 1. *The roots of the derivative of a polynomial $\phi \in \mathbb{F}[x]$ belong to the convex hull of its roots, namely, using the notation above we have:*

$$w = \sum \frac{A_i}{A} \zeta_i = \sum B_i \zeta_i, \ B_i := \frac{A_i}{A}. \tag{1}$$

Assume, for convenience, that the contact orders of w with ζ_i , that is $O_y(w-\zeta_i) = k_i$, are in order $k_1 \le k_2 \le \cdots \le k_m$; $O_y(A_i) = -2k_i$ and $O_y(A) = -2k_m$, so each $O_y(B_i) = 2(k_m - k_i)$.

We call k_m the "ϕ-departing height" of w and $\mathcal{D}_w(\phi) = \{\zeta_i | O_y(w - \zeta_i) = \sup_j O_y(w-\zeta_j) = k_m\}$ the ϕ-departing bar of w. Note that the notions ϕ-departing height and ϕ-departing bar can be similarly defined for any $\zeta \in \mathbb{F}$, allowing $k_m = \infty$ when ζ is a root of ϕ.

We write $\phi = QL$, where $Q = \prod_{i,k_i=k_m}(z-\zeta_i)$, and $L(z) = \prod_{i,k_i<k_m}(z-\zeta_i)$. It follows that

$$\frac{Q'(w)}{Q(w)} = -\frac{L'(w)}{L(w)} = -\sum_{i,k_i<k_m} \frac{1}{w-\zeta_i}.$$

A simple estimation of the y-orders gives that at least one root of the derivative of Q has contact with w bigger that k_m. This is, basically, the Kuo–Lu Theorem [2], a natural generalization of the Rolle's theorem to functions of two complex variables, having interesting applications to singularities, see for instance [1] or [3].

More precisely, for each i, such that $k_i = k_m$, $\zeta_i = \alpha + z_i y^{k_m} + HOT$, α involving all the terms in ζ_i with exponents smaller than k_m, thus clearly $w = \alpha + z_0 y^{k_m} + HOT$. In the case $k_i < k_m$, we have that $O_y(w-\zeta_i) = O_y(\alpha - \zeta_i)$ and our formula 1 becomes:

$$w = \sum B_i\zeta_i = \alpha + z_0 y^{k_m} + HOT = \sum_{i,k_i=k_m} B_i\zeta_i + \sum_{i,k_i<k_m} B_i\zeta_i =$$

$$= \sum_{i,k_i=k_m} B_i(\alpha + z_i y^{k_m} + HOT) + \sum_{i,k_i<k_m} B_i(\zeta_i - \alpha + \alpha) = \alpha + \sum_{i,k_i=k_m} B_i z_i y^{k_m}$$

$$+ \sum_{i,k_i<k_m} B_i(\zeta_i - \alpha) + HOT.$$

If we write each $B_i = b_i y^{2(k_m-k_i)} + HOT, b_i > 0$, we further have $w = \alpha + z_0 y^{k_m} + HOT = \alpha + \sum_{i,k_i=k_m} b_i z_i y^{k_m} + HOT$.

This is exactly the Gauss–Lucas Theorem at the "ϕ-departing height" k_m, namely $z_0 = \sum_{i,k_i=k_m} b_i z_i$, a genuine convex combination of the roots of the associated bar polynomial. Indeed, z_i are the roots of the polynomial associated with the bar at height k_m, see the tree below, and z_0 is a root of its derivative.

The situation is best explained, via *the notion of tree-model* [2], by an easy example. Take $\phi(\xi) = (\xi^2 - y^3)^2 - 4\xi y^5$. The Puiseux roots are

$$\zeta_i(y) = y^{3/2} \pm y^{7/4} + \cdots, 1 \le i \le 2,$$

$$\zeta_i(y) = -y^{3/2} \pm \sqrt{-1}y^{7/4} + \cdots, 3 \le i \le 4.$$

The tree-model is shown in the figure below. Tracing upward from the tree root to a tip along solid line segments amounts to identifying a Puiseux root.

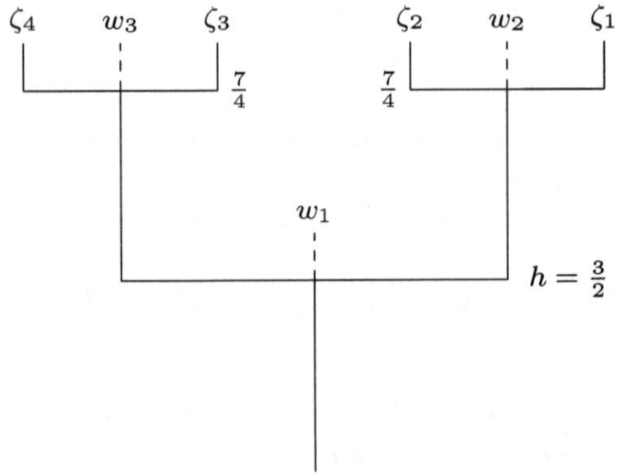

Fig. 1 Tree model

There are three essential bars of heights $3/2$, $7/4$, and $7/4$, respectively. Each is indicated by a horizontal line segment (whence the name); the associated polynomial has *at least two* distinct roots.

At these heights, the ζ_i's split away from each other. *Bars without this property are not indicated.*

The three critical points (polars) w_j, at their ϕ–departing heights, are indicated by dashed lines. The Kuo–Lu Theorem states exactly this fact (plus the counting of multiplicities), namely on each bar grows exactly $k-1$ polars, where k is the degree of the corresponding bar polynomial (Fig. 1).

In the tree below there are three essential bars and their associated bar polynomials are:

1. $(z-1)^2(z+1)^2$ at height $3/2$,
2. $(z-1)(z+1)$ and $(z-\sqrt{-1})(z+\sqrt{-1})$ at height $7/4$ respectively.

In fact for any ζ_{i_0} an arbitrary root and k an arbitrary rational number, we can consider the index family $I_k := \{i \in \{1, \cdots, m\} | O_y(\zeta_i - \zeta_{i_0}) \geq k\}$ and put $\phi_k(\zeta) := \prod_{i \in I_k}(\zeta - \zeta_i)$ (a kind of bar-wise polynomial factor of ϕ). In, inductively, counting multiplicities of the polars the following result is useful.

Theorem 2. *Let w be a root of ϕ_k' with "ϕ_k-departing height" $h \geq k$. Then there is a root of ϕ' having contact with w bigger than h.*

Proof. Indeed, if w is a multiple root of ϕ_k, so it is for ϕ, thus nothing to prove. Assume that w is not a root of ϕ_k and define $Q(\zeta) := \prod_{i \in I_k, O_y(w-\zeta_i)=h}(\zeta - \zeta_i)$, say of degree q. Note that for the roots of ϕ outside I_k, the order of contact with w is less than h. We put $L(\zeta) = \prod_{i \in I_h}(\zeta - \zeta_i)$, where $I_h := \{j \in \{1, \cdots, m\} | j \notin \{i \in I_k | O_y(w - \zeta_i) = h\}\}$, hence $O_y(w - \zeta_i) < h, i \in I_h$. By the previous calculation, $O_y(Q'(w)) > (q-1)h$ and $O_y(Q(w)) = qh$.

Accordingly, looking at the y-order of $\phi'(w)/\phi(w) = Q'(w)/Q(w) + L'(w)/L(w)$, we see that it has to be bigger than $-h$.

Assume that w has contact smaller or equal to h with all roots of ϕ' and also note that w has the ϕ-departing height equal to h as well. In this case if we look at the y-order of $\phi'(w)/\phi(w) = \sum_i \frac{1}{w-\zeta_i}$ we can see it has to be $-h$. Indeed, by construction the departing bar of w in the tree of ϕ_k is also the departing bar of w in the tree of ϕ, therefore sharing the same associated bar polynomial. This is a contradiction.

Theorem 3. *Let $\phi : \mathbb{F} \to \mathbb{F}$ be a polynomial and H a half-space containing at least one root of the derivative ϕ'. Then the restriction of ϕ to H is surjective.*

Proof. Indeed, if not, then there is $a \in \mathbb{F}$ such that all the roots of $\phi - a$ are in the complement of H, also a convex set in our sense. This contradicts the Gauss–Lucas Theorem, as the roots of $(\phi - a)' = \phi'$ are convex combinations of the roots of $\phi - a$, and they are not all in the complement of H.

We remark that for any $\zeta \in \mathbb{F}$, different from the roots of ϕ, we have

$$v := -\overline{\frac{\phi(\zeta)'}{\phi(\zeta)}} = \sum \frac{\zeta_i - \zeta}{(\zeta - \zeta_i)(\overline{\zeta} - \overline{\zeta_i})}.$$

and this implies, using our previous notation on the second page, that

$$\frac{v}{A} + \zeta = -\overline{\frac{\phi(\zeta)'}{A\phi(\zeta)}} + \zeta = \sum B_i \zeta_i,\ A_i = \frac{1}{(\zeta - \zeta_i)(\overline{\zeta} - \overline{\zeta_i})},\ A = \sum A_i,\ B_i = \frac{A_i}{A},\ \text{all in } \mathcal{R}_+^*.$$

This, an obvious generalisation of 1, is nothing but:

Theorem 4. *The $\mathcal{R}$-line passing through ζ and of direction v passes through the convex hull of the roots of ϕ.*

The last two results have been inspired by the wonderfully animated "La géométrie derrière le théorème de Gauss–Lucas. Si nous faisions danser les racines ? Un hommage à Bill Thurston", by Arnaud Chritat and Tan Lei, see http://images.math.cnrs.fr/Si-nous-faisions-danser-les.html.

References

1. Izumi, S., Koike, S., Kuo, T.-C.: Computations and stability of the Fukui invariant. Compositio Math. **130**(1), 49–73 (2002)
2. Kuo, T.-C., Lu, Y.C.: On analytic function germs of two complex variables. Topology **16**, 299–310 (1977)
3. Kuo, T.C., Paunescu, L.: Enriched Riemann sphere, Morse stability and equi-singularity in O_2. J. Lond. Math. Soc. **85**(2), 382–408 (2012)
4. Walker, R.J.: Algebraic Curves. Dover, New York (1962)

Fibonacci Numbers and Self-Dual Lattice Structures for Plane Branches

María Pe Pereira and Patrick Popescu-Pampu

Abstract Consider a plane branch, that is, an irreducible germ of curve on a smooth complex analytic surface. We define its *blow-up complexity* as the number of blow-ups of points needed to achieve its minimal embedded resolution. We show that there are F_{2n-4} topological types of blow-up complexity n, where F_n is the n-th Fibonacci number. We introduce complexity-preserving operations on topological types which increase the multiplicity and we deduce that the maximal multiplicity for a plane branch of blow-up complexity n is F_n. It is achieved by exactly two topological types, one of them being distinguished as the only type which maximizes the Milnor number. We show moreover that there exists a natural partial order relation on the set of topological types of plane branches of blow-up complexity n, making this set a *distributive lattice*. We prove that this lattice admits a *unique* order-inverting bijection. As this bijection is involutive, it defines a *duality* for topological types of plane branches. There are F_{n-2} self-dual topological types of blow-up complexity n. Our proofs are done by encoding the topological types by the associated *Enriques diagrams*.

Keywords Distributive lattices • Enriques diagrams • Fibonacci numbers • infinitely near points • Milnor number.

1 Introduction

Let C be a plane branch, that is, an irreducible germ of an analytic curve on a smooth analytic surface $\mathscr{S}$. It is a classical fact that one may get a canonical embedded

---

M. Pe Pereira (✉)
Instituto de Ciencias Matemáticas (ICMAT), Nicolás Cabrera, 13-15, Campus de Cantoblanco, UAM, 28049 Madrid, Spain
e-mail: maria.pe@icmat.es

P. Popescu-Pampu
Université Lille 1, UFR de Maths., Bâtiment M2 Cité Scientifique, 59655, Villeneuve d'Ascq Cedex, France
e-mail: patrick.popescu@math.univ-lille1.fr

D. Ibadula, W. Veys (eds.), *Bridging Algebra, Geometry, and Topology*, Springer Proceedings in Mathematics & Statistics 96,
DOI 10.1007/978-3-319-09186-0_13

resolution of it by successively blowing up the singular points of the strict transform of C. We say that the number of blow-ups needed to arrive at the minimal embedded resolution is the *blow-up complexity* of C. This notion is not to be confused with that of *resolution complexity* introduced by Lê and Oka in [7].

The blow-up complexity is a topological invariant of the pair $(\mathscr{S}, C)$. It is then natural to compare it with more common invariants, as its *multiplicity* and its *Milnor number.*

We were surprised to discover that, *if one fixes a blow-up complexity n, then the maximal multiplicity is equal to the n-th Fibonacci number* F_n, and that *there are exactly two topological types realizing this maximum* (recall that the *Fibonacci sequence* $(F_n)_{n\geq 0}$ is defined by the initial conditions $F_0 = 0$, $F_1 = 1$ and the recursive relation $F_{n+1} = F_n + F_{n-1}$, for all $n \geq 1$). These are the topological types given by $x^{F_n} - y^{F_{n+1}} = 0$ and $x^{F_n} - y^{F_n + F_{n-2}} = 0$. One may discriminate these two multiplicity-maximizing types using the Milnor number: *one of them is the unique topological type with maximal Milnor number among plane branches of blow-up complexity n* (see Theorem 3).

This motivated us to study in more detail the set $\mathscr{E}_n$ of embedded topological types of plane branches with blow-up complexity n. We discovered a second appearance of the Fibonacci numbers: *the cardinal of* $\mathscr{E}_n$ *is equal to* F_{2n-4} (see Theorem 1). But $\mathscr{E}_n$ should not be thought only as a set: we found out a natural partial order on $\mathscr{E}_n$ which makes it a *distributive lattice*, that is, any two elements have an infimum and a supremum, each one of these operations being distributive with respect to the other one (see Proposition 10). This partial order has an absolute maximum, which is the topological type of maximal Milnor number alluded to before. There is also an absolute minimum, which may be characterized as the unique topological type of blow-up complexity n with multiplicity 2 (it is the simple singularity $\mathbb{A}_{2n-4}$).

This symmetry between the minimum and the maximum extends to a *duality* on embedded topological types of plane branches: for each n, there is an order-inverting involution on $\mathscr{E}_n$ (see Definition 8). This involution is the *unique* bijection of $\mathscr{E}_n$ on itself which inverts the partial order (see Theorem 4). The Fibonacci numbers appear for a third time: *there are* F_{n-2} *self-dual topological types of complexity n* (see Proposition 7).

As far as we know, no such lattice structures or duality were known before. See Remark 6 for some comments about the classical projective duality of plane curves.

Let us explain now our way to work with embedded topological types. There are various classical ways to encode them; the most common ones are the sequence of characteristic Newton–Puiseux exponents and the weighted dual graph of the minimal embedded resolution. Nevertheless, here we describe structures on $\mathscr{E}_n$ which we discovered and we believe are most clearly understandable using what was, historically speaking, the first graphical encoding of the topological type of a plane branch: its *Enriques diagram.*

An Enriques diagram associated with a branch is a decorated graph homeomorphic to an interval, whose vertices are labeled by the infinitely near points appearing during the canonical process of embedded resolution by point blow-ups.

By Enriques' convention, the edges are either *curved* or *straight*, and moreover one tells if at the junction point of two successive straight edges the diagram is broken or not: we speak then about *breaking* versus *neutral* vertices (see Definition 3). All our results are proved by studying carefully those decorations. For instance, the duality expresses itself most easily as a symmetry between curved edges and breaking vertices: note that both ends of a curved edge are neutral and both edges adjacent to a breaking vertex are straight.

The reader who wants to get an overview of the structure of the paper may read the short introductory paragraphs of the various sections. The final Remark 7 explains how we were led to discover our results. As we intend this paper to be understandable to both singularists and combinatorialists, we wrote a rather detailed section with basic material about infinitely near points and Enriques diagrams (Sect. 2), which is standard in singularity theory, and another detailed section with basic material about partial order relations and lattices (Sect. 7), which is standard in combinatorics.

We conclude this introduction with a few words of explanation about our use of the term "complexity." Following Matveev's convention in [8], an invariant of a class of objects may be considered as a *complexity measure* if the set of isomorphism classes of objects with a given invariant is *finite*. Our *blow-up complexity* satisfies this condition, as well as the *Milnor number*, if we look at the embedded topological types of plane branches as objects. But the *multiplicity* or the *resolution complexity* of Lê and Oka do not.

2 The Enriques Diagram and the Multiplicity Sequence

In this section we recall basic vocabulary about infinitely near points, as well as the equivalent notions of *Enriques diagram* and *multiplicity sequence* associated with a plane branch. The reader who is not familiar with the process of *point blowing up* and of the way its iteration leads to resolutions of plane curve singularities, may gain a lot of intuition as well as technical skills by consulting Brieskorn and Knörrer's book [3].

Let $(\mathscr{S}, O)$ be a germ of smooth complex analytic surface. A **model** over $(\mathscr{S}, O)$ is a morphism $(\Sigma, E) \xrightarrow{\pi} (\mathscr{S}, O)$ obtained as a sequence of blowing-ups of points above O. Its **exceptional divisor** is the reduced curve $E := \pi^{-1}(O)$. If $(\Sigma_i, E_i) \xrightarrow{\pi_i} (\mathscr{S}, O)$ for $i = 1, 2$ are two models and $P_i \in E_i$, we say that the points P_1, P_2 are *equivalent* when the bimeromorphic map $\pi_2^{-1} \circ \pi_1$ is an isomorphism in a neighborhood of P_1. An **infinitely near point** of O is an equivalence class of points on various models over $(\mathscr{S}, O)$. By abuse of language, we will say also that any one of its representatives is an infinitely near point of O.

Denote by $\mathscr{C}_O$ the set of all infinitely near points of O. If $P \in \mathscr{C}_O$, we denote by $E(P)$ the smooth irreducible rational curve obtained by blowing up P. Of course, this blow up procedure has to be done in a model, but the various exceptional curves obtained like this get canonically identified by the bimeromorphic maps $\pi_2^{-1} \circ \pi_1$. We make an abuse of notation and we denote also by $E(P)$ its strict transform in further blow-ups.

If $P, Q \in \mathscr{C}_O$, we say that Q is **proximate** to P if $Q \in E(P)$, and we write $Q \mapsto P$. As the exceptional divisor on any model has normal crossings, any one of its points lies on one or two of its irreducible components, that is, it is proximate either to one or to two other points of $\mathscr{C}_O$. In the first case it is called a **free point** over O and in the second one a **satellite** over O.

Let $(C, O) \hookrightarrow (\mathscr{S}, O)$ be a **branch**, that is, a reduced irreducible germ of complex analytic curve. In the sequel, in order to insist on the fact that the surface $\mathscr{S}$ is smooth, we will say that (C, O) is a **plane branch**. A model $(\Sigma, E) \xrightarrow{\pi} (\mathscr{S}, O)$ is called an **embedded resolution** of (C, O) if the total transform $\pi^{-1}(C)$ is a normal crossings divisor. There exists a unique *minimal* embedded resolution, obtained recursively by blowing up the unique point of the last defined model where the total transform of C has not a normal crossing.

Denote by $(P_i)_{0 \leq i \leq n-1}$ the finite sequence of infinitely near points of O which are blown up in order to achieve the minimal embedded resolution of (C, O). Therefore $P_0 = O$. Moreover, either C is smooth, in which case $n = 0$, or it is singular and $n \geq 3$. In this second case, P_{n-1} is a satellite point over O and the strict transform of C passing through it is smooth and transversal to both components of the exceptional divisor.

Definition 1. Let (C, O) be a plane branch. We say that the **blow-up complexity** of C is the number of infinitely near points of O which have to be blown up in order to achieve its minimal embedded resolution.

With the previous notations, the blow-up complexity of C is equal to n.

For all $i \in \{0, \dots, n-1\}$, denote by C_i the strict transform of C passing through P_i and by $m_i(C) \in \{1, 2, \dots, m_0(C)\}$ the multiplicity of C_i at P_i. Therefore, $m_0(C)$ denotes the multiplicity of C at O. As the multiplicity of a strict transform drops (not necessarily strictly) when one does another blow-up, this sequence is decreasing. As the strict transform of C passing through P_{n-1} is smooth, one has $m_{n-1}(C) = 1$.

Definition 2. The decreasing sequence of positive integers $(m_0(C), \dots, m_{n-1}(C))$ is called the **multiplicity sequence** of the plane branch C with blow-up complexity n.

One has the following *proximity relations* between the terms of the multiplicity sequence (see [4, Prop. 3.5.3]):

Proposition 1. *For each* $i \in \{0, \dots, n-1\}$, *the multiplicity* $m_i(C)$ *is equal to the sum of multiplicities of the strict transforms of* C *which pass through points proximate to* P_i. *That is:*

$$m_i(C) = \sum_{P_j \mapsto P_i} m_j(C).$$

The Milnor number $\mu(C)$ of (C, O) (introduced first in [9] for germs of isolated complex algebraic hypersurface singularities of arbitrary dimension) may be expressed in the following way in terms of the associated multiplicity sequence (see [4, Proposition 6.4.1] or [12, Prop. 6.5.9]):

Proposition 2. *From the multiplicity sequence $(m_i(C))_{0 \leq i \leq n-1}$ associated with the plane branch (C, O), one can compute the Milnor number of (C, O) by the formula:*

$$\mu(C) = \sum_{0 \leq i \leq n-1} m_i(C) \cdot (m_i(C) - 1).$$

Note that the previous formula shows that the Milnor number is necessarily *even*. In fact, $\mu(C)/2$ is equal to the more classical δ-invariant of the branch (see for instance [12, Prop. 6.3.2]). In the sequel we will also need (see [12, Example 6.5.1]):

Proposition 3. *Let C be a plane branch defined by the equation $x^a - y^b = 0$, where a, b are coprime positive integers. Then $\mu(C) = (a-1)(b-1)$.*

Let us describe now the geometric object by which we will represent all along the paper the topological type of $(\mathcal{S}, C)$: its *Enriques diagram*, introduced first in [5, Libro Quarto, Cap. 1, Sec. 8] (see also Casas' book [4, Section 3.9]). It is a finite graph homeomorphic to an interval and enriched with *decorations* of the edges and of the vertices: the edges may be *curved/straight* and the vertices *neutral/breaking*:

Definition 3. If v, w are two vertices of a graph homeomorphic to an interval, we denote by $[vw]$ the segment joining them. The **Enriques diagram** $\epsilon(C)$ of the plane branch (C, O) with blow-up complexity n is a graph homeomorphic to an interval, whose vertices $(v_i)_{0 \leq i \leq n-1}$ correspond bijectively to the infinitely near points $(P_i)_{0 \leq i \leq n-1}$ and whose edges $(e_i)_{1 \leq i \leq n-1}$ are indexed such that $e_i = [v_{i-1} v_i]$. The edges are either **curved** or **straight** and successive straight edges make either a straight path or a broken one, according to the following rules:

1. The edge e_i is curved if and only if P_i is a free point.
2. Assume that P_i has at least two proximate points in the sequence $(P_i)_{0 \leq i \leq n-1}$. If $P_{i+1}, P_{i+2}, \ldots, P_k$ are all the points proximate to P_i (therefore $k \geq i+2$), then the path $[v_{i+1} v_k]$ is straight. Moreover, $[v_{i+1} v_k]$ is a **maximal straight path**, that is, adding any one of its adjacent edges $e_{i+1} = [v_i v_{i+1}]$ and $e_{k+1} = [v_k v_{k+1}]$, one does not get a straight path. If $[v_i v_{i+1}]$ is also straight, we say that $[v_i v_k]$ is a path which is broken at the **breaking vertex** v_{i+1}. The vertices which are not breaking ones are called **neutral**.

We say that $\epsilon(C)$ is an **Enriques diagram of complexity** n and we denote by $\mathcal{E}_n$ the set of isomorphism classes of such diagrams.

Remark 1. Note that for $n \geq 3$, the edge e_1 is always curved and the edge e_{n-1} is always straight, as explained before Definition 1. If a vertex is breaking, then the two adjacent edges are straight. In a dual manner, if an edge is curved, then its vertices are neutral. In particular, only the vertices $\{v_2, \ldots, v_{n-2}\}$ and the edges

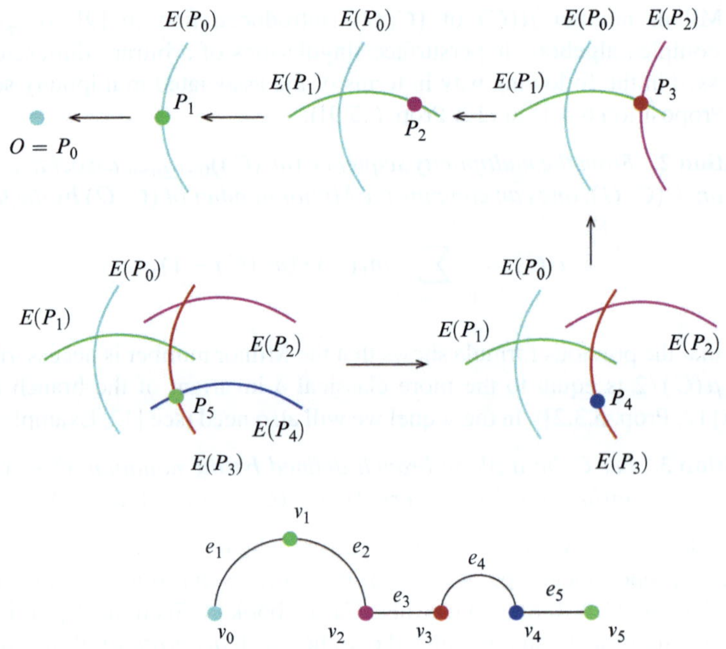

Fig. 1 A sequence of blow-ups and its Enriques diagram

$\{e_2, \ldots, e_{n-2}\}$ may have both decorations when we vary the diagram among the elements of $\mathscr{E}_n$.

In the previous definition, the attributes "curved" and "straight" associated with the edges and "breaking" or "neutral" associated with the vertices are purely combinatorial. Nevertheless, their concrete meaning gives a way to represent an Enriques diagram as a piecewise smooth embedded arc in the plane.

Example 1. In Fig. 1 are represented schematically the exceptional divisors of a sequence of point blowing-ups, as well as the associated Enriques diagram. Its complexity is 6. In order not to charge the drawing, we did not represent the strict transforms of a branch having this embedded resolution process.

Remark 2. In the original definition by Enriques, as a rule for plane representation one also supposed that any curved edge formed a C^1-smooth arc with the previous edge, be it either curved of straight. Moreover, Enriques chose to draw perpendicularly the two maximal straight segments adjacent to a breaking vertex. Here we do not keep those supplementary conventions, as they do not give more information. In this way we gain more flexibility for our drawings.

The following proposition is an immediate consequence of Proposition 1 and shows that the Enriques diagram contains the same information as the multiplicity sequence $(m_i(C))_{0 \le i \le n-1}$. Nevertheless, in the sequel it will be important to think about both of them simultaneously: the multiplicity sequence will be seen as a **multiplicity function** $\underline{m}$ defined on the set of vertices of the Enriques diagram. To simplify notations, we will write simply m_i instead of $\underline{m}(v_i)$. We say that m_0 is the **initial multiplicity** of an Enriques diagram. Proposition 2 shows that the Milnor number of a plane branch is determined by the associated Enriques diagram. Therefore, we will speak also about the **Milnor number** μ of such a diagram.

Proposition 4. *Consider an Enriques diagram of complexity* $n \geq 3$. *Then* $m_{n-1} = 1$ *and for each* $i \in \{0, \dots, n-1\}$, *the multiplicity* m_i *may be computed in the following way from the multiplicities* $(m_j)_{j>i}$:

1. *If there is no maximal straight path of the form* $[v_{i+1}v_j]$, *with* $j > i + 1$, *then:*

$$m_i = m_{i+1}. \tag{1}$$

2. *If* $[v_{i+1}v_j]$, *with* $j > i + 1$, *is a maximal straight path of the Enriques diagram, then:*

$$m_i = \sum_{k=i+1}^{j} m_k. \tag{2}$$

More precisely:

- **if** $e_{j+1} = [v_j v_{j+1}]$ **is curved**, *then* $m_k = m_j$ *for all* $k = i + 1, \dots, j$ *and:*

$$m_i = (j - i) \cdot m_{i+1}, \tag{3}$$

- **if** e_{j+1} **is straight**, *then* $m_k = m_{i+1}$ *for all* $k = i + 1, \dots, j - 1$ *and:*

$$m_i = m_j + (j - i - 1) \cdot m_{i+1}. \tag{4}$$

As explained in [12, Sections 3.5, 3.6, 5.5], the following invariants associated with a complex plane branch contain the same information:

- its multiplicity sequence;
- its sequence of generic Newton–Puiseux exponents;
- the weighted dual graph of the exceptional divisor of its minimal embedded resolution;
- its embedded topological type (i.e., the topology of the associated knot in the 3-sphere).

One may also consult [3] for the relation between the last three view-points and [10] for the relation between the third one and the Enriques diagram.

Therefore, all such objects parametrize the embedded topological types of plane branches. Nevertheless, as the notion of topological type is specific to $\mathbb{C}$, while the other ones are applicable to curves defined over arbitrary algebraically closed fields with characteristic zero, we will speak about the set of *combinatorial types* instead of *embedded topological types* of plane branches.

In general, when different logically equivalent encodings of some class of objects are available, they are not equivalent from the viewpoint of adaptability to specific situations. For instance, in this paper we will show that Enriques diagrams are especially adapted to emphasize a hidden lattice structure on the set of combinatorial types of plane branches with fixed blow-up complexity.

3 The Number of Combinatorial Types of Branches with Fixed Blow-Up Complexity

We begin the study of the sets $\mathscr{E}_n$ of combinatorial types of branches of fixed blow-up complexity. In this section we show that their cardinals are Fibonacci numbers.

Theorem 1. *The number of combinatorial types of plane branches with blow-up complexity $n \geq 3$ is the Fibonacci number F_{2n-4}.*

Proof. We decompose each set $\mathscr{E}_n$ into the disjoint union of two subsets $\mathscr{A}_n$ and $\mathscr{B}_n$, and we compute by induction the pair of cardinals $(|\mathscr{A}_n|, |\mathscr{B}_n|)$, where:

- $\mathscr{A}_n$ is the set of Enriques diagrams of complexity n whose vertex v_{n-2} is neutral.
- $\mathscr{B}_n$ is the set of Enriques diagrams of complexity n whose vertex v_{n-2} is breaking.

Each diagram of $\mathscr{E}_{n+1}$ can be obtained from a diagram of $\mathscr{E}_n$, which is determined by the decorations of the vertices $v_2, \dots, v_{n-2}$ and of the edges $e_2, \dots, e_{n-2}$, by adding the information about the decorations of v_{n-1} and e_{n-1}. We count the number of elements of $\mathscr{E}_{n+1}$ by looking at the possible ways to complete a given diagram of $\mathscr{E}_n$ (recall Remark 1). This number changes from $\mathscr{A}_n$ to $\mathscr{B}_n$:

- Each graph of $\mathscr{A}_n$ can be completed in 3 ways as an Enriques diagram of complexity $n+1$: *either* with a neutral vertex v_{n-1}, and with an edge e_{n-1} *either* curved *or* straight, *or* with a breaking vertex v_{n-1} and a straight edge.
- Each graph of $\mathscr{B}_n$ can be completed in 2 ways, *either* with a breaking vertex *or* with a neutral vertex v_{n-1}, but always with a straight edge e_{n-1}, such that v_{n-2} keeps being a breaking vertex.

We deduce that:

$$\begin{cases} |\mathscr{A}_{n+1}| = 2|\mathscr{A}_n| + |\mathscr{B}_n|, \\ |\mathscr{B}_{n+1}| = |\mathscr{A}_n| + |\mathscr{B}_n|, \end{cases} \quad \text{for all } n \geq 3.$$

As $|\mathscr{A}_3| = 1$ and $|\mathscr{B}_3| = 0$, which is seen immediately by inspection, the previous recursive relations allow to prove immediately by induction on $n \geq 3$ that:

$$\begin{cases} |\mathscr{A}_n| = F_{2n-5}, \\ |\mathscr{B}_n| = F_{2n-6}, \end{cases} \quad \text{for all } n \geq 3.$$

Therefore $|\mathscr{E}_n| = |\mathscr{A}_n| + |\mathscr{B}_n| = F_{2n-4}$. □

In Theorem 3 we will see a second appearance of Fibonacci numbers related to the sets $\mathscr{E}_n$.

4 Multiplicity Increasing Operators

In this section we introduce two types of partially defined operators on Enriques diagrams and we prove that they make strictly increase the multiplicity function and the Milnor number. Moreover, we describe the cases when also the initial multiplicity increases strictly.

Definition 4. Let $n \geq 3$. We introduce the following partially defined operators on $\mathscr{E}_n$ (see Figs. 2 and 3, in which are indicated the different possibilities for the adjacent edges):

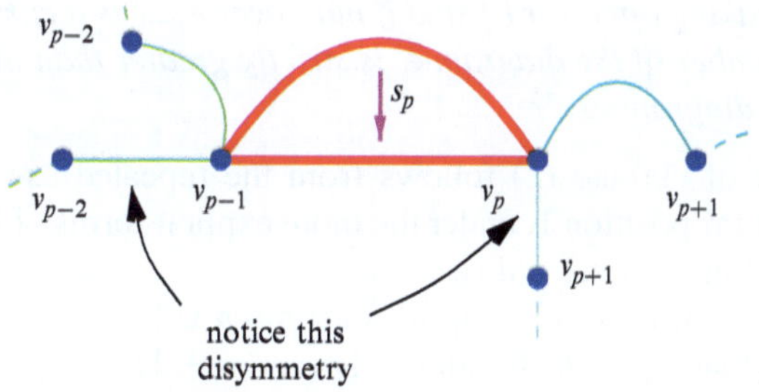

Fig. 2 The straightening operator s_p

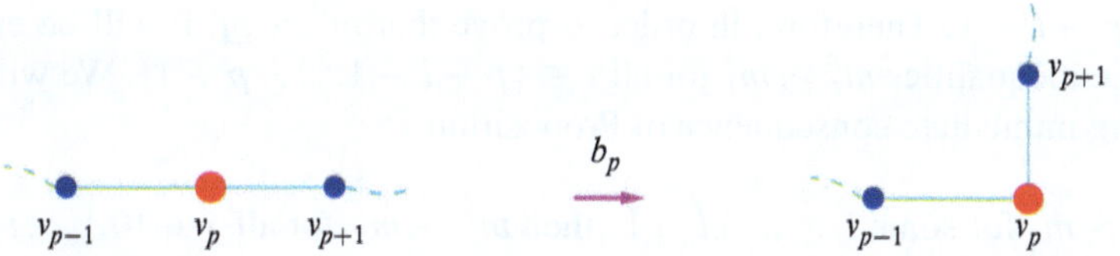

Fig. 3 The breaking operator b_p

(i) Suppose that the diagram $\epsilon \in \mathscr{E}_n$ is such that its edge e_p is curved for some $p \geq 2$. Let $s_p(\epsilon)$ be the diagram obtained from ϵ by declaring the edge e_p straight, aligned with e_{p-1} if this is straight in ϵ and declaring the vertex v_p breaking if e_{p+1} is straight. We say that s_p is the **straightening operator** at the edge e_p.
(ii) Suppose that the diagram $\epsilon \in \mathscr{E}_n$ is such that the path $[v_{p-1}v_{p+1}]$ is straight for some $p \in \{2, \dots, n-2\}$. Let $b_p(\epsilon)$ be the diagram obtained from ϵ by declaring v_p a breaking vertex. We say that b_p is the **breaking operator** at the vertex v_p.

In both cases, all the decorations of the vertices and the edges which are not mentioned are left unchanged.

One has therefore the straightening operators $s_2, \dots, s_{n-2}$ and the breaking ones $b_2, \dots, b_{n-2}$. They are *partially defined* in the sense that $s_p(\epsilon)$ is defined only if e_p is a curved edge of the diagram ϵ and $b_p(\epsilon)$ is defined only if v_p is a neutral vertex between two straight edges.

The next theorem states that the multiplicities and Milnor numbers increase when one applies either type of operator. Moreover, we describe the situations when the increase is not strict.

Theorem 2. *Let $\epsilon \in \mathscr{E}_n$ be an Enriques diagram with multiplicity function $\underline{m}$. Denote by $\epsilon' \in \mathscr{E}_n$ a diagram obtained from ϵ by applying either a straightening or a breaking operator. Denote by $\underline{m}'$ its multiplicity function. Then:*

1. *$\underline{m}' > \underline{m}$ (i.e., $m'_j \geq m_j$ for all $j \in \{0, \dots, m_{n-1}\}$ and there is at least one strict inequality).*
2. *$m'_0 \geq m_0$. The two multiplicities are equal if and only if ϵ' is obtained from ϵ by applying a breaking operator b_p and if moreover e_{p-1} is a curved edge of ϵ.*
3. *The Milnor number of the diagram ϵ' is strictly greater than the Milnor number of the starting diagram ϵ.*

Proof. The proof of (1) and (2) follows from the repeated use of the proximity relations stated in Proposition 1, under the more explicit forms of Proposition 4: the relations (1)–(4). Let us develop this.

Assume that $\epsilon' = s_p(\epsilon)$ or $\epsilon' = b_p(\epsilon)$ for some $p \in \{2, \dots, n-2\}$. This implies, by Proposition 1, that $m'_i = m_i$ for all $i \in \{p, \dots, n-1\}$.

Let $l > 0$ be maximal such that $[v_{p-l}v_{p-1}]$ is a straight path of the diagram ϵ. Notice that by Proposition 4, the rules of computation of m_i and m'_i starting from the knowledge of the multiplicities at the vertices of higher subindex are the same if $i < p - l - 1$. Therefore, in order to prove that $\underline{m}' > \underline{m}$, it will be enough to check the inequalities $m'_i \geq m_i$ for all $i \in \{p - l - 1, \dots, p - 1\}$. We will use the following immediate consequence of Proposition 1:

$$\text{if } m'_i > m_i \text{ for some } i \leq p - l - 1, \text{ then } m'_j > m_j \text{ for all } j \in \{0, \dots, i\}. \tag{5}$$

We treat now separately the two types of operators.

* **The case of a straightening operator**: assume that $\epsilon' = s_p(\epsilon)$ (see Fig. 4).

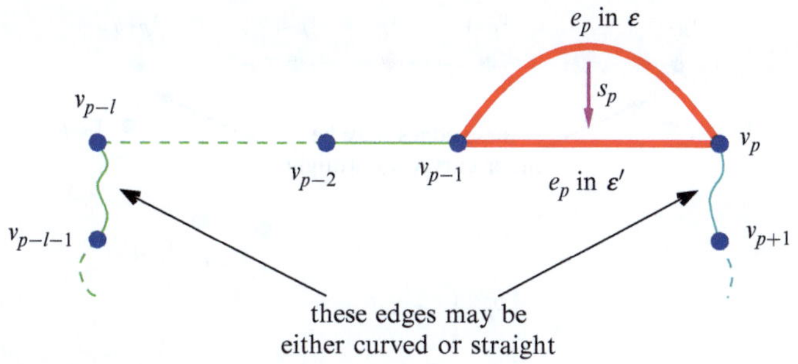

Fig. 4 The diagrams ϵ and $\epsilon' = s_p(\epsilon)$

By (2) and the fact that $m'_i = m_i$ for all $i \geq p$, we get $m'_{p-1} = m_{p-1}$. We deduce from (1) that $m'_j = m'_{p-1} = m_{p-1} = m_j$ for all $j \in \{p-l, \dots, p-1\}$. Then, by (4):

$$m'_{p-l-1} = m'_p + l \cdot m'_{p-l} > l \cdot m'_{p-l} = l \cdot m_{p-l} = m_{p-l-1}.$$

By (5), we see that $m'_j > m_j$ for all $j \in \{0, \dots, p-l-1\}$. This ends the proof for this operator.

$*$ **The case of a breaking operator**: assume that $\epsilon' = b_p(\epsilon)$ (see Fig. 5).

Let $h > 0$ be maximal such that $[v_p v_{p+h}]$ is a straight interval of the diagram ϵ (and therefore also of the diagram ϵ').

By (4), $m'_{p-1} = m'_{p+h} + h \cdot m'_p > m'_p = m_p = m_{p-1}$. From (1) we deduce that $m'_j = m'_{p-1} > m_{p-1} = m_j$ for all $j \in \{p-l, \dots, p-1\}$.

As a consequence: $m'_{p-l-1} = m'_p + l \cdot m'_{p-l} = m'_p + l \cdot m'_{p-1} = m'_p + l \cdot (m'_{p+h} + h \cdot m'_p) = (1 + lh) \cdot m'_p + l \cdot m'_{p+h} = (1 + lh) \cdot m_p + l \cdot m_{p+h} \geq (l + h) \cdot m_p + m_{p+h} = m_{p-l-1}$.

Therefore, $m'_{p-l-1} \geq m_{p-l-1}$, with an equality precisely when $1 + lh = l + h$ and $l = 1$ hold simultaneously. Therefore, $m'_{p-l-1} = m_{p-l-1}$ if and only if $l = 1$.

Let us consider now two subcases.

- **Assume that** $l > 1$. Then $m'_{p-l-1} > m_{p-l-1}$, and (5) implies that $m'_j > m_j$ for all $j \in \{0, \dots, p-l-1\}$.
- **Assume that** $l = 1$. We consider again two subcases:
  - **Assume that** e_{p-1} **is curved, that is, that** v_{p-1} **is neutral.** Therefore $m'_{p-2} = m_{p-2}$. By Proposition 4, we see that for any $j \in \{0, \dots, p-2\}$, the formulae expressing m_j and m'_j in terms of the multiplicities $m_{i>j}$ and $m'_{i>j}$ are the same and involve only subindices $i \leq p-2$. This implies, by descending induction on j, that $m'_j = m_j$ for all $j \in \{0, \dots, p-2\}$.
  - **Assume that** e_{p-1} **is straight, that is, that** v_{p-1} **is breaking** (see Fig. 6). Let $k \geq 2$ be maximal such that $[v_{p-k} v_{p-1}]$ is a straight interval of the diagram ϵ. By (1), we get that $m'_j = m_j$ for all $j \in \{p-k, \dots, p-2\}$. But then

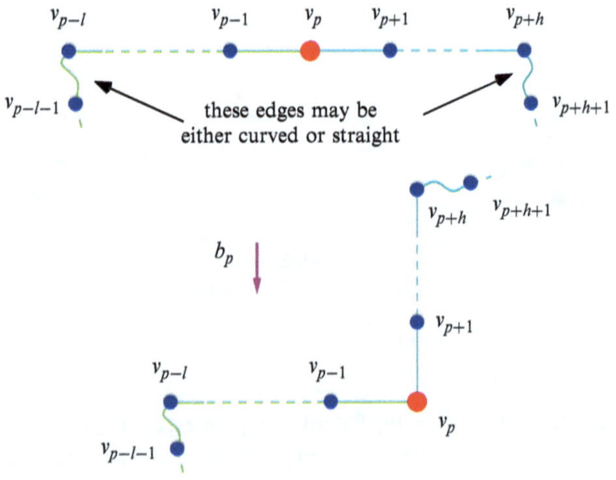

Fig. 5 The diagrams ϵ and $\epsilon' = b_p(\epsilon)$

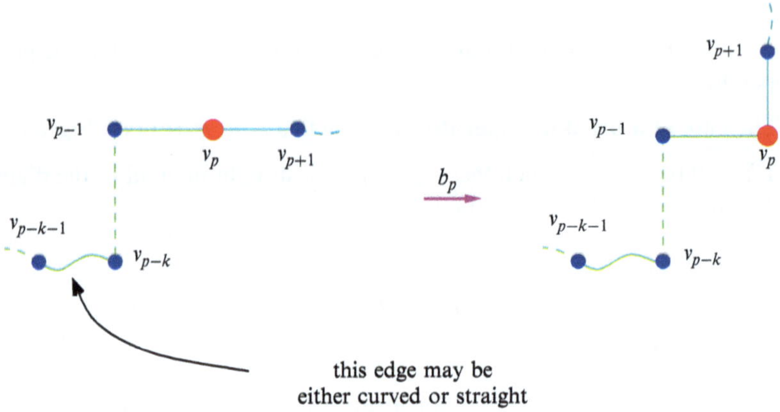

Fig. 6 The breaking operator b_p when v_{p-1} is a breaking vertex

$m'_{p-k-1} = m'_{p-1} + (k-1) \cdot m'_{p-k} > m_{p-1} + (k-1) \cdot m_{p-k}$, which by (5) implies that $m'_j > m_j$ for all $j \in \{0, \ldots, p-k-1\}$.

This finishes the proof of the points (1) and (2) of the theorem. Point (3) is then a direct consequence of them and of Proposition 2. □

As a consequence of the previous proof, one may describe the set of vertices at which the multiplicity function increases *strictly*:

Proposition 5. *Let $\epsilon, \epsilon' \in \mathscr{E}_n$ be such that either $\epsilon' = s_p(\epsilon)$ or $\epsilon' = b_p(\epsilon)$. Let $l \geq 1$ be maximal such that $[v_{p-l}v_p]$ is straight and, in the case where $l = 1$ and*

e_{p-1} is straight, $k \geq 2$ is maximal such that $[v_{p-k}v_{p-1}]$ is straight. Denote also by $J \subset \{0, \dots, n-1\}$ the set of indices i such that $m'_i > m_i$. Then:

- $J = \{0, \dots, p-l-1\}$ *if* $\epsilon' = s_p(\epsilon)$.
- $J = \{0, \dots, p-1\}$ *if* $\epsilon' = b_p(\epsilon)$ *and* $l > 1$.
- $J = \{p-1\}$ *if* $\epsilon' = b_p(\epsilon)$, $l = 1$ *and* e_{p-1} *is curved.*
- $J = \{0, \dots, p-k-1\} \cup \{p-1\}$ *if* $\epsilon' = b_p(\epsilon)$, $l = 1$ *and* e_{p-1} *is straight.*

This shows in particular that in the case when the initial multiplicity does not change, the multiplicity function changes at only one vertex.

5 Extremal Multiplicities and Milnor Numbers for Fixed Complexity

In this section we prove that for a fixed complexity $n \geq 3$, the minimal multiplicity is 2 and the maximal one is the n-th Fibonacci number F_n. The number F_n appears as the natural candidate for the biggest value m_0 attained in a sequence $(m_0, \dots, m_{n-1} = 1)$ generated in reverse order by using the rules described in Proposition 4. Anyway, the complete proof requires some meticulosity and we do it here as a consequence of Theorem 2. In particular, we show that the maximal multiplicity F_n is achieved by exactly two combinatorial types in $\mathscr{E}_n$ and that one of them is distinguished by the property of maximizing also the Milnor number. Meanwhile, the minimal multiplicity 2 is obviously achieved by only one combinatorial type.

Definition 5. For all $n \geq 3$, denote by $\alpha_n, \omega_n, \pi_n \in \mathscr{E}_n$ the diagrams represented in Fig. 7. That is:

- α_n has all its edges $e_2, \dots, e_{n-2}$ curved (therefore all its vertices $v_2, \dots, v_{n-2}$ are neutral). It is the Enriques diagram of the plane branch defined by the equation $x^{2n-3} - y^2 = 0$, that is, of the simple singularity $\mathbb{A}_{2n-4}$.
- ω_n has all its vertices $v_2, \dots, v_{n-2}$ breaking (therefore all the edges $e_2, \dots, e_{n-2}$ are straight). It is the Enriques diagram of the plane branch defined by the equation $x^{F_{n+1}} - y^{F_n} = 0$.

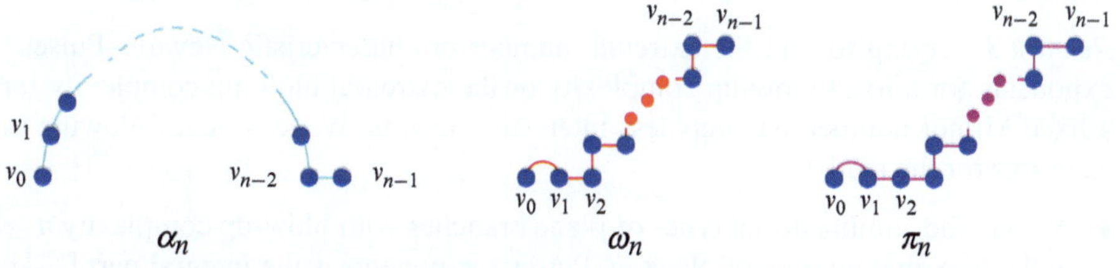

Fig. 7 The diagrams α_n, ω_n and π_n

- π_n is identical to ω_n, except that v_2 is a neutral vertex. It is the Enriques diagram of the plane branch defined by the equation $x^{F_{n-2}+F_n} - y^{F_n} = 0$.

The notations $\alpha_n, \omega_n, \pi_n$ are explained in Remark 5. In order to get the defining equations, one may use the transformation rules described for instance in [12, Section 3]. The previous Enriques diagrams may be characterized in the following way:

Theorem 3. *The diagrams $\alpha_n, \omega_n, \pi_n$ satisfy the following extremal properties among Enriques diagrams of blow-up complexity n:*

1. *α_n is the unique diagram with minimal multiplicity, which is equal to 2, and with minimal Milnor number, equal to $2n - 4$.*
2. *ω_n, π_n are the only diagrams with maximal multiplicity, equal to F_n.*
3. *ω_n is the unique diagram with maximal Milnor number, equal to the product $(F_{n+1} - 1)(F_n - 1)$.*

Proof. All the statements of this theorem are rapid consequences of Theorem 2.

- **Proof of (1):** The diagram α_n has multiplicity 2 and Milnor number $2n - 4$, as may be seen from the defining equation and Proposition 3. Any other diagram $\epsilon \in \mathcal{E}_n$ may be obtained from α_n by a sequence of straightening and breaking operators, with at least one straightening operator being applied. By Theorem 2, we deduce that $m_0(\epsilon) > m_0(\alpha_n)$ and $\mu(\epsilon) > \mu(\alpha_n)$ for all $\epsilon \in \mathcal{E}_n$.
- **Proof of (2):** We see that $\omega_n = b_2(\pi_n)$ and that e_1 is curved in π_n. By Theorem 2, we deduce that ω_n and π_n have the same multiplicity and that $\mu(\omega_n) > \mu(\pi_n)$, as may be seen also from the defining equations.

  Assume now that $\epsilon \in \mathcal{E}_n$ is a diagram different from them. Therefore, one may obtain ω_n from it by applying a sequence of at least two straightening or breaking operators. As $\epsilon \neq \pi_n$, either one of them is a straightening operator, or there is a breaking one b_i among them such that e_{i-1} is not curved. Therefore, Theorem 2 implies that $m_0(\epsilon) < m_0(\omega_n)$.
- **Proof of (3):** The reasoning is the same, but even simpler, as applying an operator makes the Milnor number increase *strictly*. The given expression of $\mu(\omega_n)$ may be deduced from the defining equation and Proposition 3. □

As an immediate consequence of the previous theorem we get:

Corollary 1. *Any plane branch of multiplicity $m \geq 2$ has blow-up complexity at least n, where the integer $n \geq 3$ is such that $F_{n-1} < m \leq F_n$.*

Remark 3. Trying to find the extremal number of characteristic Newton–Puiseux exponents for a fixed blow-up complexity or the extremal blow-up complexity for a fixed Milnor number, one gets less interesting results. We leave the following as exercises for the reader:

- Among the combinatorial types of plane branches with blow-up complexity $n \geq 3$, the maximal number of Newton–Puiseux exponents is the integral part $[\frac{n-1}{2}]$. This maximum is achieved once for n *odd* and $n - 2$ times for n *even*.

- The maximal blow-up complexity of plane branches with Milnor number $\mu \in 2\mathbb{N}^*$ is equal to $2 + \frac{\mu}{2}$. It is achieved by exactly one combinatorial type (and one analytical type), that of the simple plane branch $\mathbb{A}_\mu$, which has one Newton–Puiseux exponent equal to $\frac{\mu+1}{2}$.

6 The Self-Dual Lattice Structure on $\mathscr{E}_n$

In this section we give another way of codifying an Enriques diagram (therefore, the combinatorial type of a plane branch) as a subset of the set of symbols of vertices and edges. This produces a natural partial order relation on $\mathscr{E}_n$, coming from the inclusion relation among such subsets. We show that this relation is a *lattice structure*. Moreover, we show that there exists an order-inverting involution on $\mathscr{E}_n$. This allows to speak about the *dual* of any combinatorial type of plane branch. We present then a third occurrence of the Fibonacci numbers in our context: they appear as the cardinals of the sets of self-dual combinatorial types for each fixed complexity. The few notions about lattices which are used here are explained in the Sect. 7.

Let us fix $n \geq 3$. Denote by:

$$S_n := \{v_2, \dots, v_{n-2}, e_2, \dots, e_{n-2}\}$$

the set of *symbols* of the vertices and edges of the Enriques diagrams of complexity n whose decoration is undetermined (recall Remark 1). That is, we think about the *name* of each edge or vertex, not about the geometric object itself. We encode now each Enriques diagram by *the subset of S_n consisting of the symbols of its straight edges and of its breaking vertices*:

Definition 6. The **code** $\chi(\epsilon) \subseteq S_n$ of an Enriques diagram $\epsilon \in \mathscr{E}_n$, or of the corresponding combinatorial type of plane branch, is the set of symbols of its straight edges and of its breaking vertices. We denote by $\mathscr{K}_n \subseteq \mathscr{P}(S_n)$ the set of such codes.

Here $\mathscr{P}(S_n)$ denotes the *power set* of S_n, that is, the set of its subsets.

Example 2. For instance, if ϵ is the Enriques diagram of Example 1, which is of blow-up complexity 6, then its code is $\{e_3\}$. For any complexity $n \geq 3$, one has $\chi(\alpha_n) = \emptyset$, $\chi(\omega_n) = S_n$ and $\chi(\pi_n) = S_n \setminus \{v_2\}$.

As may be easily shown using the Definition 3, the codes of the Enriques diagrams may be characterized in the following way:

Lemma 1. *A subset χ of S_n is the code of an Enriques diagram $\epsilon \in \mathscr{E}_n$ if and only if it has one of the following equivalent properties:*

(a) for all $i = 2, \dots, n-2$, if $v_i \in \chi$ then $e_i, e_{i+1} \in \chi$;
(b) for all $j = 2, \dots, n-2$, if $e_j \notin \chi$ then $v_j, v_{j+1} \notin \chi$.

On the power set $\mathscr{P}(S_n)$, let us consider the inclusion $\subseteq$ as partial order relation. Endowed with it, $(\mathscr{P}(S_n), \subseteq)$ is a lattice (and even a Boolean algebra). Restrict this partial order to the set of codes of the Enriques diagrams of complexity n.

Lemma 2. *The subset $\mathscr{K}_n$ of $\mathscr{P}(S_n)$ is stable under the intersection $\cap$ and union $\cup$ operations. That is, it is a sublattice of $(\mathscr{P}(S_n), \subseteq)$.*

Proof. It is immediate to check that both the intersection and the union of two subsets of S_n which satisfy either one of the conditions (a) and (b) of Lemma 1 satisfy again that condition. □

Remark 4. The subset $\mathscr{K}_n \subseteq \mathscr{P}(S_n)$ is not stable by the operation of taking the complement, therefore it is not a sub-Boolean algebra of $(\mathscr{P}(S_n), \subseteq)$. For instance, $S_4 = \{v_2, e_2\}$ and $\{e_2\} \in \mathscr{K}_4$ but $S_4 \setminus \{e_2\} = \{v_2\} \notin \mathscr{K}_4$, by Lemma 1.

We are ready to define the lattice structure on the set $\mathscr{E}_n$ of Enriques diagrams of complexity n:

Definition 7. The **staircase partial order relation** $\preceq$ on the set $\mathscr{E}_n$ of Enriques diagrams of complexity n is defined by saying that for any two diagrams $\epsilon, \epsilon' \in \mathscr{E}_n$, one has $\epsilon \preceq \epsilon'$ if and only if the code of ϵ' contains the code of ϵ.

Our motivation for choosing this name comes from the fact that an Enriques diagram ϵ_2 is greater than another one ϵ_1 of the same complexity if and only if ϵ_2 has all the straight edges and breaking vertices of ϵ_1, and maybe more. That is, if and only if ϵ_2 looks more like the staircase diagram ω_n than ϵ_1.

By the Lemma 2, the staircase relation endows $\mathscr{E}_n$ with a lattice structure. As this lattice is finite, it has an absolute minimum and an absolute maximum. Their characterization is a first immediate consequence of the previous definition and of the Definition 5 of the diagrams α_n and ω_n:

Proposition 6. *The minimum of $(\mathscr{E}_n, \preceq)$ is the diagram α_n with $\chi(\alpha_n) = \emptyset$ and the maximum is the diagram ω_n with $\chi(\omega_n) = S_n$. More generally, if ϵ' is obtained from ϵ by applying a straightening or a breaking operator, then $\epsilon \prec \epsilon'$.*

Proof. The first statement results from Example 2. The second one results from the fact that $\chi(s_p(\epsilon))$ is equal either to $\chi(\epsilon) \sqcup \{e_p\}$ or to $\chi(\epsilon) \sqcup \{v_p, e_p\}$ and that $\chi(b_p(\epsilon)) = \chi(\epsilon) \sqcup \{v_p\}$ (here $\sqcup$ denotes a disjoint union). □

Remark 5. This proposition motivated us to choose the notations α_n and ω_n, as α is the first letter in the Greek alphabet and ω is the last one. Concerning the third diagram π_n appearing in the statement of Theorem 3, we chose its name as π is the Greek analog of the initial of "predecessor": indeed, π_n is a predecessor of ω_n for the relation $\preceq$.

Notice that there is a symmetry between the two equivalent characterizations of $\mathscr{K}_n$ stated in Lemma 1: one switches them by interchanging the dimensions (edges $\leftrightarrow$ vertices), the types (straight $\leftrightarrow$ neutral, curved $\leftrightarrow$ breaking), as well as the order of indexing. More precisely:

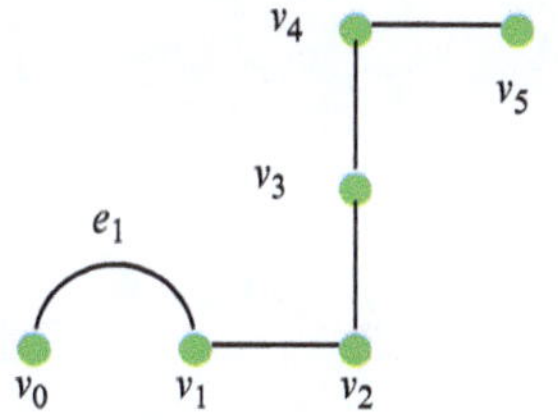

Fig. 8 The Enriques diagram dual to the diagram of Fig. 1

Definition 8. Let ϵ be an Enriques diagram of complexity $n \geq 3$. The **dual Enriques diagram** $\Delta_n(\epsilon)$ is defined by:

- v_k is a breaking vertex in $\Delta_n(\epsilon)$ if and only if e_{n-k} is a curved edge in ϵ;
- e_k is a straight edge in $\Delta_n(\epsilon)$ if and only if v_{n-k} is a neutral vertex in ϵ.

Example 3. Let us consider the Enriques diagram ϵ of Example 1. As $\chi(\epsilon) = \{e_3\}$, we get $\chi(\Delta_6(\epsilon)) = \{e_2, e_3, e_4, v_2, v_4\}$. The associated Enriques diagram $\Delta_6(\epsilon)$ is drawn in Fig. 8.

One may describe the duality map $\Delta_n : \mathcal{E}_n \longrightarrow \mathcal{E}_n$ more geometrically as follows:

- Think about $\epsilon \in \mathcal{E}_n$ as a cell decomposition of the underlying segment, which is oriented by the chosen order of its vertices.
- Take the dual cell decomposition $\Delta_n(\epsilon)$ from the topological viewpoint, endowed with the opposite orientation.
- Look only at the cells of $\Delta_n(\epsilon)$ which are dual to those of ϵ which may have both decorations. Decorate them by respecting the following associations of the decorations of dual cells: curved/straight $\leftrightarrow$ breaking/neutral.

Example 4. An example of complexity 12 is drawn in Fig. 9. The vertical lines connect cells which are dual to each other. The arrows indicate the orientations of the two cell decompositions associated with the numberings of their vertices.

It is an immediate consequence of Definition 8 that:

Corollary 2. *The duality map $\Delta_n : \mathcal{E}_n \to \mathcal{E}_n$ reverses the partial order $\preceq$.*

In fact, Δ_n *is characterized by the previous property: it is the unique bijection of the set $\mathcal{E}_n$ on itself which reverses the staircase partial order.* We will present the proof of this fact in Sect. 8 (see Theorem 4), after having recalled some generalities about partial order relations, lattices, and their Hasse diagrams in Sect. 7.

Remark 6. As proved by Wall [11] (see also [12, Section 7.4]), if one starts from a germ (C, O) of an irreducible complex analytic curve in the complex projective plane $\mathbb{P}^2$, then the germ of its projective dual $(\check{C}, t_O)$ at the tangent line t_O to C at O is strongly similar to C. More precisely, their strict transforms by the blow-ups of O and t_O respectively have the same combinatorial type. This is enough to see that

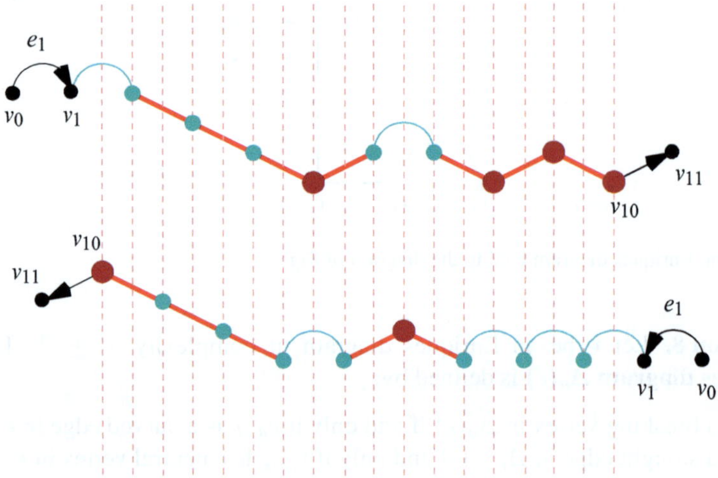

Fig. 9 Dual Enriques diagrams as dual cell complexes

our notion of duality does not correspond to this projective duality. Indeed, if the Enriques diagram $\epsilon(C)$ of (C, O) is that of Example 1, that of its strict transform after one blow-up is obtained by removing e_1 (and renumbering consequently the remaining vertices and edges). As shown by Example 3, this new diagram is not isomorphic to the one obtained by the analogous procedure from $\Delta_6(\epsilon(C))$ (here one has also to make e_1 curved). But a more fundamental difference between the two notions of duality is that the combinatorial type of the projective dual is not determined by the combinatorial type of the initial branch (see [12, Example 7.4.1]).

Let us present a third appearance of Fibonacci numbers in our context (recall from the proof of Theorem 1 that we denote by $\mathscr{A}_n$ the set of Enriques diagrams in $\mathscr{E}_{m+2}$ such that v_{n-2} is neutral):

Proposition 7. *The set $\mathscr{D}_n$ of self-dual elements of $\mathscr{E}_n$ is in a natural bijection with $\mathscr{E}_{(n+2)/2}$ if n is even and with $\mathscr{A}_{(n+3)/2}$ if n is odd. In particular, there are F_{n-2} self-dual combinatorial types of plane branches of blow-up complexity n.*

Proof. Informally speaking, the idea is that "the first half" of a self-dual Enriques diagram determines its second half. Moreover, one knows how both halves are joined. This allows to get a bijection between the set of self-dual diagrams of given complexity and a subset of the diagrams of approximately half the complexity. In order to make this argument precise, we describe it according to the parity of n.

Consider first a self-dual diagram $\epsilon \in \mathscr{D}_{2m}$. If its edge e_m were curved, then its adjacent vertex v_m would be neutral. Therefore, in the dual diagram the edge e_m

would be straight, which would contradict the self-duality. This shows that $e_m \in \chi(\epsilon)$, which implies by the same argument that $v_m \notin \chi(\epsilon)$. As a consequence, the map:

$$\begin{array}{ccc} \mathscr{D}_{2m} & \longrightarrow & \mathscr{E}_{m+1} \\ \epsilon & \longrightarrow & \text{the diagram whose code is } \chi(\epsilon) \cap S_{m+1} \end{array}$$

is bijective.

We can argue similarly for the self-dual diagrams of $\mathscr{E}_{2m+1}$. Given $\epsilon \in \mathscr{D}_{2m+1}$, one sees that its vertex v_m is neutral. Analogously to the previous case, we get a bijection:

$$\begin{array}{ccc} \mathscr{D}_{2m+1} & \longrightarrow & \mathscr{A}_{m+2} \\ \epsilon & \longrightarrow & \text{the diagram whose code is } \chi(\epsilon) \cap S_{m+2} \end{array}.$$

Using Theorem 1 as well as the computation of the cardinality of $\mathscr{A}_n$ done in its proof, we get now the cardinality of the set of self-dual diagrams of given complexity, as stated in the proposition. □

7 Basic Facts About Posets and Lattices

In this section we explain basic facts about *partially ordered sets*, *lattices*, and *Hasse diagrams*. We will apply these notions only to *finite sets*. For a more detailed introduction to lattices, one may consult Birkhoff and Bartee's book [2]. For much more details about lattices and the historical development of their theory, one may consult Birkhoff [1] and Grätzer [6].

Definition 9. A **partial order** $\preceq$ on a set S is a binary relation which is reflexive, antisymmetric and transitive. A **partially ordered set (poset)** is a set endowed with a partial order. A **hereditary subset** of a poset is such that each time it contains some element, it also contains all the elements which are less or equal to it.

As is customary for the usual partial order $\leq$ on $\mathbb{R}$, if $\preceq$ is a partial order on a set S, we denote by $\prec$ the (transitive) binary relation defined by:

$$a \prec b \iff (a \preceq b \text{ and } a \neq b).$$

Formally speaking, this second relation is not a partial order, as it is not reflexive. Nevertheless, common usage allows to speak also about *the partial order* $\prec$. Notice that the usual notation of inclusion of subsets of a given set does not respect this convention: $A \subset B$ does not imply that $A \neq B$.

Definition 10. Let $(E, \preceq)$ be a poset. The associated **successor relation**, denoted $\prec_s$, is the binary relation defined by the condition that for any $a, b \in E$, one has

$a \prec_s b$ if and only if $a \prec b$ and if there is no element $c \in E$ such that $a \prec c \prec b$. We say then that b **is a successor of** a or that a **is a predecessor of** b.

Therefore, any partial order defines canonically its associated successor relation. This relation may be empty, as illustrated by $\mathbb{Q}$ or $\mathbb{R}$ with their usual orders. But on *finite* sets, it is easy to see that the knowledge of the successor relation is enough to reconstruct the initial partial order (see [2, Section 2.4, Theorem 3]).

Given a partial order on a finite set, it is more economic to encode the associated successor relation, as it has less pairs of related elements. A visual way to do this encoding is through its associated *geometric graph*, which is the geometric realization of the successor binary relation, seen as a *directed graph*. Let us first recall this last notion.

An abstract **directed graph** is a triple (S, E, A) where S is a set of **vertices**, E is a set of **edges** and $A : E \to S \times S$ is a map. We denote $A(e) = (s(e), t(e))$ and we say that the vertex $s(e)$ is the **source**, the vertex $t(e)$ is the **target** of the edge e and that the two vertices are **connected by** e. If the map A is injective (i.e., any pair of vertices is connected by at most one edge) and its image is disjoint from the diagonal (i.e., there are no **loops**, which are edges connecting a vertex with itself), we say that we have a **simple directed graph**. A **directed path** is a finite sequence of edges such that the target of each edge is equal to the source of its successor. A **circuit** of a directed graph is a directed path starting and ending at the same vertex.

To any simple directed graph (S, E, A), one associates canonically its **geometric realization** (a **geometric simple directed graph**), which is a simplicial complex with S and E as sets of 0-simplices and 1-simplices respectively, each $e \in E$ corresponding to the oriented 1-simplex $\overrightarrow{s(e)t(e)}$.

For example, consider a binary relation R defined on a set S, such that no element of S is related to itself. The **geometric graph of** R is the geometric realization of the simple directed graph (S, E, A), where $E \subset S \times S$ is the set of pairs (a, b) such that $a \, R \, b$ and A is the inclusion map.

Let us come back to partial order relations:

Definition 11. The **Hasse diagram** of a poset $(S, \preceq)$ is the geometric graph of the associated successor relation $\prec_s$.

Example 5. Consider the poset of subsets of $\{1, 2, 3\}$, ordered by inclusion. Its Hasse diagram is represented in Fig. 10.

As stated in [2, Section 2–10] and as it may be easily verified, one may characterize the simple directed graphs associated with partial orders on finite sets in the following way:

Proposition 8. *A simple directed graph with* finite *vertex set S is the Hasse diagram of a partial order on S if and only if it has no circuits and the only directed path joining the source and the target of any edge is the edge itself.*

Among posets, we will be particularly interested in *lattices*:

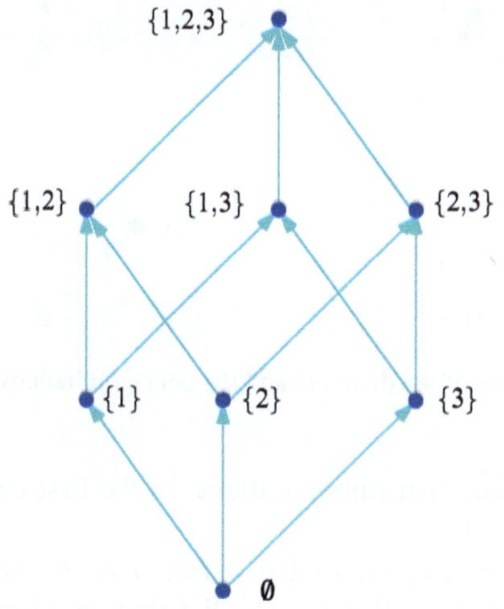

Fig. 10 The Hasse diagram of the subsets of a set with three elements

Definition 12. A **lattice** is a poset $(S, \preceq)$ such that any two elements $a, b \in S$ admit:

- a greatest lower bound denoted by $a \wedge b$ and called their **infimum**.
- a smallest upper bound denoted by $a \vee b$ and called their **supremum**.

A **sublattice** of a given lattice is a subset which is closed under the ambient infimum and supremum operations. A lattice is **distributive** if each one of the two operations is distributive with respect to the other one.

In the literature, $a \wedge b$ is also called the *meet* of a and b, and $a \vee b$ is called their *join*. The knowledge of these two operations allows to reconstruct the partial order, as:

$$a \preceq b \Leftrightarrow a = a \wedge b \Leftrightarrow b = a \vee b.$$

Example 6. The simplest examples of non-distributive lattices are the *diamond* and *pentagon* lattices, with Hasse diagrams drawn in Fig. 11. In both cases, $a \wedge (b \vee c) \neq (a \wedge b) \vee (a \wedge c)$. It is known that a lattice is distributive if and only if it does not contain any sublattice isomorphic to a diamond or a pentagon one.

As other examples of posets, let us quote:

1. the power set $\mathscr{P}(L)$ of a given set L, with the relation of inclusion;
2. the set $\mathbb{N}^*$ of positive integers, with the relation of divisibility;
3. the set of subgroups of a given group G, with the relation of inclusion; this is *not* a sublattice of $\mathscr{P}(G)$, as the union of two subgroups is not in general a group.

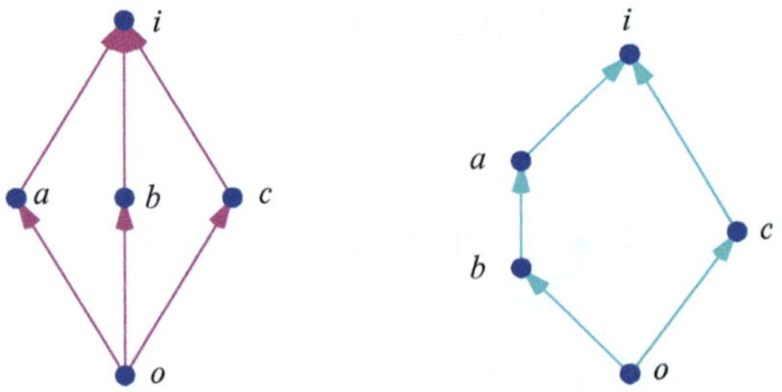

Fig. 11 The Hasse diagrams of the diamond and the pentagon lattices

The first two cases are distributive lattices. In the first case, one has moreover a structure of Boolean algebra.

Of course, *any sublattice of a distributive lattice is also distributive*. In particular, any sublattice of a power set $(\mathscr{P}(L), \subseteq)$ is distributive. Conversely, if a *finite* lattice $(S, \preceq)$ is distributive, then it may be embedded as a sublattice of a power set. Such an embedding may be obtained in a canonical way. In order to explain this, we need one more definition:

Definition 13. Let $(S, \preceq)$ be a lattice. An element $i \in S$ is called **sup-irreducible** if it cannot be written as $a \vee b$, with a and b distinct from i.

The canonical realization of a distributive lattice as a sublattice of a power set is described by the following result (see [1, Chapter IX.4] or [6, Chapter 7, Theorem 9] for the proof and Definition 9 for the notion of hereditary subset of a poset):

Proposition 9. *Let $(S, \preceq)$ be a finite distributive lattice. Let $I^{\vee}$ be its subset of sup-irreducible elements. If a is any element of S, denote by $\rho(a)$ the subset of $I^{\vee}$ consisting of the sup-irreducible elements which are less or equal to a. Then the map $\rho : S \to \mathscr{P}(I^{\vee})$ embeds $(S, \preceq)$ as a sublattice of $(\mathscr{P}(I^{\vee}), \subseteq)$. Its image consists of the hereditary subsets of the poset $(I^{\vee}, \preceq)$.*

We have now enough material to proceed to the proof of the uniqueness of the duality of $(\mathscr{E}_n, \preceq)$.

8 The Uniqueness of the Duality of Plane Branches

If a finite lattice admits an order-reversing bijection, such a bijection is not necessarily unique. For instance, as the reader may easily check, the diamond lattice of Fig. 11 has six such bijections. In this section we show that the duality Δ_n defined in Sect. 6 is the *unique* bijection of $\mathscr{E}_n$ which reverses the staircase partial order structure.

Before proving the announced uniqueness, let us draw the Hasse diagrams of a few posets $(\mathscr{E}_n, \preceq)$.

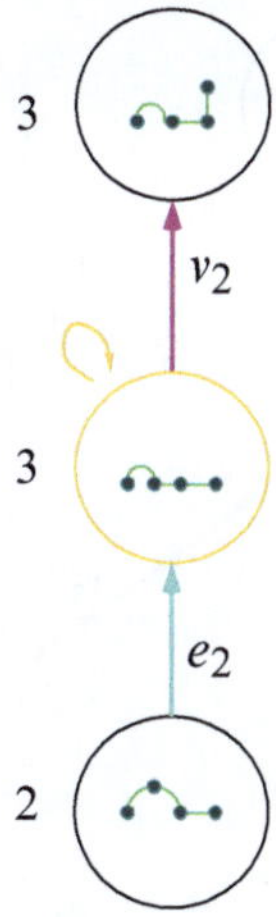

Fig. 12 The Hasse diagram of the lattice structure on $\mathscr{E}_4$

Example 7. The case $n = 3$ is trivial, as $\mathscr{E}_3$ has only one element $\alpha_3 = \omega_3$. In the Figs. 12, 13, and 14 are represented the Hasse diagrams of the lattices $(\mathscr{E}_4, \preceq)$, $(\mathscr{E}_5, \preceq)$, and $(\mathscr{E}_6, \preceq)$, respectively. If an arrow goes from a diagram ϵ to a greater diagram ϵ', it is labeled by the unique element of $\chi(\epsilon') \setminus \chi(\epsilon)$. The diagrams which have a self-referencing arrow are the self-dual ones. Near each diagram we indicate the corresponding initial multiplicity.

As a first consequence of the constructions of Sect. 6 and of the definitions of Sect. 7, one has:

Proposition 10. *The staircase partial order relation of Definition 7 endows each set $\mathscr{E}_n$ with a structure of distributive lattice.*

Proof. By Lemma 2, the set $\mathscr{K}_n$ of codes of Enriques diagrams of complexity n is a sublattice of $(\mathscr{P}(S_n), \subseteq)$. As this last lattice is distributive, we deduce that the first one has also this property. By the Definition 7 of the staircase partial order relation, we conclude that $(\mathscr{E}_n, \preceq)$ is indeed a distributive lattice. □

Note that by Proposition 9, the apparently "complicated" lattices $(\mathscr{E}_n, \preceq)$ are completely described up to lattice-isomorphisms by the much "simpler" posets $(I_n^\vee, \preceq)$. As results immediately from Lemma 1, the elements of $I_n^\vee$ are:

- $\{e_i\}$ for $i \in \{2, \dots, n-2\}$;
- $\{v_i, e_i, e_{i+1}\}$ for $i \in \{2, \dots, n-3\}$ and $\{v_{n-2}, e_{n-2}\}$.

One has the following relation between the code of an Enriques diagram (see Definition 6) and its associated hereditary subset of $(I_n^\vee, \preceq)$ (see Proposition 9):

Proposition 11. *The bijection from S_n to $I_n^\vee$ which sends each e_i to $\{e_i\}$, each v_i to $\{v_i, e_i, e_{i+1}\}$ and v_{n-2} to $\{v_{n-2}, e_{n-2}\}$, transforms the code $\chi(\epsilon) \in \mathscr{P}(S_n)$ of an*

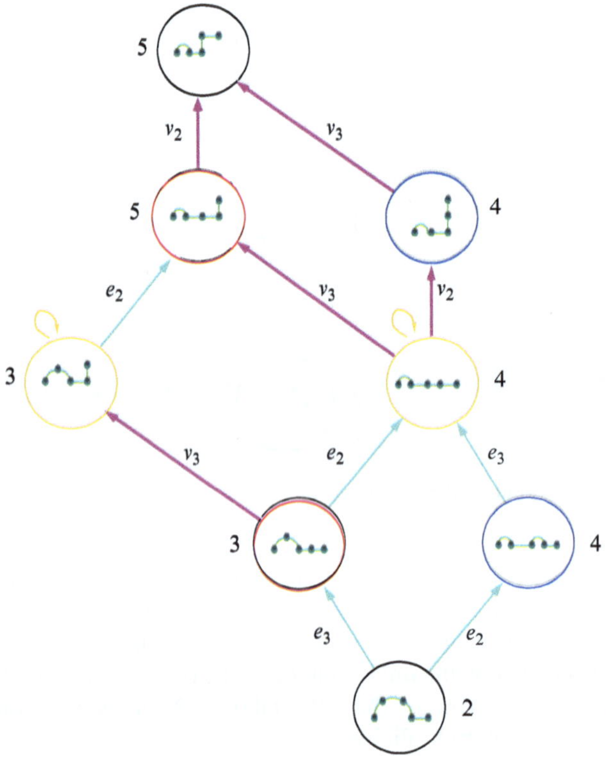

Fig. 13 The Hasse diagram of the lattice structure on $\mathcal{E}_5$

Enriques diagram $\epsilon \in \mathcal{E}_n$ *into the hereditary subset* $\rho(\epsilon) \in \mathcal{P}(I_n^{\vee})$. *Therefore, if one considers the poset structure* $(S_n, \preceq)$ *inherited from* $(I_n^{\vee}, \preceq)$ *by the previous bijection, its Hasse diagram is as drawn in Fig. 15 and* $\mathcal{K}_n$ *is exactly the set of hereditary subsets of* S_n.

Proof. The statement about the Hasse diagram is checked easily using the definition of the bijection. Then one checks using Lemma 1 the statement about the correspondence between codes and hereditary subsets of $(I_n^{\vee}, \preceq)$. □

Example 8. In Fig. 16 is represented again the lattice of Fig. 14. This time, we have represented each element of the lattice $(\mathcal{E}_6, \preceq)$ through the associated hereditary subset of $(S_6, \preceq)$. Each hereditary subset is represented as a set of discs at the vertices of the Hasse diagram of $(S_6, \preceq)$. The sup-irreducible elements of $(\mathcal{E}_6, \preceq)$, which correspond therefore to the elements of S_6 by the bijection defined in Proposition 11, are represented against a colored background.

The previous considerations allow us to prove the announced uniqueness theorem:

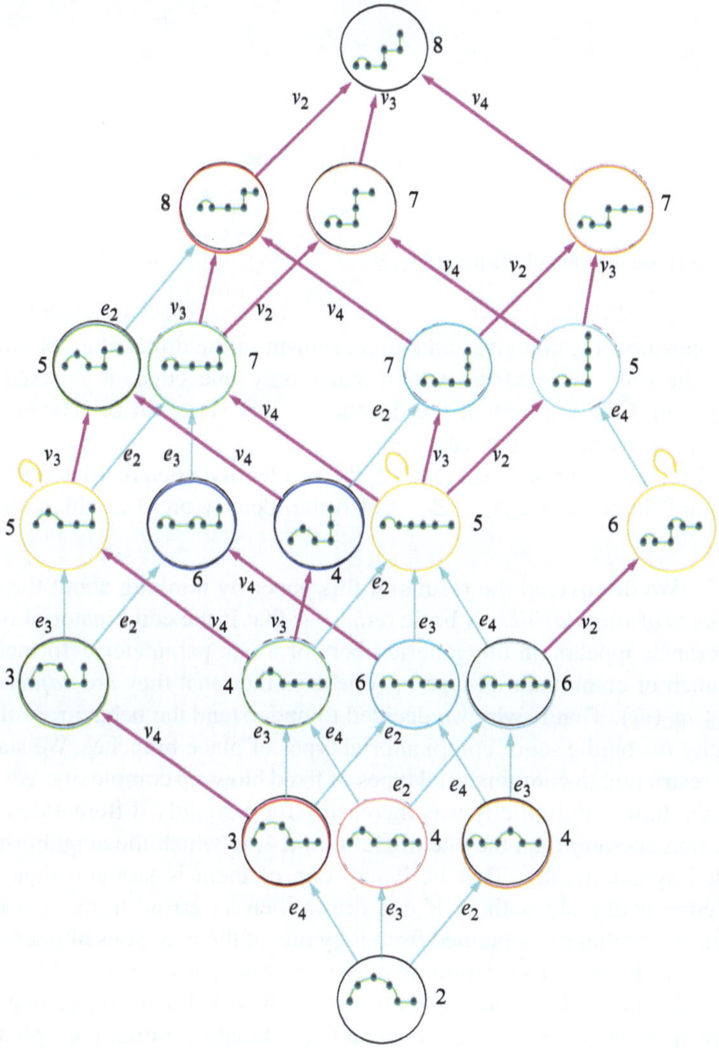

Fig. 14 The Hasse diagram of the lattice structure on $\mathscr{E}_6$

Theorem 4. *The involution Δ_n is the unique bijection of the set $\mathscr{E}_n$ on itself which reverses the staircase partial order.*

Proof. By our definition of the map Δ_n, it is equivalent to show the analogous property for the poset $(\mathscr{K}_n, \subseteq)$ of codes of the Enriques diagrams of complexity n and the corresponding involution, which we denote again by Δ_n (see Definition 8). Assume that there is another order-reversing bijection Δ'_n of $\mathscr{K}_n$. Then $\Delta_n \circ \Delta'_n$ is an automorphism of this poset. In particular, it restricts to an automorphism of the subposet $(I_n^{\vee}, \preceq)$ of its sup-irreducible elements (see Definition 13).

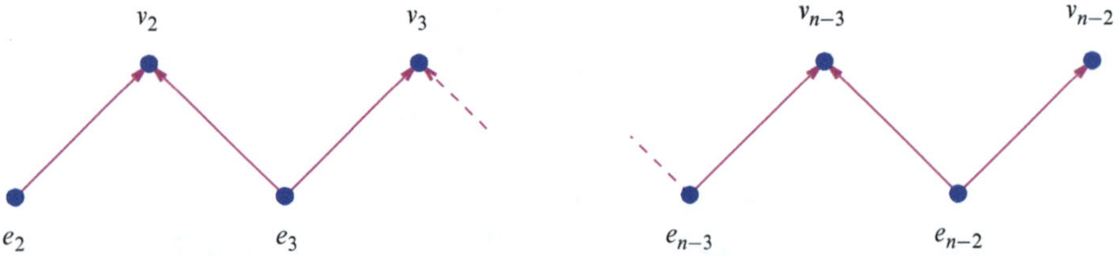

Fig. 15 The Hasse diagram of the poset $(S_n, \preceq)$

By Proposition 11, this gives an automorphism of the directed graph of Fig. 15. As e_2 is the only vertex from which starts only one edge, it is fixed by this automorphism. Looking then at the distance to this vertex in the Hasse diagram, one sees that all vertices are fixed.

Therefore, the automorphism $\Delta_n \circ \Delta'_n$ is the identity when restricted to $(I_n^\vee, \preceq)$. Proposition 9 implies that $\Delta_n \circ \Delta'_n$ is also the identity on $\mathcal{K}_n$, which shows that $\Delta_n = \Delta'_n$. □

Remark 7. We discovered the results of this paper by thinking about the problem of *adjacency of singularities*. A basic remark is that if the combinatorial type ϵ_2 of a plane branch appears on the generic fibers of a one parameter deformation of a plane branch of combinatorial type ϵ_1 (one says then that they are *adjacent*), then $m_0(\epsilon_2) \leq m_0(\epsilon_1)$. That is why we decided to understand the behavior of the initial multiplicity m_0 on the set of combinatorial types of plane branches. We started our study by restricting to combinatorial types of fixed blow-up complexity. We checked whether the initial multiplicity was increasing for a slightly different definition of straightening operator than the one of Definition 4, in which the neighboring edges are treated symmetrically. That is, if any one of them is straight, then the new straight edge is aligned with it. If one draws then an arrow from each Enriques diagram to every diagram obtained from it by one of the two types of operators, one obtains precisely the Hasse diagrams of the staircase partial orders! On that of $\mathcal{E}_6$ (see Fig. 14), the duality jumped to our eyes, which led us to prove that it was a general phenomenon. One may also see on Fig. 14 that the initial multiplicity is not necessarily increasing for the staircase partial order on $\mathcal{E}_6$. In fact, and we leave this as an exercise for the reader, it is never increasing on $(\mathcal{E}_n, \preceq)$, for $n \geq 6$.

Acknowledgements This project was supported by the Labex CEMPI (ANR-11-LABX-0007-01). The first author wants to thank the Laboratoire Painlevé for the very pleasant working conditions during her stay in Lille. The first author was also partially supported by the ERC Starting Grant project TGASS and by Spanish Contract MTM2010-21740-C02-01. The second author was also partially supported by the grant ANR-12-JS01-0002-01 SUSI.

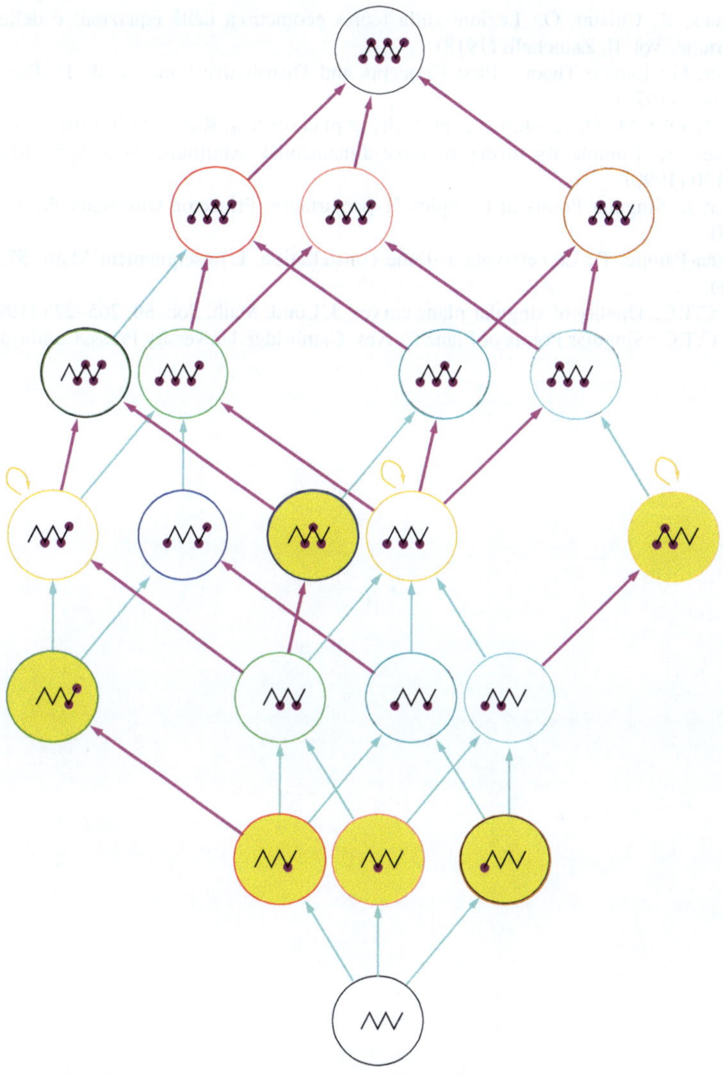

Fig. 16 The lattice structure on $\mathscr{E}_6$ encoded using its sup-irreducible elements

References

1. Birkhoff, G.: Lattice Theory. Revised edition, A.M.S. Coll. Publ. **XXV**. Amer. Math. Soc. (1948)
2. Birkhoff, G., Bartee, T.C.: Modern Applied Algebra. McGraw-Hill, New York (1970)
3. Brieskorn, E., Knörrer, H.: Plane Algebraic Curves. Translated from the German by John Stillwell. Birkhäuser Verlag, Basel (1986)
4. Casas-Alvero, E.: Singularities of Plane Curves. Cambridge University Press, Cambridge (2000)

5. Enriques, F., Chisini, O.: Lezioni sulla teoria geometrica delle equazioni e delle funzioni algebriche. Vol. II, Zanichelli (1918)
6. Grätzer, G.: Lattice Theory. First Concepts and Distributive Lattices. W. H. Freeman, San Francisco (1971)
7. Lê, D.T., Oka, M.: On resolution complexity of plane curves. Kodai Math. J. **18**(1), 1–36 (1995)
8. Matveev, S.: Complexity theory of three-dimensional manifolds. Acta Appl. Math. **19**(2), 101–130 (1990)
9. Milnor, J.: Singular Points of Complex Hypersurfaces. Princeton University Press, Princeton (1968)
10. Popescu-Pampu, P.: Le cerf-volant d'une constellation. L'Enseignement Math. **57**, 303–347 (2011)
11. Wall, C.T.C.: Duality of singular plane curves. J. Lond. Math. Soc. **50**, 265–275 (1994)
12. Wall, C.T.C.: Singular Points of Plane Curves. Cambridge University Press, Cambridge (2004)

Four Generated, Squarefree, Monomial Ideals

Adrian Popescu and Dorin Popescu

Abstract Let $I \supsetneq J$ be two squarefree monomial ideals of a polynomial algebra over a field generated in degree $\geq d$, resp. $\geq d+1$. Suppose that I is either generated by three monomials of degrees d and a set of monomials of degrees $\geq d+1$, or by four special monomials of degrees d. If the Stanley depth of I/J is $\leq d+1$ then the usual depth of I/J is $\leq d+1$ too.

Keywords Monomial ideals • Depth • Stanley depth

2010 Mathematics Subject Classification: Primary 13C15, Secondary 13F20, 13F55, 13P10.

1 Introduction

Let K be a field and $S = K[x_1, \ldots, x_n]$ be the polynomial K-algebra in n variables. Let $I \supsetneq J$ be two squarefree monomial ideals of S and suppose that I is generated by squarefree monomials of degrees $\geq d$ for some positive integer d. After a multigraded isomorphism we may assume either that $J = 0$, or J is generated in degrees $\geq d+1$. By [5, Proposition 3.1] (see [12, Lemma 1.1]) we have $\operatorname{depth}_S I/J \geq d$. Depth of I/J is a homological invariant and depends on the characteristic of the field K.

The purpose of our paper is to study upper bound conditions for $\operatorname{depth}_S I/J$. Let B (resp. C) be the set of the squarefree monomials of degrees $d+1$ (resp. $d+2$)

A. Popescu (✉)
Department of Mathematics, University of Kaiserslautern, Erwin-Schrödinger-Str., 67663 Kaiserslautern, Germany
e-mail: popescu@mathematik.uni-kl.de

D. Popescu
Simion Stoilow Institute of Mathematics of Romanian Academy, Research unit 5, P.O.Box 1-764, Bucharest 014700, Romania
e-mail: dorin.popescu@imar.ro

D. Ibadula, W. Veys (eds.), *Bridging Algebra, Geometry, and Topology*, Springer Proceedings in Mathematics & Statistics 96,
DOI 10.1007/978-3-319-09186-0_14

of $I \setminus J$. Suppose that I is generated by some squarefree monomials $f_1, \dots, f_r$ of degrees d for some $d \in \mathbb{N}$ and a set of squarefree monomials E of degree $\geq d+1$. If $d = 1$ and each monomial of $B \setminus E$ is the least common multiple of two f_i then it is easy to show that depth $_S I/J = 1$ (see Lemma 3). Trying to extend this result for $d > 1$ we find an obstruction given by Example 2. Our extension given by Lemma 4 is just a special form, but a natural condition seems to be given in terms of the Stanley depth.

More precisely, let $P_{I \setminus J}$ be the poset of all squarefree monomials of $I \setminus J$ with the order given by the divisibility. Let P be a partition of $P_{I \setminus J}$ in intervals $[u, v] = \{w \in P_{I \setminus J} : u|w, w|v\}$, let us say $P_{I \setminus J} = \cup_i [u_i, v_i]$, the union being disjoint. Define sdepth $P = \min_i \deg v_i$ and the *Stanley depth* of I/J given by sdepth $_S I/J = \max_P$ sdepth P, where P runs in the set of all partitions of $P_{I \setminus J}$ (see [5, 20]). Stanley's Conjecture says that sdepth $_S I/J \geq$ depth $_S I/J$. The Stanley depth of I/J is a combinatorial invariant and does not depend on the characteristic of the field K. Stanley's Conjecture holds when $J = 0$ and I is an intersection of four monomial prime ideals by [8, 10], or I is such that the sum of every three different of its minimal prime ideals is a constant ideal by [11] (see also [14]), or I is an intersection of three monomial primary ideals by [22], or a monomial almost complete intersection by [4].

Theorem 1 (D. Popescu [12, Theorem 4.3]). *If* sdepth $_S I/J = d$ *then* depth $_S I/J = d$, *that is Stanley's Conjecture holds in this case.*

Next step in the study of Stanley's Conjecture is to show the following weaker conjecture.

Conjecture 1. Suppose that $I \subset S$ is minimally generated by some squarefree monomials $f_1, \dots, f_r$ of degrees d, and a set E of squarefree monomials of degrees $\geq d+1$. If sdepth $_S I/J = d+1$ then depth $_S I/J \leq d+1$.

Set $s = |B|$, $q = |C|$. In the study of the above conjecture very useful seem to be the following two particular results of [13, Theorem 1.3] and [19, Theorem 2.4].

Theorem 2 (D. Popescu). *If* $s > q + r$ *then* depth $_S I/J \leq d+1$.

Theorem 3 (Y. Shen). *If* $s < 2r$ *then* depth $_S I/J \leq d+1$.

These results were hinted by Stanley's Conjecture since it is obvious that $s > q + r$, or $s < 2r$ imply sdepth $_S I/J \leq d+1$. The proof of Theorem 2 uses Koszul homology (see [1, Section 1.6]). Shen's proof of the above theorem as well of Theorem 2 is easy and uses the Hilbert depth considered by Bruns–Krattenhaler–Uliczka [2] (see also [6, 21]).

An equivalent definition for the Stanley depth is:

$$\text{sdepth}\,(M) = \max\{\text{sdepth}\,\mathcal{D} \mid \mathcal{D} \text{ is a Stanley decomposition of } M\},$$

where a Stanley decomposition of a $\mathbb{Z}$-graded (resp. $\mathbb{Z}^n$-graded) S-*module* M is $\mathcal{D} = (S_i, u_i)_{i \in I}$, where u_i are homogenous elements of M and S_i are graded (resp.

$\mathbb{Z}^n$-graded) K-algebra retracts of S and $S_i \cap Ann(u_i) = 0$ such that $M = \oplus_i S_i u_i$; and sdepth $\mathcal{D}$ is the depth of the S-module $\oplus_i S_i u_i$. A more general concept is the one of Hilbert depth of a $\mathbb{Z}$-graded module M, denoted by hdepth $_1(M)$. Instead of considering equality, we only assume that $M \cong \oplus S_i(-s_i)$, where $s_i \in \mathbb{Z}$. One can also construct hdepth$_n$ analogously if M is a multigraded (that is $\mathbb{Z}^n$) module.

In [9] is presented (and implemented) an algorithm that computes hdepth $_1(M)$ based on a Theorem of Uliczka [21]; and in [7] was presented an algorithm that computes hdepth $_n(M)$. Meanwhile, another algorithm that computes hdepth $_1$ and more was given in [3]. Popescu [9, Proposition 1.9] gives a partial answer to a question of Herzog asking whether sdepth m = sdepth $(S \oplus m)$, where m is the graded maximal ideal of S. More precisely, for $n \in \{1,2,3,4,5,7,9,11\}$ one obtains hdepth $_1 m$ = hdepth $_1(S \oplus m)$, which gives sdepth m = sdepth $(S \oplus m)$ (again Hilbert depth helps the study of Stanley depth). For $n = 6$ we have hdepth $_1 m \neq$ hdepth $_1(S \oplus m)$, which means that in general Herzog's question could have a negative answer. Later Ichim and Zarojanu checked the case $n = 6$ and found indeed a counterexample to Herzog's question, which will be included in the new version of [7].

An important step in proving Conjecture 1 is the following theorem.

Theorem 4 (D. Popescu-A. Zarojanu [16], [17, Theorem 1.5]). *Conjecture 1 holds in each of the following three cases:*

1. $r = 1$,
2. $1 < r \leq 3$, $E = \emptyset$.

Next theorem is the main result of this paper.

Theorem 5. *Conjecture 1 holds in each of the following two cases:*

1. $r \leq 3$,
2. $r = 4$, $E = \emptyset$ *and there exists* $c \in C$ *such that* supp $c \not\subset \cup_{i \in [4]}$supp f_i.

This follows from our Theorems 6, 8. The proof of 6 extends the proof of [17, Theorem 2.3].

We owe thanks to A. Zarojanu, who noticed some small mistakes in a previous version of this paper and gave us the bad Example 5.

2 Depth and Stanley Depth

Let $I \supsetneq J$ be two squarefree monomial ideals of S. We assume that I is generated by squarefree monomials $f_1, \ldots, f_r$ of degrees d for some $d \in \mathbb{N}$ and a set of squarefree monomials E of degree $\geq d + 1$. We may suppose that either $J = 0$, or is generated by some squarefree monomials of degrees $\geq d + 1$. B (resp. C) denotes the set of the squarefree monomials of degrees $d + 1$ (resp. $d + 2$) of $I \setminus J$.

Lemma 1. *Let $J \subset I$ be square free monomial ideals and $j \in [n]$ be such that $(J : x_j) \neq (I : x_j)$. Then* depth $_S(I : x_j)/(J : x_j) \geq$ depth $_S I/J$.

Proof. We have

$$pd_S I/J \geq pd_{S_{x_j}}(I/J) \otimes S_{x_j} = pd_{S_{x_j}}((I : x_j)/(J : x_j)) \otimes S_{x_j}$$
$$= pd_S((I : x_j)/(J : x_j))$$

the last equality holds since x_j does not appear among the generators of $(I : x_j)$ and $(J : x_j)$. Now it is enough to apply the Auslander–Buchsbaum Theorem.

Lemma 2. *Let $t \in [n]$. Suppose that $I \neq J + I \cap (x_t)$ and* depth $_S I/(J + I \cap (x_t)) = d$. *If* depth $_S I/J \geq d + 1$ *then* depth $_S I/J = d + 1$.

Proof. In the following exact sequence

$$0 \to (I : x_t)/(J : x_t) \xrightarrow{x_t} I/J \to I/(J + I \cap (x_t)) \to 0$$

the first term has depth $d + 1$ by the Depth Lemma. Now it is enough to apply the above lemma.

Let w_{ij} be the least common multiple of f_i and f_j and set W to be the set of all $w_{ij} \in B$.

Lemma 3. *If $d = 1$ and $B \subset E \cup W$ then* depth $_S I/J = 1$.

Proof. First suppose that $E = \emptyset$, let us say $I = (x_1, \ldots, x_r)$. Set $S' = K[x_1, \ldots, x_r]$, $I' = I \cap S'$, $J' = J \cap S'$. By hypothesis $B \subset S'$ and it follows that $(x_{r+1}, \ldots, x_n)I \subset J$ and so depth $_S I =$ depth $_{S'} I' = 1$. But depth $_S J \geq 2$, if $J \neq 0$, and so depth $_S I/J = 1$ by the Depth Lemma.

Now, suppose that $E \neq \emptyset$. In the following exact sequence

$$0 \to (x_1, \ldots, x_r)/J \cap (x_1, \ldots, x_r) \to I/J \to I/(J, x_1, \ldots, x_r) \to 0$$

the first term has depth 1 as above and the last term has depth $\geq d + 1$ since it is generated by squarefree monomials of degrees ≥ 2 from E. Again the Depth Lemma gives depth $_S I/J = 1$.

Lemma 4. *Suppose that $I \subset S$ is generated by some squarefree monomials $f_1, \ldots f_r$ of degree d. Assume that for all $b \in B$ all divisors of b of degree d are among $\{f_1, \ldots, f_r\}$. Then* depth $_S I/J = d$.

Proof. Apply induction on $d \geq 1$. If $d = 1$ then apply the above lemma. Assume $d > 1$. We may suppose that $n \in$ supp f_1. $(I : x_n)$ is an extension of a squarefree monomial ideal I' of $S' = K[x_1, \ldots, x_{n-1}]$ which is generated in degree $\geq d - 1$. Similarly $(J : x_n)$ is generated by a squarefree monomial ideal J' of S'. Note that the generators of I' of degree $d - 1$ have the form $f_i' = f_i/x_n$ for $f_i \in (x_n)$, and the squarefree monomials B' of degrees d from $I' \setminus J'$ have the form $b' = b/x_n$

for some $b \in (B \cap (x_n))$. Certainly we must consider also the case when $f_j \notin (x_n)$. If $x_n f_j \in J$ then $f_j \in (J : x_n)$ is not in B'. Otherwise, $f_j = (x_n f_j)/x_n \in B'$. Note that all divisors of degree $d-1$ of each $b' \in B'$ are among f_i'. By induction hypothesis we have depth $_{S'} I'/J' = d-1$ and so depth $_S (I : x_n)/(J : x_n) = d$. Now it is enough to apply Lemma 1.

An obstruction to improve Lemma 3 and the above lemma is given by the following example.

Example 1. Let $n = 5, d = 2, r = 5, I = (x_1x_2, x_1x_3, x_2x_3, x_1x_4, x_3x_5)$, $J = (x_1x_2x_5, x_1x_4x_5, x_2x_3x_4, x_3x_4x_5)$, $B = \{x_1x_2x_3, x_1x_2x_4, x_1x_3x_4, x_1x_3x_5, x_2x_3x_5\}$. We have depth $_S I/J = 3$ because depth $_S S/J = 3$, depth $_S S/I = 2$ and with the help of Depth Lemma. Note that each $b \in B$ is the least common multiple of two generators of I, but for example $b = x_1x_2x_4$ has $x_2x_4 \notin I$ as a divisor of degree 2.

Let $C_2 = C \cap W$ and C_3 be the set of all $c \in C$ having all divisors from $B \setminus E$ in W. In particular each monomial of C_3 is the least common multiple of three of f_i. The converse is not true as shows the following example.

Example 2. Let $n = 4, d = 2, r = 3, f_1 = x_1x_2, f_2 = x_2x_3, f_3 = x_3x_4, I = (f_1, f_2, f_3)$ and $J = 0$. Then $c = x_1x_2x_3x_4$ is the least common multiple of f_1, f_2, f_3 but has a divisor $b = x_1x_2x_4 \in B$ which is not the least common multiple of two f_i.

Next theorem is our key result, its proof is based on [17, Theorem 2.1] and will be given in the last section. The main reason that this proof works for $r \leq 3$ but not for $r = 4$ is that in the first case $|C_3| \leq 1$ but in the second one we may have $|C_3| = 4$, which makes the things harder. However, for $r \geq 5$ will appear a new problem since we may have $B \subset W$ and $s \geq 2r$ (for example when $r = 5, d = 2$ we may have $s = 10 = 2r$). We remind that by Theorem 3 we had to check Stanley's Conjecture only when $s \geq 2r$.

Theorem 6. *Conjecture 1 holds for $r \leq 3$, the case $r = 1$ being given in Theorem 4.*

Example 3. Let $n = 5, f_1 = x_1x_2, f_2 = x_1x_3, f_3 = x_1x_4, a = x_2x_3x_5$, E = {a}, $I = (f_1, f_2, f_3, a)$, $J = (x_4a)$. We have $w_{12} = f_1x_3, w_{13} = f_1x_4, w_{23} = f_2x_4$. Set $c = w_{12}x_4, c_1 = w_{12}x_5, c_2 = w_{23}x_5, c_3 = w_{13}x_5$. Then $C = \{c, c_1, c_2, c_3\}$ and $B \setminus E = B \cap (\cup_i [f_i, c_i])$. Thus $s = 7, q = 4, r = 3$. It is easy to see that sdepth $_S I/J = 3$. Indeed, note that c_1 is the only $c' \in C$ which is multiple of a. Suppose that there exists a partition P on $P_{I/J}$ with sdepth 4. Then we have necessarily in P the interval $[a, c_1]$. If P contains the interval $[f_1, c]$ then it must contain also the intervals $[f_2, c_2]$ and so $[f_3, c_3]$, but then $w_{13} \in [f_1, c] \cap [f_3, c_3]$, that is the union is not disjoint. If P contains the interval $[f_1, c_3]$ then P contains either $[f_3, c], [f_2, c_2]$, or $[f_2, c], [f_3, c_2]$, in both cases the intersection of these two intervals contains w_{23}, which is false. By Theorem 6 we get depth $_S I/J \leq 3$, this inequality being in fact an equality.

3 A Special Case of $r = 4$

Theorem 7. *Suppose that $I \subset S$ is minimally generated by some squarefree monomials $\{f_1, \dots, f_r\}$ of degrees d such that there exists $c \in C$ with* $\text{supp}\, c \not\subset \cup_{i\in[r]}\text{supp}\, f_i$. *If Conjecture 1 holds for $r' < r$ and* $\text{sdepth}_S I/J = d + 1$, *then* $\text{depth}_S I/J \leq d + 1$.

Proof. By [17, Lemma 1.1] we may assume that $C \subset (W)$. By hypothesis, choose $t \in \text{supp}\, c$ such that $t \notin \cup_{i\in[r]}\text{supp}\, f_i$. We may suppose that $B_t = B \cap (x_t) = \{x_t f_1, \dots x_t f_e\}$ for some $1 \leq e \leq r$. Set $I_t = I \cap (x_t)$, $J_t = J \cap (x_t)$ and $U_t = I_t/J_t$. Then B_t generates I_t.

First assume that $\text{sdepth}_S U_t \leq d + 1$. It follows that $\text{depth}_S U_t \leq d + 1$ by [12, Theorem 4.3]. But $U_t \cong (I : x_t)/(J : x_t)$ and so $\text{depth}_S U_t \geq \text{depth}_S I/J$ by Lemma 1, which is enough.

Now assume that U_t has sdepth $\geq d+2$. Let P_{U_t} be a partition on U_t with sdepth $d+2$ and let $[b_i, c_i]$ be the disjoint intervals starting with $b_i = x_t f_i$, $i \in [e]$. We may suppose that $c_i \in C$ for $i \in [e]$. We have $c_i = x_t w_{ik_i}$ for some $1 \leq k_i \leq r$, $k_i \neq i$ because $C \subset (W)$. Note that $x_t f_{k_i} \in B$ and so $k_i \leq e$. We consider the intervals $[f_i, c_i]$. These intervals contain $x_t f_i$ and w_{ik_i}. If $w_{ik_i} = w_{jk_j}$ for $i \neq j$ then we get $c_i = c_j$ which is false. Thus these intervals are disjoint.

Let I_e be the ideal generated by f_j for $e < j \leq r$ and $B \setminus (\cup_{i=1}^{e}[f_i, c_i])$. Set $J_e = I_e \cap J$. Note that $c_i \notin I_e$ for any $i \in [e]$. In the following exact sequence

$$0 \to I_e/J_e \to I/J \to I/J + I_e \to 0$$

the last term has a partition of sdepth $d+2$ given by the intervals $[f_i, c_i]$ for $1 \leq i \leq e$. It follows that $I_e \neq J_e$ because $\text{sdepth}_S I/J = d + 1$. Then $\text{sdepth}_S I_e/J_e \leq d+1$ using [18, Lemma 2.2] and so $\text{depth}_S I_e/J_e \leq d+1$ by Conjecture 1 applied for $r' < r$. But the last term of the above sequence has depth $> d$ because x_t does not annihilate f_i for $i \in [e]$. With the Depth Lemma we get $\text{depth}_S I/J \leq d + 1$.

Example 4. Let $n = 5$, $r = 4$, $f_1 = x_2x_3$, $f_2 = x_1x_2$, $f_3 = x_3x_4$, $f_4 = x_3x_5$ and $J = (x_1x_2x_4x_5)$. We have $w_{12} = x_1x_2x_3$, $w_{13} = x_2x_3x_4$, $w_{14} = x_2x_3x_5$, $w_{34} = x_3x_4x_5$, $w_{23} = x_1x_2x_3x_4$, $w_{24} = x_1x_2x_3x_5$, $C_2 = \{w_{23}, w_{24}\}$, $C = C_2 \cup \{x_1w_{34}, x_2x_3x_4x_5\}$, $\tilde{I}_1 = \{x_1f_3, x_1f_4, f_2\} \supset J$, $\tilde{I}_4 = \{f_3, x_4f_2\} \supset J$ and $B \cap (x_1) = \{x_1f_3, x_1f_4, x_4f_2, x_5f_2, w_{12}\}$, $B \cap (x_4) = \{w_{13}, w_{14}, w_{34}, x_4f_2, x_1f_3\}$. Note that $\text{sdepth}_S U_1 \leq d + 1 = 3$, $\text{sdepth}_S U_4 \leq 3$ because $|B \cap (x_1)| = |B \cap (x_4)| = 5 > |C \cap (x_1)| + 1 = |C \cap (x_4)| + 1 = 4$. Thus $\text{depth}_S U_1 = \text{depth}_S U_4 \leq 3$ and so we get $\text{depth}_S I/J \leq 3$ using two different t.

Theorem 8. *Suppose that $I \subset S$ is minimally generated by four squarefree monomials $\{f_1, \dots, f_4\}$ of degrees d such that there exists $c \in C$ such that* $\text{supp}\, c \not\subset \cup_{i\in[4]}\text{supp}\, f_i$. *If* $\text{sdepth}_S I/J = d + 1$ *then* $\text{depth}_S I/J \leq d + 1$.

Proof. Apply Theorem 7, since Conjecture 1 holds for $r < 4$ by Theorem 6.

4 Proof of Theorem 6

Suppose that $E \neq \emptyset$ and $s \leq q + r$. For $b = f_1x_i \in B$ set $I_b = (f_2, \ldots, f_r, B \setminus \{b\})$, $J_b = J \cap I_b$. If $\text{sdepth}_S I_b/J_b \geq d + 2$ then let P_b be a partition on I_b/J_b with sdepth $d + 2$. We may choose P_b such that each interval starting with a squarefree monomial of degree d, $d + 1$ ends with a monomial of C. In P_b we have some intervals $[f_k, f_kx_{i_k}x_{j_k}]$, $1 < k \leq r$ and for all $b' \in B \setminus \{b, f_2x_{i_2}, f_2x_{j_2}, \ldots, f_rx_{i_r}, f_rx_{j_r}\}$ an interval $[b', c_{b'}]$. We define $h : [[\{f_2, \ldots, f_r\} \cup B] \setminus \{b, f_2x_{i_2}, f_2x_{j_2}, \ldots, f_rx_{i_r}, f_rx_{j_r}\}] \to C$ by $f_k \to f_kx_{i_k}x_{j_k}$ and $b' \to c_{b'}$. Then h is an injection and $|\text{Im}\, h| = s - r \leq q$ (if $s = r + q$ then h is a bijection).

Lemma 5. *Suppose that the following conditions hold:*

1. $r = 2$, $4 \leq s \leq q + 2$,
2. $C \subset ((f_1) \cap (f_2)) \cup ((E) \cap (f_1, f_2)) \cup (\cup_{a,a' \in E, a \neq a'} (a) \cap (a'))$,
3. $\text{sdepth}_S I_b/J_b \geq d + 2$ *for a* $b \in (B \cap (f_1)) \setminus (f_2)$.

Then either $\text{sdepth}_S I/J \geq d + 2$, *or there exists a nonzero ideal* $I' \subsetneq I$ *generated by a subset of* $\{f_1, f_2\} \cup B$ *such that* $\text{sdepth}_S I'/J' \leq d + 1$ *for* $J' = J \cap I'$ *and* $\text{depth}_S I/(J, I') \geq d + 1$.

Proof. Since $\text{sdepth}_S I_b/J_b \geq d + 2$ we consider h as above for a partition P_b with sdepth $d + 2$ of I_b/J_b. We have an interval $[f_2, c'_2]$ in P_b. Suppose that $B \cap [f_2, c'_2] = \{u, u'\}$. A sequence $a_1, \ldots, a_k$ is called a *path* from a_1 to a_k if the following statements hold:

(i) $a_l \in B \setminus \{b, u, u'\}$, $l \in [k]$,
(ii) $a_l \neq a_j$ for $1 \leq l < j \leq k$,
(iii) $a_{l+1} | h(a_l)$ and $h(a_l) \notin (b)$ for all $1 \leq l < k$.

This path is *weak* if $h(a_j) \in (u, u')$ for some $j \in [k]$. It is *bad* if $h(a_k) \in (b)$ and it is *maximal* if either $h(a_k) \in (b)$, or all divisors from B of $h(a_k)$ are in $\{u, u', a_1, \ldots, a_k\}$. If $a = a_1$ we say that the above path *starts with* a.

By hypothesis $s \geq 4$ and so there exists $a_1 \in B \setminus \{b, u, u'\}$. Set $c_1 = h(a_1)$. If $c_1 \in (b)$ then the path $\{a_1\}$ is maximal and bad. We construct below, as an example, a path with $k > 1$. By recurrence choose if possible a_{p+1} to be a divisor from B of c_p, which is not in $\{b, u, u', a_1, \ldots, a_p\}$ and set $c_p = h(a_p)$, $p \geq 1$. This construction ends at step $p = e$ if all divisors from B of c_e are in $\{b, u, u', a_1, \ldots, a_e\}$. If $c_i \notin (b)$ for $1 \leq i < e$ then $\{a_1, \ldots, a_e\}$ is a maximal path. If one $c_p \in (u, u')$ then the constructed path is weak. If $c_e \in (b)$ then this path is bad. We have three cases:

1) there exist no weak path and no bad path starting with a_1,
2) there exists a weak path starting with a_1 but no bad path starts with a_1,
3) there exists a bad path starting with a_1.

In **the first case**, set $T_1 = \{b' \in B :$ there exists a path $a_1, \ldots, a_k$ with $a_k = b'\}$, $G_1 = B \setminus T_1$, and $I'_1 = (f_1, G_1)$, $I'_2 = (f_2, G_1)$, $I'_{12} = (f_1, f_2, G_1)$, $I'' = (G_1)$,

$J_1' = I_1' \cap J$, $J_2' = I_2' \cap J$, $J_{12}' = I_{12}' \cap J$, $J'' = I'' \cap J$. Note that $I'' \neq 0$ because $b \in I''$ and all divisors from B of a monomial $c \in U_1 = h(T_1)$ belong to T_1. Consider the following exact sequence

$$0 \to I_{12}'/J_{12}' \to I/J \to I/(J, I_{12}') \to 0.$$

If $U_1 \cap (f_1, f_2) = \emptyset$ then the last term has depth $\geq d+1$ and sdepth $\geq d+2$ using the restriction of P_b to $(T_1) \setminus (J, I_{12}')$ since $h(b') \notin I_{12}'$, for all $b' \in T_1$. When the first term has sdepth $\geq d+2$ then by [18, Lemma 2.2] the middle term has sdepth $\geq d+2$. Otherwise, the first term has sdepth $\leq d+1$ and we may take $I' = I_{12}'$.

If $U_1 \cap (f_1) = \emptyset$, but $b_2 \in T_1 \cap (f_2)$, then in the following exact sequence

$$0 \to I_1'/J_1' \to I/J \to I/(J, I_1') \to 0$$

the last term has sdepth $\geq d+2$ since $h(b') \notin I_1'$, for all $b' \in T_1$ and we may substitute the interval $[b_2, h(b_2)]$ from the restriction of P_b by $[f_2, h(b_2)]$, the second monomial from $[f_2, h(b_2)] \cap B$ being also in T_1. As above we get either sdepth $_S I/J \geq d+2$, or sdepth $_S I_1'/J_1' \leq d+1$, depth $_S I/(J, I_1') \geq d+1$. Similarly, we do when $U_1 \cap (f_2) = \emptyset$ but $U_1 \cap (f_1) \neq \emptyset$.

Now, suppose that $b_1 \in T_1 \cap (f_1)$ and $b_2 \in T_1 \cap (f_2)$. We claim to choose $b_1 \neq b_2$ and such that one from $h(b_1), h(b_2)$ is not in (w_{12}), let us say $h(b_1) \notin (w_{12})$. Indeed, if $w_{12} \notin B$ and $h(b_1), h(b_2) \in (w_{12})$ then necessarily $h(b_1) = h(b_2)$ and it follows $b_1 = b_2 = w_{12}$ which is false. Suppose that $w_{12} \in B$ and $h(b_2) = x_j w_{12}$. Then choose $b_1 = x_j f_1 \in T_1$. If $h(b_1) \in (w_{12})$ then we get $h(b_1) = h(b_2)$ and so $b_1 = b_2 = w_{12}$ which is impossible.

In the following exact sequence

$$0 \to I''/J'' \to I/J \to I/(J, I'') \to 0$$

the last term has sdepth $\geq d+2$ since we may replace the intervals $[b_1, h(b_1)]$, $[b_2, h(b_2)]$ of the restriction of P_b to $(T_1) \setminus (J, I'')$ with the disjoint intervals $[f_1, h(b_1)]$, $[f_2, h(b_2)]$. Also the last term has depth $\geq d+1$ because in the exact sequence

$$0 \to (f_2)/(J, I'') \cap (f_2) \to I/(J, I'') \to I/(J, I'', f_2) \to 0$$

the end terms have depth $\geq d+1$ since $h(b_1) \notin (f_2)$, otherwise $h(b_1) \in (w_{12})$, which is false. As above we get either sdepth $_S I/J \geq d+2$, or sdepth $_S I''/J'' \leq d+1$, depth $_S I/(J, I'') \geq d+1$.

In **the second case**, let $a_1, \ldots, a_{t_1}$ be a weak path and set $c_j = h(a_j)$ for $j \in [t_1]$. We may suppose that $c_{t_1} \in (u)$, otherwise take a shorter path. Denote T_1, U_1 as in the first case, which we keep it fix even we will change a little h. Suppose that $a_{t_1} \in (f_2)$. Then change in P_b the intervals $[a_{t_1}, c_{t_1}]$, $[f_2, c_2']$ by $[f_2, c_{t_1}]$, $[u', c_2']$. Thus the new c_2' is among $\{c_1, \ldots, c_{t_1}\} \subset U_1$, though the old $c_2' \notin U_1$. Also the new u' is in T_1. However, if the old u' is not a divisor of a c from U_1, then the proof goes as in the first case using $T_1' = T_1 \cup \{u\}$, $G_1' = B \setminus T_1'$ with $I' = I_2'$, or $I' = I_{12}'$. Otherwise, T_1' should be completed because u' is not now in $[f_2, c_2']$ and we may consider some paths starting with u'. Note that there exists a path from a_1 to u' since u' is a divisor of a monomial from U_1. It follows that there exist no bad path starting with u'. Take $\tilde{T}_1 = T_1' \cup \{b' \in B$: there exists a path from u' to $b'\}$ and the proof goes as above with $\tilde{T}_1$ instead T_1', that is with I' generated by a subset of $\{f_1, f_2\} \cup \tilde{G}_1$ for $\tilde{G}_1 = B \setminus \tilde{T}_1$.

Now suppose that $a_{t_1} \notin (f_2)$ but there exists $1 \leq v < t_1$ such that $a_v \in (f_2)$ and $a_v | c_{t_1}$. Then we may replace in P_b the intervals $[a_p, c_p], v \leq p \leq t_1$ with the intervals $[a_v, c_{t_1}], [a_{p+1}, c_p], v \leq p < t_1$. The old c_{t_1} becomes the new c_v, that is we reduce to the case when u divides c_v and $a_v \in (f_2)$, subcase solved above.

Remains to study the subcase when there exist no $a_v \in (f_2)$, $1 \leq v \leq t_1$ with $a_v | c_{t_1}$. Then there exists an $a_{t_1+1} \in B \cap (f_2)$, $a_{t_1+1} \neq u$ such that $a_{t_1+1} | c_{t_1}$. Clearly, $a_{t_1+1} \neq u'$ because otherwise $c_2' = c_{t_1}$. We have two subcases:

1′) there exists a path $a_{t_1+1}, \ldots, a_l$ such that $h(a_l) \in (a_{v'})$ for some $1 \leq v' \leq t_1$,

2′) for any path $a_{t_1+1}, \ldots, a_p$, any $h(a_j)$, $t_1 < j \leq p$ does not belong to $(a_1, \ldots, a_{t_1})$.

In the first subcase, we replace in P_b the intervals $[a_j, c_j], v' \leq j \leq l$ with the intervals $[a_{v'}, c_l], [a_{j+1}, c_j], v' \leq j < l$. The new $h(a_{t_1+1})$ is the old c_{t_1} and we may proceed as above. In the second case, we set

$$T_2 = \{b' \in B : \text{there exists a path from } a_{t_1+1} \text{ to } b'\}.$$

Note that any path starting from a_{t_1+1} can be completed to a path from a_1 by adding the monomials $a_1, \ldots, a_{t_1}$. Thus there exists no bad path starting with a_{t_1+1}, otherwise we can get one starting from a_1, which is false.

If there exists no weak path starting with a_{t_1+1} then we proceed as in the first case with T_2 instead T_1. If there exists a weak path starting with a_{t_1+1} then we proceed as above in case 2) with T_2', or $\tilde{T}_2$ instead T_1', or $\tilde{T}_1$, except in the subcase 2′) when we will define similarly a T_3 given by the paths starting with a certain a_{t_2+1}. Note that the whole set $\{a_1, \ldots, a_{t_2}\}$ has different monomials. After several such steps we must arrive in the case $p = t_m$ when $\{a_1, \ldots, a_{t_m}\}$ has different monomials and the subcase 2′) does not appear. We end this case using T_m, or T_m', or $\tilde{T}_m$ instead T_1, or T_1', or $\tilde{T}_1$.

In **the third case,** let $a_1, \ldots, a_{t_1}$ be a bad path starting with a_1. Set $c_j = h(a_j)$, $j \in [t_1]$. Then $c_{t_1} = bx_{l_1}$ and let us say $b = f_1 x_i$. If $a_{t_1} \in (f_1)$ then changing in P_b the interval $[a_{t_1}, c_{t_1}]$ by $[f_1, c_{t_1}]$ we get a partition on I/J with sdepth $d + 2$. Thus we may assume that $a_{t_1} \notin (f_1)$. If $f_1 x_{l_1} \in \{a_1, \ldots, a_{t_1 - 1}\}$, let us say $f x_{l_1} = a_v$, $1 \leq v < t_1$ then we may replace in P_b the intervals $[a_p, c_p], v \leq p \leq t_1$ with the intervals $[a_v, c_{t_1}], [a_{p+1}, c_p], v \leq p < t_1$. Now we see that we have in P_b the interval $[f_1 x_{l_1}, f_1 x_i x_{l_1}]$ and switching it with the interval $[f_1, f_1 x_i x_{l_1}]$ we get a partition with sdepth $\geq d + 2$ for I/J.

Thus we may assume that $f_1 x_{l_1} \notin \{a_1, \ldots, a_{t_1}\}$. Now set $a_{t_1+1} = f x_{l_1}$. Let $a_{t_1+1}, \ldots, a_k$ be a path starting with a_{t_1+1} and set $c_j = h(a_j)$, $t_1 < j \leq k$. If $a_p = a_v$ for $v \leq t_1$, $p > t_1$ then change in P_b the intervals $[a_j, c_j], v \leq j \leq p$ with the intervals $[a_v, c_p], [a_{j+1}, c_j], v \leq j < p$. We have in P_b an interval $[f_1 x_{l_1}, f_1 x_i x_{l_1}]$ and switching it to $[f_1, f_1 x_i x_{l_1}]$ we get a partition with sdepth $\geq d + 2$ for I/J. Thus we may suppose that in fact $a_p \notin \{b, a_1, \ldots, a_{p-1}\}$ for any $p > t_1$ (with respect to any path starting with a_{t_1+1}). We have three subcases:

1″) there exist no weak path and no bad path starting with a_{t_1+1},
2″) there exists a weak path starting with a_{t_1+1} but no bad path starts with a_{t_1+1},
3″) there exists a bad path starting with a_{t_1+1}.

Set $T_2 = \{b' \in B$: there exists a path $a_{t_1+1}, \ldots, a_k$ with $a_k = b'\}$, $G_2 = B \setminus T_2$, $U_2 = h(T_2)$ in the first subcase, and see that I' generated by a subset of $\{f_1, f_2\} \cup G_2$ chosen as above works.

In the second subcase, let $a_{t_1+1}, \ldots, a_k$ be a weak path and set $c_j = h(a_j)$ for $t_1 < j \leq k$. We may suppose that $c_k \in (u)$. Changing P_b we may suppose that the new c'_2 is in U_2 as above. If the old u' was not a divisor of a $c \in U_2$ then the proof goes as in the first case with $T'_2 = T_2 \cup \{u\}$, $I' = I'_2$. Otherwise, T'_2 should be completed to a $\tilde{T}_2$ similar to $\tilde{T}_1$. The proof goes as above with $\tilde{T}_2$ instead T'_2.

In the third subcase, let $a_{t_1+1}, \ldots, a_{t_2}$ be a bad path starting with a_{t_1+1} and set $c_j = h(a_j)$ for $j > t_1$. We saw that the whole set $\{a_1, \ldots, a_{t_2}\}$ has different monomials. As above $c_{t_2} = b x_{l_2}$ and we may reduce to the case when $f_1 x_{l_2} \notin \{a_1, \ldots, a_{t_1}\}$. Set $a_{t_2+1} = f_1 x_{l_2}$ and again we consider three subcases, which we treat as above. Anyway after several such steps we must arrive in the case $p = t_m$ when $b | c_{t_m}$ and again a certain $f_1 x_{l_m}$ is not among $\{a_1, \ldots, a_{t_m}\}$ and there exist no bad path starting with $a_{t_m+1} = f_1 x_{l_m}$. This follows since we may reduce to the case when the set $\{a_1, \ldots, a_{t_m}\}$ has different monomials and so the procedures should stop for some m. Finally, using

$$T_m = \{b' \in B : \text{there exists a path } a_{t_m+1}, \ldots, a_k \text{ with } a_k = b'\}$$

(resp. T'_m, or $\tilde{T}_m$) as T_1 (resp. T'_1, or $\tilde{T}_1$) above we are done.

Lemma 6. *Suppose that the following conditions hold:*

1. $r = 3$, $6 \leq s \leq q + 3$,
2. $C \subset (\cup_{i,j \in [3], i \neq j} (f_i) \cap (f_j)) \cup ((E) \cap (f_1, f_2, f_3)) \cup (\cup_{a,a' \in E, a \neq a'} (a) \cap (a'))$,
3. There exists $b \in (B \cap (f_1)) \setminus (f_2, f_3)$ *such that* sdepth $_S I_b / J_b \geq d + 2$.

Then either sdepth $_S I/J \geq d + 2$, *or there exists a nonzero ideal* $I' \subsetneq I$ *generated by a subset of* $\{f_1, f_2, f_3\} \cup B$ *such that* sdepth $_S I'/J' \leq d + 1$ *for* $J' = J \cap I'$ *and* depth $_S I/(J, I') \geq d + 1$.

Proof. We follow the proof of Lemma 5. Since sdepth $_S I_b/J_b \geq d + 2$ we consider h as above for a partition P_b with sdepth $d + 2$ of I_b/J_b. We have two intervals $[f_2, c'_2]$, $[f_3, c'_3]$ in P_b. Suppose that $B \cap [f_i, c'_i] = \{u_i, u'_i\}$, $1 < i \leq 3$. As in Lemma 5 we define a path $a_1, \ldots, a_k$ from a_1 to a_k and a bad path. The above path is *weak* if $h(a_j) \in (u_2, u_2', u_3, u_3')$ for some $j \in [k]$. It is *maximal* if either $h(a_k) \in (b)$, or all divisors from B of $h(a_k)$ are in $\{b, u_2, u_2', u_3, u_3', a_1, \ldots, a_k\}$.

By hypothesis $s \geq 6$ and there exists $a_1 \in B \setminus \{b, u_2, u_2', u_3, u_3'\}$. Set $c_1 = h(a_1)$. If $c_1 \in (b)$ then the path $\{a_1\}$ is maximal and bad. We construct below a path with $k > 1$. By recurrence choose if possible a_{p+1} to be a divisor from B of c_p, which is not in $\{b, u_2, u_2', u_3, u_3', a_1, \ldots, a_p\}$ and set $c_p = h(a_p)$, $p \geq 1$. This construction ends at step $p = e$ if all divisors from B of c_e are in $\{b, u_2, u_2', u_3, u_3', a_1, \ldots, a_e\}$. If $c_j \notin (b)$ for $1 \leq j < e$ then $\{a_1, \ldots, a_e\}$ is a maximal path. If one $c_p \in (u_2, u_2', u_3, u_3')$ then the constructed path is weak. If $c_e \in (b)$ then this path is bad.

We may reduce to the situation when P_b satisfies the following property:

(∗) For all $1 \leq i < j \leq 3$ if $w_{ij} \in B \setminus \{u_2, u_2', u_3, u_3'\}$, $1 \leq i < j \leq 3$ then $h(w_{ij}) \notin (u_i, u_i', u_j, u_j')$ if $i > 1$.

Indeed, suppose that $w_{ij} \in B \setminus \{u_2, u_2', u_3, u_3'\}$ and $h(w_{ij}) \in (u_j)$. Then $h(w_{ij}) = x_l w_{ij}$ for some $l \notin \operatorname{supp} w_{ij}$ and we must have let us say $u_j = x_l f_j$. Changing in P_b the intervals $[f_j, c_j']$, $[w_{ij}, h(w_{ij})]$ with $[f_j, h(w_{ij})]$, $[u_j', c_j']$ we see that we may assume $u_j = w_{ij}$. Suppose that (∗) holds. We have three cases:

1) there exist no weak path and no bad path starting with a_1,
2) there exists a weak path starting with a_1 but no bad path starts with a_1,
3) there exists a bad path starting with a_1.

In **the first case**, set $T_1 = \{b' \in B$: there exists a path $a_1, \ldots, a_t$ with $a_t = b'\}$, $G_1 = B \setminus T_1$, and for $k = (k_1, \ldots, k_m)$, $1 \leq k_1 < \ldots k_m \leq 3, 0 \leq m \leq 3$ set $I_k' = (f_{k_1}, \ldots, f_{k_m}, G_1)$, $J_k' = I_k' \cap J$, and $I_0' = (G_1)$, $J_0' = I_0' \cap J$ for $m = 0$. Note that all divisors from B of a monomial $c \in U_1 = h(T_1)$ belong to T_1, and $I_0' \neq 0$ because $b \in I_0'$. Consider the following exact sequence

$$0 \to I_k'/J_k' \to I/J \to I/(J, I_k') \to 0.$$

If $U_1 \cap (f_1, f_2, f_3) = \emptyset$ then the last term of the above exact sequence given for $k = (1, 2, 3)$ has depth $\geq d + 1$ and sdepth $\geq d + 2$ using the restriction of P_b to $(T_1) \setminus (J, I_k')$ since $h(b') \notin I_k'$, for all $b' \in T_1$. When the first term has sdepth $\geq d + 2$ then by [18, Lemma 2.2] the middle term has sdepth $\geq d + 2$ which is enough.

If $U_1 \cap (f_1, f_2) = \emptyset$, but there exists $b_3 \in T_1 \cap (f_3)$, then set $k = (1, 2)$. In the following exact sequence

$$0 \to I_k'/J_k' \to I/J \to I/(J, I_k') \to 0$$

the last term has sdepth $\geq d + 2$ since $h(b') \notin I_k'$, for all $b' \in T_1$ and we may substitute the interval $[b_3, h(b_3)]$ from the restriction of P_b by $[f_3, h(b_3)]$, the second monomial from $[f_3, h(b_3)] \cap B$ being also in T_1. As above we get either $\operatorname{sdepth}_S I/J \geq d + 2$, or $\operatorname{sdepth}_S I_1'/J_1' \leq d + 1$, $\operatorname{depth}_S I/(J, I_1') \geq d + 1$.

Now, we omit other subcases considering only the worst subcase $m = 0$. Let $b_1 \in T_1 \cap (f_1)$, $b_2 \in T_1 \cap (f_2)$ and $b_3 \in T_1 \cap (f_3)$. For $1 \leq l < j \leq 3$ we claim that we may choose $b_l \neq b_j$ and such that one from $h(b_l), h(b_j)$ is not in (w_{lj}). Indeed, if $w_{lj} \notin B$ and $h(b_l), h(b_j) \in (w_{lj})$ then necessarily $h(b_l) = h(b_j)$ and it follows $b_l = b_j = w_{lj}$, which is false. Suppose that $w_{lj} \in B$ and $h(b_j) = x_p w_{lj}$. Then choose $b_l = x_p f_l \in T_1$. If $h(b_l) \in (w_{lj})$ then we get $h(b_l) = h(b_j)$ and so $b_l = b_j = w_{lj}$, which is impossible.

Therefore, we may choose b_j such that $h(b_1) \notin (w_{12}), h(b_2) \notin (w_{23})$. Note that it is possible that $f_1|c$ for some $c \in h(T_1)$ even $b \not| c$ for any $c \in U_1$. If $h(b_1) \in (w_{13})$ then we may also choose $h(b_3) \notin (w_{13})$. In the case when $h(b_1) \notin (w_{13})$, choose any $b_3 \in T_1 \cap (f_3)$ different from the others b_j. We conclude that the possible intervals $[f_j, h(b_j)]$, $j \in [3]$ are disjoint. Next we change the intervals $[b_j, h(b_j)]$, $j \in [3]$ from the restriction of P_b to $(T_1) \setminus (J, I_0')$ by $[f_j, h(b_j)]$, the second monomial from $[f_j, h(b_j)] \cap B$ being also in T_1. We claim that $I/(J, I_0')$ has depth $\geq d+1$. Indeed, in the following exact sequence

$$0 \to (f_2)/(f_2) \cap (J, I_0', f_3) \to I/(J, I_0', f_3) \to I/(J, I_0', f_2, f_3) \to 0$$

the first term has depth $\geq d+1$ because $h(b_2) \notin (f_2) \cap (f_3)$. If $h(b_1) \notin (f_3)$ then $h(b_1) \notin (f_2, f_3) \cap (f_1)$ and so the last term has depth $\geq d+1$. If $h(b_1) \in (w_{13})$ then we may find a $b' \in (B \cap (f_1)) \setminus (f_3)$ dividing $h(b_1)$. It follows that $b' \in T_1$ and $b' \notin (f_2, f_3) \cap (f_1)$, which implies that the last term has again depth $\geq d+1$. Thus depth $_S I/(J, I_0', f_3) \geq d+1$ by the Depth Lemma. Our claim follows from the exact sequence

$$0 \to (f_3)/(f_3) \cap (J, I_0') \to I/(J, I_0') \to I/(J, I_0', f_3) \to 0$$

because the first term has depth $\geq d+1$. Therefore, as above we get either sdepth $_S I/J \geq d+2$, or sdepth $_S I_0'/J_0' \leq d+1$, depth $_S I/(J, I_0') \geq d+1$.

In **the second case**, let $a_1, \ldots, a_{t_1}$ be a weak path and set $c_j = h(a_j)$ for $j \in [t_1]$. We may suppose that $c_{t_1} \in (u_2)$, otherwise take a shorter path. Denote T_1, U_1 as in the first case. First **consider the subcase when** $U_1 \cap (f_3) = \emptyset$. Suppose that $a_{t_1} \in (f_2)$. Then change in P_b the intervals $[a_{t_1}, c_{t_1}]$, $[f_2, c_2']$ by $[f_2, c_{t_1}]$, $[u_2', c_2']$. Thus the new c_2' is among $\{c_1, \ldots, c_{t_1}\} \subset U_1$, though the old $c_2' \notin U_1$. If the old u_2' is not a divisor of any $c \in U_1$ then the proof goes as in the first case with $T_1' = T_1 \cup \{u_2\}$. If the old u_2' is a divisor of a monomial from U_1 then T_1' should be completed because the old u_2' is not now in $[f_2, c_2']$. Note that there exists a path from a_1 to u_2' since u_2' is a divisor of a monomial from U_1. It follows that there exist no bad path starting with u_2'. It is worth to mention that the old c_2' is now in U_1 and we should consider all paths starting with divisors of c_2' from B. Take $\tilde{T}_1 = T_1' \cup \{b' \in B :$ there exists a path from u_2' to $b'\}$ and the proof goes as above with $\tilde{T}_1$ instead T_1', that is with I' generated by a subset of $\{f_1, f_2, f_3\} \cup \tilde{G}_1$, where $\tilde{G}_1 = B \setminus \tilde{T}_1$.

Now suppose that $a_{t_1} \notin (f_2)$ but there exists $1 \leq v < t_1$ such that $a_v \in (f_2)$ and $a_v | c_{t_1}$. Then we may replace in P_b the intervals $[a_p, c_p], v \leq p \leq t_1$ with the intervals $[a_v, c_{t_1}], [a_{p+1}, c_p], v \leq p < t_1$. The old c_{t_1} becomes the new c_v, that is we reduce to the case when u_2 divides c_v and $a_v \in (f_2)$, subcase solved above.

Remains to study the subcase when there exist no $a_v \in (f_2)$, $1 \leq v \leq t_1$ with $a_v | c_{t_1}$. Then there exists an $a_{t_1+1} \in B \cap (f_2)$, $a_{t_1+1} \neq u_2$ such that $a_{t_1+1} | c_{t_1}$. Clearly, $a_{t_1+1} \neq u_2'$ because otherwise $c_2' = c_{t_1}$. We have two subcases:

1′) there exists a path $a_{t_1+1}, \ldots, a_l$ such that $h(a_l) \in (a_{v'})$ for some $1 \leq v' \leq t_1$,
2′) for any path $a_{t_1+1}, \ldots, a_p$, any $h(a_j)$, $t_1 < j \leq p$ does not belong to $(a_1, \ldots, a_{t_1})$.

In the first subcase, we replace in P_b the intervals $[a_j, c_j], v' \leq j \leq l$ with the intervals $[a_{v'}, c_l], [a_{j+1}, c_j], v' \leq j < l$. The new $h(a_{t_1+1})$ is the old c_{t_1} and we may proceed as above. In the second subcase we set

$$T_2 = \{b' \in B : \text{there exists a path from } a_{t_1+1} \text{ to } b'\}.$$

Note that any path starting from a_{t_1+1} can be completed to a path from a_1 by adding the monomials $a_1, \ldots, a_{t_1}$. Thus there exists no bad path starting with a_{t_1+1}, otherwise we can get one starting from a_1, which is false.

If there exists no weak path starting with a_{t_1+1} then we proceed as in the first case with T_2 instead T_1. If there exists a weak path starting with a_{t_1+1} then we proceed as above in case 2) with T_2', or $\tilde{T}_2$ instead T_1', or $\tilde{T}_1$, except in the subcase 2′) when we will define similarly a T_3 given by the paths starting with a certain a_{t_2+1}. Note that the whole set $\{a_1, \ldots, a_{t_2}\}$ has different monomials. After several such steps we must arrive in the case $p = t_m$ when $\{a_1, \ldots, a_{t_m}\}$ has different monomials and the subcase 2′) does not appear. We end this case using T_m, or T_m', or $\tilde{T}_m$ instead T_1, T_1', or $\tilde{T}_1$. We should mention that if there exists $b_1 \in T_m$ (or in T_m', $\tilde{T}_m$) such that $h(b_1) \in (f_1)$ then changing P_b as in case 1) we may suppose that $h(b_1) \notin (w_{12})$ and $b_1 \in T_m \cap (f_1)$. Thus we may consider the interval $[f_1, h(b_1)]$ disjoint of $[f_2, c_2']$.

Consider **the subcase when there exist** $b_j \in T_1$, $j = 2, 3$ such that $h(b_2) \in (u_2)$ and $h(b_3) \in (f_3)$ **but** $h(T_1) \cap (u_3, u_3') = \emptyset$. As above we may suppose that after several procedures we changed P_b such that $b_j \in (f_j)$ and the new c_2' is the old $h(b_2)$. If $h(b_2) \notin C_3 \cup C_2$ then we may suppose that $h(b_2) \notin (w_{12})$. As in the first case we may change b_3 such that $h(b_3) \notin (w_{23})$. Indeed, the only problem could be if the old $h(b_3) \in \{u_3, u_3'\}$, which is not the case. We have no obstruction to change as usual b_1 such that $h(b_1) \notin (w_{13})$ and so note that the interval $[f_2, h(b_2)]$ (resp. $[f_3, h(b_3)]$, or $[f_1, h(b_1)]$) has at most w_{23} (resp. w_{13}, or w_{12}) from W. Thus the intervals $[f_j, h(b_j)]$, $j \in [3]$ are disjoint.

If $h(b_2) \in C_3$ then either $b_2 = w_{23}$, or w_{12}. But $b_2 \neq w_{23}$ because otherwise $h(w_{23}) \in (u_2)$ contradicting (∗). Similarly, $b_2 \neq w_{12}$. If $h(b_2) = w_{12}$ (resp. $h(b_2) = w_{23}$) then $b_2 \neq w_{23}$ (resp. $b_2 \neq w_{12}$) because otherwise we get a contradiction with (∗). Thus w_{12} (resp. w_{23}) is the only monomial of W which belongs to $[f_2, h(b_2)]$. Choosing b_3 such that $h(b_3) \notin (w_{23})$ (resp. $h(b_3) \notin (w_{12})$) and b_1 such that $h(b_1) \notin (w_{12})$ (resp. $h(b_1) \notin (w_{13})$) we get disjoint the corresponding intervals.

Now consider **the subcase when there exist** $b_j \in T_1$, $j = 2, 3$ such that $h(b_2) \in (u_2)$ **and** $h(b_3) \in (u_3)$. If $h(b_2) \notin (f_3)$ and $h(b_3) \notin (f_2)$ then as above we may assume that with a different P_b, if necessary, we may reduce to the subcase when $b_j \in (f_j)$, $j = 2, 3$. In general this is not simple because $h(b_2)$ as in

Example 5 can have no divisors from $B \cap (f_2)$, which are not in $\{u_2, u_2', u_3, u_3'\}$ and there exist no other $c \in U_1$ multiple of u_2. In such situation we are force to remain on the old c_2' taking $T_1' = T_1 \cup \{u_2, u_2'\}$ and $U_1' = U_1 \cup \{c_2'\}$. If there exists a bad path starting on a divisor from $B \setminus \{u_2, u_2', u_3, u_3'\}$ of c_2' then we go to case 3). Otherwise, we should consider also the paths starting with the divisors of c_2' from $B \setminus \{u_2, u_2', u_3, u_3'\}$ completing T_1' to $\tilde{T}_1$. Note that because of 2′) we may speak now about T_m instead T_1.

Changing in P_b the intervals $[f_j, c_j']$, $[b_j, h(b_j)]$, $j = 2, 3$ with $[f_j, h(b_j)]$, $[u_j', c_j']$, $j = 2, 3$ we may assume the new c_2', c_3' are in $U_m = h(T_m)$ for some m and the proof goes as above. If let us say $h(b_2) \in (f_3)$ then we must be carefully since it is possible that the new intervals $[f_j, c_j']$ could be not disjoint. A nice subcase is for example when $h(b_2)$ is a least common multiple of u_2, u_3, which we study below.

If $w_{23} \in B$ then we **we may suppose** $u_2 = w_{23}$. Indeed, if $w_{23} \notin \{u_2, u_2', u_3, u_3'\}$ and $u_2 = x_p f_2$ for some $p \notin \operatorname{supp} w_{23}$ then $h(b_2) = x_p w_{23}$. Since $b_2 \neq u_2$ and $b_2 \in (f_2)$ it follows that $b_2 = w_{23}$. But this contradicts the property $(*)$. Suppose that $a_{t_1} = b_2$. Then note that $h(b_2) = c_{t_1} = x_p w_{23}$ for some p and it follows that $a_{t_1} = b_2 = x_p f_2$ since $a_{t_1} \neq w_{23} = u_2$. Changing in P_b the intervals $[f_2, c_2']$, $[b_2, h(b_2)]$ with $[f_2, h(b_2)]$, $[u_2', c_2']$, we may assume the new c_2' is in U_m. We claim that w_{23} is the only monomial from $B \cap W$ which is in $[f_2, c_2']$. Indeed, w_{12} could be another monomial from $B \cap W$ which is present in the new $[f_2, c_2']$. This could be true only if $a_{t_1} = w_{12}$. Thus $h(w_{12}) = c_{t_1} \in (u_2)$ which is not possible again by $(*)$. The same procedure we use to include a new c_3' in U_m. Since $u_2 = w_{23}$ cannot be among u_3, u_3' we see that only w_{13} could be among them. Suppose that $u_3 = w_{13}$. Clearly the new $[f_3, c_3']$ cannot contain w_{23}. Choose as in the first case $b_1 \in (f_1)$ such that $h(b_1) \notin (w_{13})$ and the new intervals $[f_j, c_j']$, $j \in [3]$ are disjoint. If $w_{13} \notin \{u_3, u_3'\}$ then we might have only $b_3 = w_{13}$ and we may repeat the argument.

A problem could appear when the new $[f_j, c_j']$, $j = 2, 3$ contain w_{12}, w_{13} because then we may not find b_1 as before. Note that this problem could appear only when $w_{12}, w_{13} \in \{u_2, u_2', u_3, u_3'\}$ because of $(*)$. We will change the new c_2' such that will not belong to (f_1). Changing P_b we may suppose that $b_j \in (f_j)$, $j = 2, 3$ (again this change is not so simple as we saw above). We have $h(b_2) = c_{t_1} = x_p w_{12}$ for some p and it follows that $a_{t_1} = b_2 = x_p f_2$ since $a_{t_1} \neq w_{12} = u_2$. Suppose that $t_1 > 1$. Thus $a_{t_1} | c_{t_1 - 1}$ and we see that c_{t_1-1} is not in (f_1) because otherwise we get $c_{t_1-1} = x_p w_{12} = c_{t_1}$, which is false. **If** $a_{t_1-1} \in (f_2)$ then changing in P_b the intervals $[a_{t_1}, c_{t_1}]$, $[a_{t_1-1}, c_{t_1-1}]$, $[f_2, c_2']$ by $[f_2, c_{t_1-1}]$, $[u_2, c_{t_1}]$, $[u_2', c_2']$ we see that the new c_2' is not in (f_1) and belongs to U_m. **If** $w_{12} \in C$ then **we get** $h(b_2) = w_{12}$ and the above argument works again, c_{t_1-1} being the new c_2'.

When $a_{t_1-1} \notin (f_2)$ **but** $u_2' | c_{t_1-1}$ we reduce the problem to the subcase when the path $\{a_1, \ldots, a_{t_1-1}\}$ goes from a_1 to u_2' and now $u_2' \notin (f_1)$. As above we may change P_b such that the new $b_2 = a_{t_1-1} \in (f_2)$ and the new c_2', that is the old c_{t_1-1} is not in (f_1).

If $a_{t_1-1} \notin (f_2)$, $u_2' \not| c_{t_1-1}$ but **there exists** $\tilde{a} \in B \cap (f_2)$ a divisor of c_{t_1-1} then $\tilde{a} \neq u_2$ because otherwise we get $c_{t_1-1} = c_{t_1}$. Now we repeat the first part of the case 2). If $\tilde{a} = a_v$ for some $1 \le v < t_1 - 1$ then changing in P_b the intervals $[a_p, c_p]$,

$v \le p < t_1$ by $[a_v, c_{t_1-1}]$, $[a_{p+1}, c_p]$, $v \le p < t_1 - 1$ we see that the new c_v (resp. c_{v+1}) is the old c_{t_1-1} (resp. c_{t_1})). Now changing the intervals $[a_v, c_v]$, $[a_{v+1}, c_{v+1}]$, $[f_2, c_2']$ by $[f_2, c_v]$, $[w_{12}, c_{v+1}]$, $[u_2', c_2']$ we see that the new $c_2' \notin (f_1)$ and belongs to U_m. If $\tilde{a} \notin \{a_1, \dots, a_{t_1}\}$ then we are in one of the above subcases 1′), 2′) solved already.

We may use this argument to change c_j', $j = 2, 3$ such that it is not in (f_1) anymore, but as long as $h(b_j) \neq c_1$, that is the corresponding $t_1 > 1$. However, we may have $h(b_j) = c_1$ only for one $j > 1$, because if for instance $h(b_3) = c_1$ then $c_1 \in C_2 \cup C_3$. If $c_1 \in C_3$ then we see that $w_{23} \in B$ and $a_1 = w_{23}$. But this contradicts (∗) because $h(w_{23}) \in (u_2)$. If $c_2' \in C_2$ then $c_2' = w_{23}$ and either $a_1 \in (f_2)$, or $a_1 \in (f_3)$, that is a_1 cannot be b_2 and b_3 in the same time. Thus at least one of the new c_j', $j = 2, 3$ could be taken $\notin (f_1)$. If let us say only $c_3' \in (f_1)$ then choose $b_1 \in T_1 \cap (f_1)$ such that $h(b_1) \notin (w_{13})$ as before. The interval $[f_1, h(b_1)]$ is disjoint from the other new constructed intervals, which is enough as we saw in case 1).

In **the third case**, let $a_1, \dots, a_{t_1}$ be a bad path starting with a_1. Set $c_j = h(a_j)$, $j \in [t_1]$. Then $c_{t_1} = bx_{l_1}$ and let us say $b = f_1 x_i$. If $a_{t_1} \in (f_1)$ then changing in P_b the interval $[a_{t_1}, c_{t_1}]$ by $[f_1, c_{t_1}]$ we get a partition on I/J with sdepth $d + 2$. Thus we may assume that $a_{t_1} \notin (f_1)$. If $f_1 x_{l_1} \in \{a_1, \dots, a_{t_1-1}\}$, let us say $f x_{l_1} = a_v$, $1 \le v < t_1$ then we may replace in P_b the intervals $[a_p, c_p], v \le p \le t_1$ with the intervals $[a_v, c_{t_1}]$, $[a_{p+1}, c_p]$, $v \le p < t_1$. Now we see that we have in P_b the interval $[f_1 x_{l_1}, f_1 x_i x_{l_1}]$ and switching it with the interval $[f_1, f_1 x_i x_{l_1}]$ we get a partition with sdepth $\ge d + 2$ for I/J.

Thus we may assume that $f_1 x_{l_1} \notin \{a_1, \dots, a_{t_1}\}$. Now set $a_{t_1+1} = f x_{l_1}$. Let $a_{t_1+1}, \dots, a_k$ be a path starting with a_{t_1+1} and set $c_j = h(a_j)$, $t_1 < j \le k$. If $a_p = a_v$ for $v \le t_1$, $p > t_1$ then change in P_b the intervals $[a_j, c_j], v \le j \le p$ with the intervals $[a_v, c_p]$, $[a_{j+1}, c_j], v \le j < p$. We have in P_b an interval $[f_1 x_{l_1}, f_1 x_i x_{l_1}]$ and switching it to $[f_1, f_1 x_i x_{l_1}]$ we get a partition with sdepth $\ge d + 2$ for I/J. Thus we may suppose that in fact $a_p \notin \{b, a_1, \dots, a_{p-1}\}$ for any $p > t_1$ (with respect to any path starting with a_{t_1+1}). We have three subcases:

1″) there exist no weak path and no bad path starting with a_{t_1+1},
2″) there exists a weak path starting with a_{t_1+1} but no bad path starts with a_{t_1+1},
3″) there exists a bad path starting with a_{t_1+1}.

Set $T_2 = \{b' \in B$: there exists a path $a_{t_1+1}, \dots, a_k$ with $a_k = b'\}$. We treat the subcases 1″), 2″) as the cases 1), 2) and find I' generated by a subset of $\{f_1, f_2, f_3\} \cup G_2$, or $\{f_1, f_2, f_3\} \cup G_2'$, or $\{f_1, f_2, f_3\} \cup \tilde{G}_2$, where $G_2, G_2', \tilde{G}_2$, are obtained from T_2 and as above T_2', or $\tilde{T}_2$.

In the subcase 3″), let $a_{t_1+1}, \dots, a_{t_2}$ be a bad path starting with a_{t_1+1} and set $c_j = h(a_j)$ for $j > t_1$. We saw that the whole set $\{a_1, \dots, a_{t_2}\}$ has different monomials. As above $c_{t_2} = bx_{l_2}$ and we may reduce to the case when $f_1 x_{l_2} \notin \{a_1, \dots, a_{t_1}\}$. Set $a_{t_2+1} = f_1 x_{l_2}$ and again we consider three subcases, which we treat as above. Anyway after several such steps we must arrive in the case $p = t_m$ when either we may proceed as in the subcases 1″), 2″), or $b | c_{t_m}$ and again a certain $f_1 x_{l_m}$ is not among $\{a_1, \dots, a_{t_m}\}$ and taking $a_{t_m+1} = f_1 x_{l_m}$ there exist no bad path starting with

a_{t_m+1}. This follows since we may reduce to the subcase when the set $\{a_1, \dots, a_{t_m}\}$ has different monomials and so the procedures should stop for some m. Finally, using

$$T_m = \{b' \in B : \text{there exists a path } a_{t_m+1}, \dots, a_k \text{ with } a_k = b'\}$$

(resp. T'_m, or $\tilde{T}_m$) as T_1 (resp. T'_1, or $\tilde{T}_1$) above we are done.

Proof of Theorem 6. By Theorems 2, 3 we may suppose that $2r \le s \le q+r$ and we may assume that E contains only monomials of degrees $d+1$ by [15, Lemma 1.6]. Apply induction on $|E|$. If $E = \emptyset$ we may apply Theorem 4. Suppose that $|E| > 0$ and $B \cap (f_1, \dots, f_r) \neq \emptyset$, $r = 2, 3$, otherwise we get depth $_S I/J \le d+1$ using [15, Lemma 1.5] applied to any f_i. We may choose $b \in B \cap (f_1, f_2, f_3)$ which is not in W if $r = 2, 3$ and $|B \cap (f_1, \dots, f_r)| > 3 \ge |B \cap W|$. However, $|B \cap (f_1, \dots, f_r)| < 2r$ gives depth $_S (f_1, \dots, f_r)/J \cap (f_1, \dots, f_r) \le d+1$ by Theorem 3 and it follows that depth $_S I/J \le d+1$ using the Depth Lemma applied to the exact sequence

$$0 \to (f_1, \dots, f_r)/J \cap (f_1, \dots, f_r) \to I/J \to (E)/(J, f_1, \dots, f_r) \cap (E) \to 0.$$

Thus if $r = 2, 3$ we may suppose to find $b \in B \cap (f_1, \dots, f_r) \setminus W$. Renumbering f_i we may suppose that $b \in (f_1) \setminus (f_2, \dots, f_3)$.

Apply induction on $r \le 3$. Using Theorem 4 and induction hypothesis on $|E|$ and r apply [17, Lemma 1.1]. Thus we may suppose that $C \subset ((f_1) \cap (f_2)) \cup ((E) \cap (f_1, f_2)) \cup (\cup_{a,a' \in E, a \neq a'} (a) \cap (a'))$, if $r = 2$, or $C \subset (\cup_{i,j \in [3], i \neq j} (f_i) \cap (f_j)) \cup ((E) \cap (f_1, f_2, f_3)) \cup (\cup_{a,a' \in E, a \neq a'} (a) \cap (a'))$ if $r = 3$.

Set $I'_b = (f_2, \dots, f_r, B \setminus \{b\})$, $J'_b = I'_b \cap J$. Clearly $b \notin I'_b$ and so in the following exact sequence

$$0 \to I'_b/J'_b \to I/J \to I/(J, I'_b) \to 0$$

the last term has depth $\ge d+1$. If the first term has sdepth $\le d+1$ then it has depth $\le d+1$ by induction hypothesis on r, case $r = 1$ being done in Theorem 4. Thus we may suppose that sdepth $_S I'_b/J'_b \ge d+2$ and we may apply Lemmas 5, 6. Then we get either sdepth $_S I/J \ge d+2$ contradicting our assumption, or there exists a nonzero ideal $I' \subsetneq I$ generated by a subset G of B, or by G and a subset of $\{f_1, f_2, f_3\}$ such that sdepth $_S I'/J' \le d+1$ for $J' = J \cap I'$ and depth $_S I/(J, I') \ge d+1$. In the last case we see that depth $_S I'/J' \le d+1$ by induction hypothesis on r, $|E|$, or by Theorem 4 and so depth $_S I/J \le d+1$ by the Depth Lemma applied to the following exact sequence

$$0 \to I'/J' \to I/J \to I/(J, I') \to 0.$$

□

The following bad example is useful to illustrate somehow our proof.

Example 5. Let $n = 6, r = 3, d = 1, f_i = x_i$ for $i \in [3], E = \{x_4x_5, x_5x_6\}, I = (x_1, x_2, x_3, E)$ and $J = (x_2x_4, x_3x_4, x_1x_2x_6, x_1x_3x_6, x_1x_4x_6, x_1x_5x_6, x_2x_3x_6, x_2x_5x_6, x_3x_5x_6)$. Then $B = \{x_1x_2, x_1x_3, x_2x_3, x_1x_4, x_1x_5, x_2x_5, x_3x_5\} \cup E$ and

$$C = \{x_1x_2x_3, x_1x_2x_5, x_2x_3x_5, x_1x_3x_5, x_1x_4x_5, x_4x_5x_6\}.$$

Take $b = x_1x_4$ and $I_b = (x_2, x_3, x_1x_2, x_1x_3, x_1x_5, E)$, $J_b = I_b \cap J$. There exists a partition P_b with sdepth 3 on I_b/J_b given by the intervals $[x_2, x_1x_2x_3]$, $[x_3, x_1x_3x_5]$, $[x_1x_5, x_1x_2x_5]$, $[x_2x_5, x_2x_3x_5]$, $[x_4x_5, x_1x_4x_5]$, $[x_5x_6, x_4x_5x_6]$. We have $c'_2 = x_1x_2x_3, c'_3 = x_1x_3x_5$ and $u_2 = x_2x_3, u'_2 = x_1x_2, u_3 = x_3x_5, u'_3 = x_1x_3$. Clearly, $u_2 = w_{23}$.

Take $a_1 = x_1x_5, c_1 = x_1x_2x_5, a_2 = x_2x_5, c_2 = x_2x_3x_5$. The path $\{a_1, a_2\}$ is maximal weak because the divisors from B of c_2 are a_2, u_2, u_3. Then $T_1 = \{a_1, a_2\}$ and we change in P_b as in the proof the intervals $[x_2, c'_2]$, $[a_2, c_2]$ by $[x_2, c_2]$, $[u'_2, c'_2]$. Thus the new c'_2 is the old c_2. Now note that this new c'_2 is a multiple of u_3 and it is the only monomial from $h(T_1)$, which is a such multiple. Thus we had to take u_3 in the new T'_1, and u'_3 as well and certainly c'_3 is added to $h(T_1)$. Clearly, all divisors from B of c'_3 are in $T'_1 = T_1 \cup \{u_3, u'_3\}$. But the former u'_2 divides c_1 and so should be added to T'_1. Thus we have $I' = (b, E)$, $J' = J \cap I'$ and $I/(J, I')$ has a partition of sdepth 3 given by the intervals $[x_2, c_2]$, $[x_3, c'_3]$, $[x_1, x_1x_2x_5]$. If sdepth $_S I'/J' \geq 3$ then we get sdepth $_S I/J \geq 3$, which is false. Otherwise, sdepth $_S I'/J' \leq 2$ and we get depth $_S I'/J' \leq 2$ by [12, Theorem 4.3] and so depth $_S I/J \leq 2$ using the Depth Lemma.

Acknowledgements The support of the first author from the Department of Mathematics of the University of Kaiserslautern and the support of the second author from grant PN-II-RU-TE-2012-3-0161 of Romanian Ministry of Education, Research and Innovation are gratefully acknowledged.

References

1. Bruns, W., Herzog, J.: Cohen-Macaulay Rings, Revised edition. Cambridge University Press, Cambridge (1998)
2. Bruns, W., Krattenthaler, C., Uliczka, J.: Stanley decompositions and Hilbert depth in the Koszul complex. J. Commutative Algebra **2**, 327–357 (2010)
3. Bruns, W., Moyano-Fernández, J.J., Uliczka, J.: Hilbert regularity of ZZ-graded modules over polynomial rings. arXiv:AC/1308.2917 (2013)
4. Cimpoeas, M.: The Stanley conjecture on monomial almost complete intersection ideals, Bull. Math. Soc. Sci. Math. Roumanie **55**(103), 35–39 (2012)
5. Herzog, J., Vladoiu, M., Zheng, X.: How to compute the Stanley depth of a monomial ideal. J. Algebra **322**, 3151–3169 (2009)
6. Ichim, B., Moyano-Fernández, J.J.: How to compute the multigraded Hilbert depth of a module. To appear in Math. Nachr. arXiv:AC/1209.0084
7. Ichim, B., Zarojanu, A.: An algorithm for computing the multigraded Hilbert depth of a module. to appear in Exp. Math., arXiv:AC/1304.7215v2
8. Popescu, A.: Special Stanley decompositions, Bull. Math. Soc. Sc. Math. Roumanie **53**(101), no 4, 363–372 (2010). arXiv:AC/1008.3680

9. Popescu, A.: An algorithm to compute the Hilbert depth. J. Symb. Comput. **66**, 1–7 (2015). arXiv:AC/1307.6084
10. Popescu, D.: Stanley conjecture on intersections of four monomial prime ideals. Commun. Algebra **41**, 4351–4362 (2013). arXiv:AC/1009.5646
11. Popescu, D.: Graph and depth of a square free monomial ideal. Proc. AMS 140, 3813–3822 (2012). arXiv:1104.5596v1, 2011
12. Popescu, D.: Depth of factors of square free monomial ideals. Proc. AMS **142**, 1965–1972 (2014). arXiv:AC/1110.1963
13. Popescu, D.: Upper bounds of depth of monomial ideals. J. Commutative Algebra **5**, 323–327 (2013). arXiv:AC/1206.3977
14. Popescu, D.: Depth of a monomial ideal. In: Beznea et al. (eds.) Invited Contributions of the Seventh Congress of Romanian Mathematicians, Brasov, 2011 (2013), 253–257
15. Popescu, D., Zarojanu, A.: Depth of some square free monomial ideals. Bull. Math. Soc. Sci. Math. Roumanie **56**(104), 117–124 (2013)
16. Popescu, D., Zarojanu, A.: Depth of some special monomial ideals. Bull. Math. Soc. Sci. Math. Roumanie **56**(104), 365–368 (2013). arxiv:AC/1301.5171v1
17. Popescu, D., Zarojanu, A.: Three generated, squarefree, monomial ideals. Bull. Math. Soc. Sci. Math. Roumanie **58**(106), no. 3 (2015) (to appear). arXiv:AC/1307.8292v6
18. Rauf, A.: Depth and Stanley depth of multigraded modules. Comm. Algebra **38**, 773–784 (2010)
19. Shen, Y.H.: Lexsegment ideals of Hilbert depth 1 (2012), arXiv:AC/1208.1822v1
20. Stanley, R.P.: Linear diophantine equations and local cohomology. Invent. Math. **68**, 175–193 (1982)
21. Uliczka, J.: Remarks on Hilbert series of graded modules over polynomial rings. Manuscripta Math. **132**, 159–168 (2010)
22. Zarojanu, A.: Stanley conjecture on intersection of three monomial primary ideals. Bull. Math. Soc. Sci. Math. Roumanie **55**(103), 335–338 (2012)

The Connected Components of the Space of Alexandrov Surfaces

Joël Rouyer and Costin Vîlcu

Abstract Denote by $\mathcal{A}(\kappa)$ the set of all compact Alexandrov surfaces with curvature bounded below by κ, without boundary, endowed with the topology induced by the Gromov–Hausdorff metric. We determine the connected components of $\mathcal{A}(\kappa)$ and of its closure.

Keywords Space of Alexandrov surfaces

Math. Subj. Classification(2010): 53C45

1 Introduction and Results

In this note, by an *Alexandrov surface* we understand a compact 2-dimensional Alexandrov space with curvature bounded below by κ, without boundary. Roughly speaking, an Alexandrov surface is a closed topological surface endowed with an intrinsic geodesic distance satisfying Toponogov's angle comparison condition. See [5] or [11] for definitions and basic facts about such spaces.

Denote by $\mathcal{A}(\kappa)$ the set of all Alexandrov surfaces. Endowed with the Gromov–Hausdorff metric d_{GH}, $\mathcal{A}(\kappa)$ becomes a Baire space in which Riemannian surfaces form a dense subset [6].

Perelman's stability theorem (see [7,9]) states, in our case, that close Alexandrov surfaces are homeomorphic, so Alexandrov surfaces with different topology are in different connected components of $\mathcal{A}(\kappa)$. Here we show that homeomorphic Alexandrov surfaces are in the same component of $\mathcal{A}(\kappa)$.

J. Rouyer (✉) • C. Vîlcu
"Simion Stoilow" Institute of Mathematics of the Romanian Academy,
P.O. Box 1-764, Bucharest 70700, Romania
e-mail: Joel.Rouyer@ymail.com; Joel.Rouyer@imar.ro; Costin.Vilcu@imar.ro

D. Ibadula, W. Veys (eds.), *Bridging Algebra, Geometry, and Topology*,
Springer Proceedings in Mathematics & Statistics 96,
DOI 10.1007/978-3-319-09186-0_15

Let $\mathcal{A}(\kappa, \chi, o)$ denote the set of all (classes of isometric) surfaces in $\mathcal{A}(\kappa)$ of Euler–Poincaré characteristic χ and orientability o, where $o = 1$ if the surface is orientable and $o = -1$ otherwise.

Theorem 1. *If nonempty, $\mathcal{A}(\kappa, \chi, o)$ is a connected component of $\mathcal{A}(\kappa)$, for $\kappa \in \mathbb{R}$, $\chi \leq 2$ and $o = \pm 1$.*

A special motivation for this result comes from the study of *most* (in the sense of Baire category) Alexandrov surfaces. For example, we prove in [10] that most Alexandrov surfaces have either infinitely many simple closed geodesics, or no such geodesic, depending on the value of κ and the connected component of $\mathcal{A}(\kappa)$ to which they belong. Moreover, for descriptions of most Alexandrov surfaces given in [1] and [6], one has to exclude from the whole space $\mathcal{A}(0)$ its components consisting of flat surfaces.

Denote by $\bar{\mathcal{A}}(\kappa)$ (respectively $\bar{\mathcal{A}}(\kappa, \chi, o)$) the closure with respect to d_{GH} of $\mathcal{A}(\kappa)$ (respectively $\mathcal{A}(\kappa, \chi, o)$) in the space of all compact metric spaces. Using Theorem 1, we can also give the connected components of $\bar{\mathcal{A}}(\kappa)$.

Theorem 2. *If $\kappa \geq 0$, $\bar{\mathcal{A}}(\kappa)$ is connected. If $\kappa < 0$, the connected components of $\bar{\mathcal{A}}(\kappa)$ are $\bigcup_{\chi \geq 0, o = \pm 1} \bar{\mathcal{A}}(\kappa, \chi, o)$ and $\mathcal{A}(\kappa, \chi, o)$ ($\chi = -1, -2, \ldots, o = \pm 1$).*

2 Proofs

Perelman's stability theorem can be found, for example, in [9] or [7]; we only need a particular form of it.

Lemma 1. *Each Alexandrov surface A has a neighborhood in $\mathcal{A}(\kappa)$ whose elements are all homeomorphic to A.*

Let $\mathbb{M}_\kappa^d$ stand for the simply-connected and complete Riemannian manifold of dimension d and constant curvature κ.

Denote by $\mathcal{R}(\kappa)$ the set of all closed Riemannian surfaces with Gauss curvature at least κ, and by $\mathcal{P}(\kappa)$ the set of all κ-polyhedra. Recall that a *κ-polyhedron* is an Alexandrov surface obtained by naturally gluing finitely many geodesic polygons from $\mathbb{M}_\kappa^2$.

A formal proof for the following result can be found in [6].

Lemma 2. *The sets $\mathcal{R}(\kappa)$ and $\mathcal{P}(\kappa)$ are dense in $\mathcal{A}(\kappa)$.*

A *convex surface* in $\mathbb{M}_\kappa^3$ is the boundary of a compact convex subset of $\mathbb{M}_\kappa^3$ with nonempty interior. Such a surface is endowed with the so-called intrinsic metric: the distance between two points is the length (measured with the metric of $\mathbb{M}_\kappa^3$) of a shortest curve joining them and lying on the surface.

Lemma 3 ([2]). *Every convex surface in $\mathbb{M}_\kappa^3$ belongs to $\mathcal{A}(\kappa, 2, 1)$. Conversely, every surface $A \in \mathcal{A}(\kappa, 2, 1)$ is isometric to some convex surface in $\mathbb{M}_\kappa^3$.*

In order to settle the case of $\mathcal{A}(0, 0, o)$, we need the following lemma.

Lemma 4. $\mathcal{A}(0,0,1)$ *contains only flat tori, and* $\mathcal{A}(0,0,-1)$ *contains only flat Klein bottles.*

Proof. Recall that geodesic triangulations with arbitrarily small triangles exist for any Alexandrov surface [2].

Consider $A \in \mathcal{A}(0,0,1)$ and a geodesic triangulation $T = \{\Delta_i\}$ of A. For each Δ_i, consider a comparison triangle $\tilde{\Delta}_i$ (i.e., a triangle with the same edge lengths) in $\mathbb{M}_0^2$. Glue together the triangles $\tilde{\Delta}_i$ to obtain a surface P, in the same way the triangles Δ_i are glued together to compose A. By the definition of Alexandrov surfaces, the angles of $\tilde{\Delta}_i$ are lower than or equal to the angles of Δ_i. It follows that the total angles $\theta_1, \ldots, \theta_n$ of P around its (combinatorial) vertices are at most 2π, hence P is a 0-polyhedron. By the Gauss-Bonnet formula for polyhedra,

$$0 = 2\pi\chi = \sum_{i=1}^{n} (2\pi - \theta_i),$$

whence $\theta_i = 2\pi$ and P is indeed a flat torus.

Now consider a sequence of finer and finer triangulations T_m of A and denote by P_m the corresponding flat tori ($m \in \mathbb{N}$). A result of Alexandrov and Zalgaller (Theorem 10 in [3, p. 90]) assures that P_m converges to A, which is therefore flat.

The same argument holds for $\mathcal{A}(0,0,-1)$. □

Notice that for $\kappa' > \kappa$, $\mathcal{A}(\kappa')$ is a nowhere dense subset of $\mathcal{A}(\kappa)$; indeed, $\mathcal{A}(\kappa')$ is closed and its complement contains the κ-polyhedra, which are dense in $\mathcal{A}(\kappa)$. Therefore, there is no direct relationship between the connected components of $\mathcal{A}(\kappa)$ and those of $\mathcal{A}(\kappa')$.

Now we are in a position to prove Theorem 1.

Proof. By Lemma 1, each set $\mathcal{A}(\kappa, \chi, o)$ is open in $\mathcal{A}(\kappa)$, so we just need to prove that it is connected.

Each Alexandrov surface A is in particular a metric space. Multiplying all distances in $A \in \mathcal{A}(\kappa)$ with the same constant $\delta > 0$ provides another Alexandrov surface, denoted by δA, which belongs to $\mathcal{A}(\frac{\kappa}{\delta^2})$. Moreover, it is easy to see that for any metric spaces M, N we have $d_{GH}(\delta M, \delta N) = \delta d(M, N)$. So there is a natural homothety between $\mathcal{A}(\kappa)$ and $\mathcal{A}(\frac{\kappa}{\delta^2})$, and therefore we may assume that

$$\kappa \in \{-1, 0, 1\}.$$

We consider several cases.

Case 1. The sets $\mathcal{A}(-1, \chi, o)$ are connected in $\mathcal{A}(-1)$.

Choose $A_0, A_1 \in \mathcal{A}(-1, \chi, o) \cap \mathcal{R}(-1)$. There exist a differentiable surface S of Euler–Poincaré characteristic χ and orientability o, and Riemannian metrics g_0, g_1 on S such that A_i is isometric to (S, g_i) ($i = 0,1$). For $\lambda \in [0, 1]$ we set

$$\tilde{g}_\lambda = \lambda g_1 + (1 - \lambda) g_0.$$

Denote by κ_λ the minimal value of the Gauss curvature of $\tilde{g}_\lambda$, and define the Riemannian metric g_λ on S by

$$g_\lambda = \begin{cases} \tilde{g}_\lambda & \text{if } \kappa_\lambda \geq -1, \\ \frac{\tilde{g}_\lambda}{\sqrt{-\kappa_\lambda}} & \text{if } \kappa_\lambda < -1. \end{cases}$$

A straightforward computation shows that the Gauss curvature K_λ of g_λ verifies $K_\lambda \geq -1$.

Denote by γ the (obviously continuous) canonical map from the set of Riemannian structures on S to $\mathcal{A}(-1,\chi,o)$, which maps g to (S,g). Then $A_\lambda \stackrel{\text{def}}{=} \gamma(g_\lambda)$ defines a path from A_0 to A_1. Hence $\mathcal{A}(-1,\chi,o) \cap \mathcal{R}(-1)$ is connected and, by the density of $\mathcal{R}(-1)$, so is $\mathcal{A}(-1,\chi,o)$.

Next we treat the connected components of $\mathcal{A}(0)$.

Case 2. The sets $\mathcal{A}(0,0,1)$ and $\mathcal{A}(0,0,-1)$ are connected in $\mathcal{A}(0)$.

By Lemma 4, the set $\mathcal{A}(0,0,1)$ contains only flat tori, hence it is continuously parametrized by the parameters describing the fundamental domains. Similarly for $\mathcal{A}(0,0,-1)$, which consists of flat Klein bottles.

Case 3. The set $\mathcal{A}(0,2,1)$ is connected in $\mathcal{A}(0)$.

Denote by $\mathcal{S}$ the space of all convex surfaces in $\mathbb{R}^3$, endowed with the Pompeiu–Hausdorff metric. Lemma 3 shows that any surface $A \in \mathcal{A}(0,2,1)$ can be realized as a convex surface in $\mathbb{R}^3$.

Given two convex surfaces S_0, S_1, define for $\lambda \in [0,1]$

$$S_\lambda = \partial\left(\lambda \text{conv}(S_1) + (1-\lambda)\text{conv}(S_0)\right), \tag{1}$$

where ∂C stands for the boundary of C, $\text{conv}(S)$ for the convex hull of S, and $+$ for the Minkowski sum. Then $S_\lambda \in \mathcal{S}$, and we have a path in $\mathcal{S}$ joining S_0 to S_1. Since the canonical map σ from $\mathcal{S}$ to $\mathcal{A}(0,2,1)$ is continuous [2, Theorem 1 in Chapter 4], we obtain a path in $\mathcal{A}(0,2,1)$.

Case 4. The set $\mathcal{A}(0,1,-1)$ is connected in $\mathcal{A}(0)$.

Consider surfaces A_0, A_1 in $\mathcal{A}(0,1,-1)$ as quotients of centrally-symmetric convex surfaces S_0, S_1 via antipodal identification, $A_i = \sigma(S_i)/\mathbb{Z}_2$ $(i = 0,1)$. Then the surface S_λ defined by (1) is also centrally-symmetric, and therefore $A_\lambda = \sigma(S_\lambda)/\mathbb{Z}_2$ defines a path in $\mathcal{A}(0,1,-1)$ from A_0 to A_1.

We finally treat the two connected components of $\mathcal{A}(1)$.

Case 5. The set $\mathcal{A}(1,2,1)$ is connected in $\mathcal{A}(1)$.

Consider in $\mathbb{R}^4$ the subspace $\mathbb{R}^3 = \mathbb{R}^3 \times \{0\}$, and the open half-sphere H of center $c = (0,0,0,1)$ and radius 1 included in $\mathbb{R}^3 \times [0,1[$.

Let $q : \mathbb{R}^3 \to H$ be the homeomorphism associating to each $x \in \mathbb{R}^3$ the intersection point of the line-segment $[xc]$ with H. Clearly, q maps line-segments of $\mathbb{R}^3$ to geodesic segments of H and vice-versa, and thus it maps bijectively convex sets in $\mathbb{R}^3$ to convex sets in H. Denote by $\mathcal{S}_H$ the set of convex surfaces in H. We can define $Q : \mathcal{S} \to \mathcal{S}_H$ by $Q(S) \stackrel{\text{def}}{=} q(S)$. Hence $\mathcal{S}_H$ is homeomorphic to $\mathcal{S}$, which is connected by Case 3.

Consider now two surfaces $A_0, A_1 \in \mathcal{A}(1,2,1)$ and choose

$$\mu < \min\left\{\frac{\pi}{2\text{diam}\,(A_0)}, \frac{\pi}{2\text{diam}\,(A_1)}, 1\right\}.$$

Obviously, A_i is path-connected to μA_i in $\mathcal{A}(1,2,1)$, and the diameter of μA_i is less than $\pi/2$ ($i = 0, 1$). By Lemma 3, μA_i is isometric to a surface S_i in $\mathbb{M}_1^3$; moreover, the smallness of μ easily implies that S_i is isometric to a surface in $\mathcal{S}_H$, and $\mathcal{S}_H$ is connected.

Case 6. The set $\mathcal{A}(1,1,-1)$ is connected in $\mathcal{A}(1)$.

This follows directly from the previous argument, because the universal covering of any surface $\tilde{A} \in \mathcal{A}(1,1,-1)$ is a surface $A \in \mathcal{A}(1,2,1)$ endowed with an isometric involution without fixed points, $\tilde{A} = A/\mathbb{Z}_2$.

The proof of Theorem 1 is complete. □

Recall that the 2-dimensional Hausdorff measure $\mu(A)$ is always finite and positive for $A \in \mathcal{A}(\kappa)$. The following lemma is Corollary 10.10.11 in [4, p. 401], stated in our framework.

Lemma 5. *Let $A_n \in \mathcal{A}(\kappa)$ converge to a compact space X. Then* $\dim(X) < 2$ *if and only if $\mu(A_n) \to 0$.*

Now we are in a position to prove Theorem 2.

Proof. We may assume, as in the proof of Theorem 1, that $\kappa \in \{-1, 0, 1\}$.

To prove that $\bar{\mathcal{A}}(\kappa)$ is connected for $\kappa \geq 0$, it suffices to show that the space consisting of a single point belongs to the closure of any connected component of $\mathcal{A}(\kappa)$. This is indeed the case, because for any $A \in \mathcal{A}(\kappa, \chi, o)$ and $0 < \delta \leq 1$ we have $\delta A \in \mathcal{A}(\kappa, \chi, o)$, and $\lim_{\delta \to 0} \delta A$ is a point.

This also implies that

$$\bigcup_{\substack{o=\pm 1\\ \chi=0,1,2}} \bar{\mathcal{A}}(-1, \chi, o)$$

is connected.

Consider now $A \in \mathcal{A}(-1, \chi, o)$ with $\chi < 0$. Let ω be the curvature measure on A (see [3]). Y. Machigashira [8] proved that $\omega \geq \kappa\mu$ holds for any Alexandrov surface of curvature bounded below by κ. Therefore, by a variant of the Gauss-Bonnet theorem,

$$2\pi\chi = \omega(A) \geq \kappa\mu\,(A) = -\mu\,(A),$$

hence $\mu(A) \geq 2\pi|\chi|$. Lemma 5 shows now that $\mathcal{A}(-1, \chi, o)$ is closed in the space of all compact metric spaces ($o = \pm 1, \chi < 0$). □

Acknowledgements The authors were supported by the grant PN-II-ID-PCE-2011-3-0533 from the Romanian National Authority for Scientific Research, CNCS-UEFISCDI.

References

1. Adiprasito, K., Zamfirescu, T.: Few Alexandrov spaces are Riemannian, J. Nonlinear Convex Anal., (to appear)
2. Alexandrov, A.D.: Die innere Geometrie der konvexen Flächen. Akademie-Verlag, Berlin (1955)
3. Aleksandrov, A.D., Zalgaller, V.A.: Intrinsic Geometry of Surfaces. Transl. Math. Monographs. American Mathematical Society, Providence (1967)
4. Burago, D., Burago, Y., Ivanov, S.: A Course in Metric Geometry. American Mathematical Society, Providence (2001)
5. Burago, Y., Gromov, M., Perel'man, G.: A. D. Alexandrov spaces with curvature bounded below. Russ. Math. Surv. **47**, 1–58 (1992). English. Russian original
6. Itoh, J., Rouyer, J., Vîlcu, C.: Moderate smoothness of most Alexandrov surfaces. arXiv:1308.3862 [math.MG]
7. Kapovitch, V.: Perelman's stability theorem. In: Cheeger, J., et al. (eds.) Metric and Comparison Geometry. International Press. Surveys in Differential Geometry 11, 103–136 (2007)
8. Machigashira, Y.: The Gaussian curvature of Alexandrov surfaces. J. Math. Soc. Japan. **50**, 859–878 (1998)
9. Perel'man, G.: A.D. Alexandrov spaces with curvatures bounded from below II. preprint 1991
10. Rouyer, J., Vîlcu, C.: Simple closed geodesics on most Alexandrov surfaces. arXiv:1311.4873 [math.MG]
11. Shiohama, K.: An Introduction to the Geometry of Alexandrov Spaces. Lecture Notes Series. Seoul National University, Seoul (1992)

Motivic Milnor Fibre for Nondegenerate Function Germs on Toric Singularities

J.H.M. Steenbrink

Abstract We study function germs on toric varieties which are nondegenerate for their Newton diagram. We express their motivic Milnor fibre in terms of their Newton diagram. We extend a formula for the motivic nearby fibre to the case of a toroidal degeneration. We illustrate this by some examples.

Keywords Motivic milnor fibre • Toric singularity • Newton diagram

1 Introduction

In the calculation of invariants of isolated hypersurface singularities, Newton diagram methods have always been an important tool. These methods are closely connected with the theory of torus embeddings. They were first used for the computation of resolutions of cusp singularities by Ehlers [8]. Kouchnirenko in [12] used them to compute the Milnor number for nondegenerate functions, and Varchenko in [18] to compute the zeta function of their monodromy. A conjecture for the Hodge spectrum of nondegenerate hypersurface singularities in terms of their Newton diagram was formulated in [17] and proved by Danilov in [6]. It was reformulated by Saito in terms of the Newton filtration on the jacobian module and proved in [16]. Guibert [10] determined the motivic Igusa zeta function for nondegenerate curve singularities and Bories and Veys [4] proved the monodromy conjecture for the p-adic and motivic zeta function of a nondegenerate surface singularity.

In this paper we deal with the motivic Milnor fibre of nondegenerate function germs. We generalize the concepts of Newton diagram and nondegeneracy to germs defined on a toric variety at a zero-dimensional torus orbit, thereby widening the range of examples of singularities where Newton diagram methods can be applied.

J.H.M. Steenbrink (✉)
IMAPP, Radboud University Nijmegen, The Netherlands

Mathematical Sciences Center, Tsinghua University, Beijing, China
e-mail: J.Steenbrink@math.ru.nl

D. Ibadula, W. Veys (eds.), *Bridging Algebra, Geometry, and Topology*,
Springer Proceedings in Mathematics & Statistics 96,
DOI 10.1007/978-3-319-09186-0_16

We use the Newton diagram to define a toric modification of the ambient space which is a toric variety and such that the zero fibre of our germ defines a toroidal divisor on it. The proof of our formula relies then on the extension of a formula for the motivic nearby fibre from the normal crossing case to the toroidal case.

The importance of the motivic Milnor fibre lies in the fact that several additive invariants like Euler characteristic, Zeta function of monodromy and certain invariants connected with mixed Hodge structures are determined by this invariant.

Each of the terms occurring in our formula for the motivic Milnor fibre is a hypersurface in a torus given by an equation $g(z) = 1$ with g weighted homogeneous; we will calculate the Hodge spectrum of such hypersurfaces in a subsequent paper with Sander Rieken.

More friends I made during my mathematical career who deserve a birthday present than I am able to produce papers of the quality they deserve. Therefore I dedicate this paper to the sixtieth birthdays of Sabir Gusein-Zade, Wolfgang Ebeling and Alexandru Dimca. I hope they do not mind to share this present.

2 Motivic Nearby Fibre

We briefly recall the notions of motivic nearby fibre and motivic Milnor fibre as defined by Denef and Loeser [7]. We prove a formula for them in a toroidal setting.

In this section, k is an algebraically closed field of characteristic zero. By a k-variety we mean a reduced scheme of finite type over k.

2.1 Grothendieck Groups of Varieties

Let us recall the Grothendieck group $K_0(\mathrm{Var}_k)$ of varieties over k. It is defined as an abelian group by the following generators and relations. Generators are the isomorphism classes $[X]$ of algebraic k-varieties and relations are of the form $[X] = [Y] + [X \setminus Y]$ for any pair $Y \subset X$ where Y is a closed subvariety. By defining the product of $[Z]$ and $[W]$ to be $[Z \times W]$ we obtain a commutative ring with unit $[\mathrm{Spec}(k)]$.

Let X be a smooth projective variety and $Y \subset X$ a smooth closed subvariety. Let $\pi : X' \to X$ be the blowing-up with center Y and let $Y' = \pi^{-1}(Y)$. Then π defines an isomorphism between $X' \setminus Y'$ and $X \setminus Y$, hence

$$[X'] - [Y'] = [X] - [Y] \text{ in } K_0(\mathrm{Var}_k)\ . \tag{1}$$

Bittner [2] showed that this type of relation suffices to describe the group $K_0(\mathrm{Var}_k)$:

Theorem 1. *The group $K_0(\mathrm{Var}_k)$ is isomorphic to the abelian group with generators $[X]$ where X is smooth projective and relations (1).*

If S is a k-variety, we have a relative Grothendieck group $K_0(\mathrm{Var}_S)$ with generators $[X]$ the S-isomorphism classes of varieties over S and again relations $[X] = [Y] + [X \setminus Y]$ for any pair $Y \subset X$ where Y is a closed subvariety. Fibre product over S equips $K_0(\mathrm{Var}_S)$ with the structure of a ring. We let $\mathbb{L} = [\mathbb{A}^1 \times S \to S] \in K_0(\mathrm{Var}_S)$, the map to S being the projection to the second factor, and $\mathcal{M}_S = K_0(\mathrm{Var}_S)[\mathbb{L}^{-1}]$.

We also need equivariant versions of these constructions. For $n \in \mathbb{N}$ we let μ_n denote the group of n-th roots of unity in k. By mapping μ_{nd} to μ_n by $x \mapsto x^d$ we obtain a projective system and we let $\hat{\mu} = \varprojlim \mu_n$.

A *good* $\hat{\mu}$*-action* on an S-variety X is given by an action of μ_n on X by S-morphisms for some n with the property that each orbit is contained in an affine open subset. The group $K_0(\mathrm{Var}_S^{\hat{\mu}})$ has generators $[X, \hat{\mu}]$ where X is an S-variety with good $\hat{\mu}$-action and relations $[X, \hat{\mu}] = [Y, \hat{\mu}] + [X \setminus Y, \hat{\mu}]$ for $Y \subset X$ closed and $[V, \hat{\mu}] = [X, \hat{\mu}] \cdot \mathbb{L}^m$ when V is an affine space bundle of rank m over X with any good $\hat{\mu}$-action lifting the action on X. Finally $\mathcal{M}_S^{\hat{\mu}} = K_0(\mathrm{Var}_S^{\hat{\mu}})[\mathbb{L}^{-1}]$. It becomes a ring when one equips $X \times Y$ with the diagonal group action.

If $f : S \to T$ is a k-morphism, then every S-variety $h : X \to S$ becomes also a T-variety $fh : X \to T$. This defines group homomorphisms $f_! : K_0(\mathrm{Var}_S) \to K_0(\mathrm{Var}_T)$ and $f_! : \mathcal{M}_S \to \mathcal{M}_T$. In the equivariant setting we obtain

$$f_! : \mathcal{M}_S^{\hat{\mu}} \to \mathcal{M}_T^{\hat{\mu}}. \tag{2}$$

On the other hand, if Y is a T-variety, then $Y \times_T S$ is an S-variety. This defines ring homomorphisms

$$f^* : \mathcal{M}_T^{\hat{\mu}} \to \mathcal{M}_S^{\hat{\mu}}. \tag{3}$$

2.2 Nearby and Milnor Fibre

Let X be a smooth connected quasi-projective variety over k and let $f : X \to k$ be a non-constant regular function. Assume that $E = f^{-1}(0)$ is a divisor with strict normal crossings on X. Let E_i, $i \in I$, be the irreducible components of E and let e_i denote the multiplicity of f along E_i.

Choose a common multiple e of all e_i, $i \in I$. Let $\tau : k \to k$ be defined by $\tau(z) = z^e$, and let $\tau^* X$ denote the pull-back of X via τ. Finally let $\tilde{X}$ be the normalization of $\tau^* X$. Then we have a diagram

$$\begin{array}{ccccc} D & \hookrightarrow & \tilde{X} & \overset{\rho}{\to} & X \\ \downarrow & & \tilde{f} \downarrow & & f \downarrow \\ 0 & \in & k & \overset{\tau}{\to} & k \end{array} \tag{4}$$

Let $\zeta \in \mu_e$. The covering transformation $z \mapsto \zeta z$ of τ extends to an automorphism $\gamma(\zeta)$ of order e of $\tilde{X}$ and in this way we obtain a good $\hat{\mu}$-action on $\tilde{X}$ and D.

We have decompositions of the zero fibres E and D into nonsingular locally closed subsets as follows. Let $J \subset I$ be a non-empty subset. We define $E_J = \bigcap_{i\in J} E_i$ and $E_J^0 = E_J \setminus \bigcup_{i\notin J} E_i$. Moreover we let $D_J^0 = \rho^{-1}(E_J^0)$, and γ_J the $\hat{\mu}$-action on D_J^0 induced by γ. Then $\rho_J : D_J^0 \to E_J^0$ is an etale covering of degree equal to the greatest common divisor of the multiplicities $e_i, i \in J$.

The *motivic nearby fibre* of f is defined as

$$\psi_f = \sum_J [(D_J^0, \gamma_J)](1 - \mathbb{L})^{\sharp J - 1} \in K_0(\mathrm{Var}_E^{\hat{\mu}}) \tag{5}$$

where the sum runs over the non-empty subsets J of I. See [13].

Remarks

1. The pairs (D_J, γ_J) do not depend on the choice of the integer e (as long as $e_i | e$ for each $i \in I$).
2. The element ψ_f does not change when the zero fibre E is modified by blowing-up. See e.g. [15, Lemma 12.2.4].

In [7] the motivic nearby fibre ψ_f has been defined for any $f : X \to k$ with X nonsingular via the theory of arc spaces. Their definition leads to formula (5) when the zero fibre X_0 of f has strict normal crossings. Under this hypothesis ψ_f is characterized by the following properties:

1. if $\pi : X' \to X$ is a proper birational map which induces an isomorphism $X' \setminus X'_0 \to X \setminus X_0$ and $\pi_0 : X'_0 \to X_0$ is the induced map, then $\psi_f = \pi_{0!}\psi_{f\pi}$;
2. if X_0 is a divisor with strict normal crossings, then ψ_f is given by formula (5).

These two properties enable one to extend the definition of $\psi_f \in K_0(\mathrm{Var}_{X_0}^{\hat{\mu}})$ to the case where X is singular but $\mathrm{Sing}(X) \subset X_0$: one chooses an embedded resolution $\pi : (X', E) \to (X, X_0)$ such that E is a divisor with strict normal crossings and defines $\psi_f = \pi_{0!}\psi_{f\pi}$.

In the sequel we will need a definition of ψ_f in the general setting where X may have singularities not contained in X_0. Here we use Bittner's *nearby cycle functor*

$$\Psi_f : \mathcal{M}_X \to \mathcal{M}_{X_0}^{\hat{\mu}} \tag{6}$$

as defined in [3]. Bittner shows that $\mathcal{M}_X$ is generated by $[Y]_X$ with Y proper over X and nonsingular, and for such $[Y]$ she defines for $f_{|Y}$ non-constant:

$$\Psi_f([Y]) = \pi_{0!}\psi_{f\pi}, \tag{7}$$

where $\pi : Y \to X$ is the given morphism and $\pi_0 : Y_0 \to X_0$ its restriction to the zero fibres. Moreover $\Psi_f([Y]) = 0$ if f is constant on Y.

This formula respects the relations between generators by [3, Claim 8.2]. Hence we obtain an $\mathcal{M}_k$-linear map $\Psi_f : \mathcal{M}_X \to \mathcal{M}_{X_0}^{\hat{\mu}}$ and may define

$$\psi_f := \Psi_f([X]). \tag{8}$$

Following [7] for $x \in X_0$ we define the *motivic Milnor fibre of f at x* by

$$\psi_{f,x} = i_x^* \psi_f \tag{9}$$

where $i_x : \{x\} \hookrightarrow X_0$ is the inclusion.

2.3 Toroidal Embeddings

We first recall some basic notions concerning toric varieties. Details can be found in [5,9] or [14].

An n-dimensional *toric variety* is a normal variety X which contains an algebraic torus $T \simeq (\mathbb{G}_m)^n$ as a dense Zariski open subset such that the action of T on itself by translation extends to an action of T on X.

We let M denote the character group of T; it is free abelian of rank n. The dual group $N = \mathrm{Hom}(M, \mathbb{Z})$ can be identified with the group of one-parameter subgroups of T. We define $M_{\mathbb{R}} = M \otimes \mathbb{R}$ and $N_{\mathbb{R}} = N \otimes \mathbb{R}$.

Each T-invariant affine open subset U of X corresponds to a cone in $N_{\mathbb{R}}$ which is the convex hull of all one-parameter subgroups of T which extend to a morphism $\mathbb{A}_k^1 \to U$. Conversely, the affine variety U is determined by its cone $\sigma \subset N_{\mathbb{R}}$ as follows. Let $\sigma^\vee = \{m \in M_{\mathbb{R}} \mid \forall p \in \sigma : \langle m, p\rangle \geq 0\}$. Then

$$U = X_\sigma := \mathrm{Spec}\, k[M \cap \sigma^\vee]. \tag{10}$$

The (finite) collection Σ of these cones is called the *fan* in $N_{\mathbb{R}}$ associated with X, and X can be recovered from its fan. The variety X is complete if and only if $\bigcup_{\sigma \in \Sigma} \sigma = N_{\mathbb{R}}$.

Definition 1. A *toroidal embedding (without self-intersection)* is an open embedding $U \hookrightarrow X$ of varieties with the following property: for every $x \in X$ there exists an open neighborhood U_x and an etale morphism $\phi : U_x \to X_{\sigma_x}$ to a toric variety X_{σ_x} with torus T_x such that $U_x \cap U = \phi^{-1}(T_x)$. Such a morphism is called a *chart at x*.

If $U \hookrightarrow X$ is a toroidal embedding such that X is nonsingular, then $X \setminus U$ is a divisor on X with strict normal crossings. For every toroidal embedding $U \hookrightarrow X$ we have a canonical stratification with the following properties: the strata are connected locally closed subvarieties of X and for every chart $\phi : U_x \to X_{\sigma_x}$ the intersection of the strata with U_x are the preimages of the torus orbits in X_{σ_x} under ϕ.

Definition 2. A *proper modification* of toroidal embeddings $[U \hookrightarrow X'] \to [U \hookrightarrow X]$ is a proper morphism $\pi : X' \to X$ which is the identity on U (hence birational) and which maps strata of X' onto strata of X.

Theorem 2. *For every toroidal embedding $U \hookrightarrow X$ there exists a proper modification of toroidal embeddings $[U \hookrightarrow X'] \to [U \hookrightarrow X]$ such that X' is nonsingular.*

Proof. This follows from [11, Theorem 11*].

2.4 Motivic Nearby Fibre in Toroidal Setting

We first explain what is meant by "toroidal setting". We consider a toroidal embedding $U \subset X$ and a regular function $f : X \to k$ with the property that $U \cap X_0 = \emptyset$. As an important example we have the case where X is nonsingular and X_0 is a divisor with strict normal crossings on X; here we take $U = X \setminus X_0$. In general, $\mathrm{Sing}(X)$ will not be contained in X_0; then U has to be strictly smaller than $X \setminus X_0$.

We are going to show that in the toroidal setting an explicit formula for ψ_f analogous to (5) is valid. The strata E_J^0 in that formula will be replaced by the strata of the toroidal boundary $X \setminus U$ which are contained in X_0.

We start our computation by analyzing the strata of the canonical toroidal stratification of X. Let S denote the set of strata, and S_h the subset of S consisting of all strata which are not contained in X_0. For $s \in S_h$ let $\bar{s}$ be its closure in X and $i_s : \bar{s} \hookrightarrow X$ the inclusion.

We have

$$[X] = [X_0] + \sum_{s \in S_h} [s] \tag{11}$$

in $\mathcal{M}_X$ and hence

$$\psi_f := \Psi_f([X]) = \sum_{s \in S_h} \Psi_f([s]) \tag{12}$$

in $\mathcal{M}_{X_0}^{\hat{\mu}}$, because Ψ_f is $\mathcal{M}_k$-linear and $\Psi_f([X_0]) = 0$ by [3, Properties 8.4]. As i_s is a closed embedding, we also have

$$\Psi_f([s]) = \Psi_f i_{s!}[s] = i_{s0!} \Psi_{f i_s}[s] \tag{13}$$

where i_{s0} is the inclusion of $\bar{s} \cap X_0$ in X_0. Note that $s \subset \bar{s}$ is also a toroidal embedding. So we have reduced the computation of ψ_f to the case of $\Psi_f([U])$.

Lemma 1. *Let $U \subset X$ be a toroidal embedding and $f : X \to k$ a regular function such that $U \cap X_0 = \emptyset$. Let $\pi : X' \to X$ be a toroidal modification such that X' is smooth. Let $f' = f\pi$. Then*

$$\Psi_f([U]) = \pi_{0!} \Psi_{f'}([U]). \tag{14}$$

Proof. By [3, Properties 8.4] we have $\Psi_f \pi_! = \pi_{0!} \Psi_{f\pi}$ because π is proper. Moreover $\pi_!([U]_{X'}) = [U]_X$.

Hence to compute $\Psi_f([U])$ we may assume that X is smooth and $X \setminus U$ is a divisor with strict normal crossings.

Theorem 3. *Let $f : X \to k$ be a regular function on a smooth variety X. Let $Y \subset X$ be a closed subset such that $Y \cup X_0$ is a divisor with strict normal crossings on X and f is non-constant on every irreducible component of Y. Let $U = X \setminus (Y \cup X_0)$ and $\tilde{U} = X \setminus Y \overset{j}{\hookrightarrow} X$ and $\tilde{f} = fj$. Let $j_0 : \tilde{U} \cap X_0 \hookrightarrow U$. Then*

$$\Psi_f([U]) = \Psi_f([\tilde{U}]) = j_{0!} \psi_{\tilde{f}}. \tag{15}$$

Moreover, $\psi_{\tilde{f}}$ is given by formula (5) with $E = X_0 \setminus Y$.

Proof. We have $[\tilde{U}] = \sum_{J \subset I} (-1)^{|J|} [Y_J]$ in $K_0(\mathrm{Var}_X)$, where I is the set of irreducible components of Y and $Y_J = \bigcap_{j \in J} Y_j$ (so $X = Y_\emptyset$). The inclusion $i_J : Y_J \to X$ is proper so $\Psi_f([Y_J]) = (i_J)_{0!} \psi_{f|_{Y_J}}$, which is computed using the formula (5) with $E = Y_J \cap X_0$. Putting all these terms together we obtain the result.

Theorem 4. *Let $\pi : X' \to X$ be a proper equivariant modification of n-dimensional toroidal embeddings with good $\hat{\mu}$-action. Let s be a stratum of X and let I'_s denote the set of strata of $\pi^{-1}(s)$. Let $c(t) = n - \dim t$ for any stratum t. Then*

$$(1 - \mathbb{L})^{c(s)} [s] = \pi_! \sum_{t \in I'_s} (1 - \mathbb{L})^{c(t)} [t] \text{ in } K_0(\mathrm{Var}_X^{\hat{\mu}}). \tag{16}$$

Proof. Each stratum $t \in I'_s$ is a trivial $\mathbb{G}_m^k$-bundle over s with $k = c(s) - c(t)$. Hence

$$\pi_![t] = (\mathbb{L} - 1)^{c(s)-c(t)} [s] \tag{17}$$

so

$$\pi_! (1 - \mathbb{L})^{c(t)} [t] = (-1)^{c(s)-c(t)} (1 - \mathbb{L})^{c(s)}. \tag{18}$$

We claim that $\sum_{t \in I'_s} (-1)^{c(t)} = (-1)^{c(s)}$. By restricting to a chart intersecting s we obtain the situation where all strata are pulled back from a toric variety. So for the proof of (16) we may assume that π is a modification of torus embeddings. Let Σ be the fan corresponding to X and Σ' the subdivision of Σ corresponding to π. For $\sigma \in \Sigma$ let σ^0 denote the relative interior of σ; it is homeomorphic to a cell of dimension $c(s)$, where s is unique closed torus orbit in X_σ. The orbits $t \subset \pi^{-1}(s)$ correspond to the cones $\tau \in \Sigma'$ for which $\tau^0 \subset \sigma^0$ and the interiors of these cones form a cell decomposition of σ^0. Then

$$\sum_{t \in I_s} (-1)^{c(t)} = (-1)^{c(s)} \tag{19}$$

is a consequence of the additivity the Euler characteristic with compact supports, because an open cell of dimension k has $\chi_c = (-1)^k$.

3 Nondegenerate Laurent Polynomials

In this section we consider function germs on toric singularities. We compute their motivic Milnor fibre.

3.1 Toric and Toroidal Singularities

As before, k is an algebraically closed field of characteristic zero and T is an n-dimensional algebraic torus over k with character group M as in Sect. 2.3. We keep the notations of that section.

Definition 3. A *toric singularity* is a germ (X, x) where X is a toric variety and x is a fixed point of the torus action on X. A *toroidal singularity* is a germ which is analytically isomorphic to a toric singularity.

In studying a toric singularity (X, x) we may suppose that the representative X is affine (else replace X by the star of x). Then $X = X_\sigma = \operatorname{Spec} k[M \cap \sigma^\vee]$ for a strictly convex rational polyhedral cone σ in $N_\mathbb{R}$. We let $A_\sigma = k[M \cap \sigma^\vee]$.

3.2 Newton Polyhedron and Nondegeneracy

A regular k-valued function on T is called a *Laurent polynomial.* Any character m of the torus can be considered as a regular function $e(m) : T \to k$, and the $e(m)$ with $m \in M$ form a k-basis of the coordinate ring $k[T]$.

Definition 4. Given a Laurent polynomial $f = \sum_{m \in M} a_m e(m)$ we define its *support* by $\operatorname{supp}(f) := \{m \in M \mid a_m \neq 0\}$. For any $V \subset M$ we let $f^V := \sum_{m \in M \cap V} a_m e(m)$

Definition 5. Consider a toric singularity (X_σ, x) and the corresponding k-algebra $A_\sigma \subset k[T]$. The *Newton polyhedron* Δ of f is the convex hull of $\bigcup_{m\in \mathrm{supp}(f)}(m + \sigma^\vee)$ in $M_\mathbb{R}$.

The function $f \in A_\sigma$ is called *convenient* if $\sigma^\vee \setminus \Delta$ is bounded. Equivalently: supp(f) has non-empty intersection with each one-dimensional face of $\sigma^\vee$.

The function f is called *Newton nondegenerate* if for each compact face Γ of Δ the functions $x_1\frac{\partial f^\Gamma}{\partial x_1}, \dots, x_n\frac{\partial f^\Gamma}{\partial x_n}$ have no common zero in the torus T.

This concept is similar to Δ-regularity as in [1].

Let P be an integral convex polytope in $M_\mathbb{R}$, i.e. the convex hull of a finite subset of M. We let $L(P)$ denote the k-linear span of all monomials $e(m)$ with $m \in M \cap P$. If P has dimension n, these monomials embed the torus T in projective space, and the associated toric variety $\mathbb{P}(P)$ is the closure of the image $T(P)$ of T under this mapping. It is also equal to Proj $S(P)$ where $S(P) = \bigoplus_{k\in\mathbb{N}} L(kP)$.

The space $L(P)$ is in a natural way the space of global sections in a very ample line bundle $\mathcal{O}(1)$ on $\mathbb{P}(P)$. Any nonzero element $g \in L(P)$ therefore determines a hyperplane section $\bar{Z}_g$ of $\mathbb{P}(P)$. Such g is called *P-regular* if P is the convex hull of supp(g) and for each face Q of P the polynomial equations

$$g^Q = g_1^Q = \cdots = g_n^Q = 0$$

where $g_j = x_j\frac{\partial g}{\partial x_j}$ have no common solution on the torus T.

By [1, Prop. 4.16] this is equivalent with the condition that $\bar{Z}_g$ has smooth intersection with all strata $T(Q)$ of the toric variety $\mathbb{P}(P)$, where Q runs over the faces of P.

Lemma 2. *Let $U \subset X$ be a toroidal embedding and let $D \subset X$ be a codimension one subvariety which intersects all strata of $X \setminus U$ transversely. Then D has toroidal singularities and $U \cap D \hookrightarrow D$ is a toroidal embedding.*

Proof. See [5, Sect. 13.2].

3.3 Newton Modification

Consider the inclusion $\Delta \subset \sigma^\vee$ for a convenient function f. Though Δ is not a polytope, it corresponds to a toric modification of X_σ as follows (see [6, Sect. 2.1]). For each face Γ of Δ we let $\sigma_\Gamma^\vee \subset \sigma^\vee$ be the cone spanned by all $u' - u$ with $u \in \Gamma$ and $u' \in \Delta$. Then the dual cones σ_Γ form a subdivision Σ_Δ of σ, called the *polar fan* of Δ. Hence the corresponding map $\pi : \mathbb{P}_\Delta := X_{\Sigma_\Delta} \to X_\sigma$ is a toric modification. It has been considered by Varchenko [18] and Danilov [6]. Note that $\pi : \mathbb{P}_\Delta \setminus \pi^{-1}(x) \to X_\sigma \setminus \{x\}$ is an isomorphism.

Theorem 5. *Let $f \in A_\sigma$ be convenient and Newton nondegenerate. Then there exists a closed subset $B \in X_\sigma$ with $x \notin B$ such that the inclusion*

$$T \setminus (f^{-1}(0) \cup B) \hookrightarrow \mathbb{P}_\Delta \setminus B \qquad (20)$$

is a toroidal embedding.

Proof. Let $\theta_1, \ldots, \theta_n$ be a basis of invariant vector fields on T and let $Z_i \subset T$ be given by $\theta_i(f) \neq 0$. Further, let $\bar{Z}_i$ be the closure of Z_i in $\mathbb{P}_\Delta$. Then the fact that f is Newton nondegenerate implies that

$$\bigcap_{i=1}^{n} \bar{Z}_i \cap \pi^{-1}(x) = \emptyset \tag{21}$$

by [1, Prop. 4.3]. We take $B = \bigcap_{i=1}^{n} \bar{Z}_i$. Then B is closed in $\mathbb{P}_\Delta$. As $T \hookrightarrow \mathbb{P}_\Delta$ is a torus embedding, evidently $T \setminus B \hookrightarrow \mathbb{P}_\Delta \setminus B$ is a toroidal embedding.

Consider the toroidal embedding $(T \setminus B) \times \mathbb{G}_m \hookrightarrow (\mathbb{P}_\Delta \setminus B) \times \mathbb{A}^1_k$. The graph D of $f\pi : \mathbb{P}_\Delta \setminus B \to \mathbb{A}^1_k$ satisfies the requirements of Lemma 2. Hence $D \cap ((T \setminus B) \times \mathbb{G}_m) \hookrightarrow D \cap ((\mathbb{P}_\Delta \setminus B) \times \mathbb{A}^1_k)$ is a toroidal embedding. Pulling this back via the embedding of the graph we obtain the result.

3.4 Motivic Milnor Fibre

It is our aim to give an explicit formula for the motivic Milnor fibre $\psi_{f,x}$ for $f \in A_\sigma$ Newton nondegenerate. We use the Newton modification $\pi : \mathbb{P}_\Delta \to X_\sigma$.

Lemma 3. *Let e be a sufficiently divisible positive integer. Let $\tilde{\Delta}$ be the Newton polyhedron of the function $F = f(z) - t^e \in A_\sigma[t]$. Then F is $\tilde{\Delta}$ nondegenerate. Let $\tau : k \to k$ be defined by $\tau(z) = z^e$ and let $\tau^*\mathbb{P}_\Delta$ be the pull-back of $\mathbb{P}_\Delta$ by τ. Then the normalization $\tilde{\mathbb{P}}_\Delta$ of $\tau^*\mathbb{P}_\Delta$ is isomorphic to the strict transform of the divisor of the function F in $\mathbb{P}_{\tilde{\Delta}}$. The μ_e-action on these varieties is induced by multiplication of the coordinate t by roots of unity.*

Observe that Theorems 5 and 3 imply that

$$\psi_f = \sum_s [D_s^0](1 - \mathbb{L})^{\sharp J - 1} \in K_0(\mathrm{Var}_k^{\hat{\mu}}) \tag{22}$$

where for each stratum s of $(f\pi)^{-1}(0)$ we let D_s^0 denote the corresponding subvariety of $\tilde{\mathbb{P}}_\Delta$ with its $\hat{\mu}$-action. We conclude that $\psi_{f,x}$ is given by the same formula, but where s runs only over the strata contained in $\pi^{-1}(x)$. The only remaining difficulty is to describe D_s^0 in simple terms. This however has essentially been done already in [6, Sect. 3]. We briefly recall the result.

First note that each stratum of $\pi^{-1}(x)$ in $\mathbb{P}_\Delta$ is isomorphic to an algebraic torus T_Γ. In fact, the closure of such a stratum is of the form $\mathbb{P}_\Gamma$ for a compact face Γ of Δ. The strata of $\pi^{-1}(x)$ in the toroidal embedding (5) are then of two types: either the zero set Z_s of f^Γ in T_Γ or $T_\Gamma \setminus Z_{f^\Gamma}$. It remains to describe the corresponding strata of the variety $\tilde{\mathbb{P}}_\Delta$.

Lemma 4. *For any compact face Γ of Δ let $\hat{\Gamma}$ be the convex hull of Γ and $\{0\}$. Then the corresponding strata of $\widetilde{\mathbb{P}}_\Delta$ are either the $U'_\Gamma := Z_{f^\Gamma} \subset T_\Gamma$ or the $U_\Gamma := Z_{f^\Gamma - 1} \subset T_{\hat{\Gamma}}$.*

Note that for each face Γ of Δ there exists a unique face τ of $\sigma^\vee$ with $\Gamma^0 \subset \tau^0$ (recall that τ^0 is the relative interior of τ). Let I_τ denote the set of these compact faces of Δ. The toric stratification of $X_\sigma \setminus B$ induces a decomposition $\psi_f = \sum \psi_{f^\tau}$ and U_Γ and U'_Γ contribute to $\psi_{f^\tau,x}$ exactly when $\Gamma \in I_\tau$. Let $c_\Gamma = \dim \tau - \dim \Gamma$ for $\Gamma \in I_\tau$. Then we find by Theorem 3:

Theorem 6.

$$\psi_{f,x} = \sum_\tau \sum_{\Gamma \in I_\tau} \left((1-\mathbb{L})^{c_\Gamma - 1}[U_\Gamma] + (1-\mathbb{L})^{c_\Gamma}[U'_\Gamma]\right). \tag{23}$$

4 Example: Weighted Homogeneous Laurent Polynomials

Definition 6. Let $T = T_N$ be an n-dimensional algebraic torus over k with coordinate ring $k[M]$. A Laurent polynomial $f \in k[M]$ is called *weighted homogeneous* if there exists a nonzero linear form $\ell \in \mathrm{Hom}(M, \mathbb{Q})$ such that the support of f is a subset of $\ell^{-1}(1)$.

Let $f \in k[M]$ be weighted homogeneous with respect to the linear form ℓ. This linear form determines a positive integer e by setting $\ell(M) = \frac{1}{e}\mathbb{Z}$. It also determines an action γ^* of μ_e on $k[M]$ and hence a dual action γ on T_N by

$$\gamma(\zeta)^*(X^m) = \zeta^{e\ell(m)} X^m.$$

Here we write X^m for the monomial corresponding to $m \in M$ in $k[M]$. Note that f is invariant under this action.

Let Γ_f denote the convex hull $\mathrm{Conv}(\mathrm{Supp}(f))$ in $M \otimes \mathbb{R}$ of the support of f. Then f is weighted homogeneous if and only if Γ_f is contained in an affine hyperplane not passing through 0. We will assume that $\dim \Gamma_f = n - 1$ and that f is Γ_f-regular. In that case the linear form ℓ is uniquely determined by f.

We have an n-dimensional cone $\sigma^\vee \subset M_\mathbb{R}$ given by

$$\sigma^\vee = \bigcup_{k \geq 0} \mathrm{Conv}(0, k\Delta_f).$$

We let A_σ denote the semigroup algebra $k[M \cap \sigma^\vee]$ and $X_\sigma := \mathrm{Spec}(A_\sigma)$. Let $\hat{\Delta}_f := \mathrm{Conv}(O, \Delta_f)$. It is a compact n-dimensional polyhedron and corresponds to a projective n-dimensional toric variety $\mathbb{P}_{\hat{\Delta}_f}$, which contains X_σ as a dense

Zariski-open subset. Clearly, the function $f - 1$ is $\hat{\Delta}$-regular and invariant under the action of μ_e. We put V the zero set of $f - 1$ in $\mathbb{P}_{\hat{\Delta}}$ with its μ_e-action and V_∞ the zero set of V in $\mathbb{P}_\Delta$. Then

$$\psi_f = [V] - [V_\infty] \text{ and } \psi_{f,x} = [V] - \mathbb{L}[V_\infty]. \tag{24}$$

Indeed, the first formula follows from Theorem 22 as follows. To each face of $\sigma^\vee$ correspond only three strata of $(f\pi)^{-1}(0)$, namely the intersection of the exceptional divisor with the strict transform of $f^{-1}(0)$ and its complement in these two divisors. Adding corresponding terms for all faces of $\sigma^\vee$ and passing to the e-fold ramified cover we obtain V_∞, $f^{-1}(0) \setminus \{0\}$ and $V \setminus V_\infty$ respectively. Hence

$$\psi_f = [V \setminus V_\infty] + [f^{-1}(0) \setminus \{0\}] + (1 - \mathbb{L})[V_\infty] = [V] - [V_\infty] \tag{25}$$

because $f^{-1}(0) \setminus \{0\}$ is a $\mathbb{G}_m$- bundle over V_∞, hence

$$[f^{-1}(0) \setminus \{0\}] = (\mathbb{L} - 1)[V_\infty].$$

In the formula for $\psi_{f,x}$ we have to subtract the term $[f^{-1}(0) \setminus \{0\}]$ from this, so

$$\psi_{f,x} = [V] - \mathbb{L}[V_\infty].$$

References

1. Batyrev, V.V.: Variations of the mixed Hodge structure of affine hypersurfaces in algebraic tori. Duke Math. J. **69**, 349–409 (1993)
2. Bittner, F.: The universal Euler characteristic for varieties of characteristic zero. Comp. Math. **140**, 1011–1032 (2004)
3. Bittner, F.: On motivic zeta functions and the motivic nearby fibre. Math. Z **249**, 63–83 (2005)
4. Bories, B., Veys, W.: Igusa's p-adic local zeta function and the monodromy conjecture for non-degenerate surface singularities. Preprint (2013) arXiv:1306.6012
5. Danilov, V.I.: The geometry of toric varieties. Russ. Math. Surv. **33**, 97–154 (1978)
6. Danilov, V.I.: Newton polyhedra and vanishing cohomology (Russian). Funktsional Anal i Prilozhen **13**(2), 32–47 (1979)
7. Denef, J., Loeser, F.: Geometry on arc spaces of algebraic varieties. European Congress of Mathematics, vol I (Barcelona, 2000). Progr. Math. vol. 201, pp. 327–348. Birkhäuser, Basel (2001)
8. Ehlers, F.: Eine Klasse komplexer Mannigfaltigkeiten und die Auflösung einiger isolierter Singularitäten. Math. Ann. **218**(2), 127–156 (1975)
9. Fulton, W.: Introduction to Toric Varieties. Annals of Mathematics Studies, vol. 131. Princeton University Press, Princeton (1993)
10. Guibert, G.: Espaces d'arcs et invariants d'Alexander. Comment. Math. Helv. **77**(4), 783–820 (2002)

11. Kempf, G., Knudsen, F., Mumford, D., Saint-Donat, B.: Toroidal embeddings I. Lecture Notes in Math., vol. 139. Springer, Berlin (1974)
12. Kouchnirenko, A.G.: Polyèdres de Newton et nombres de Milnor. Invent. Math. **32**, 1–31 (1976)
13. Looijenga, E.: Motivic measures. Séminaire Bourbaki 52ème année 1999–2000, no. 874. Astérisque **276**, 267–297 (2002)
14. Oda, T.: Convex Bodies and Algebraic Geometry. Springer, Berlin (1988)
15. Peters, C., Steenbrink, J.: Hodge number polynomials for nearby and vanishing cohomology. In: Nagel, J., Peters, C. (eds.) Algebraic Cycles and Motives, Volume 2. London Math Soc Lect Note Series, vol. 344, p. 289–303. Cambridge University Press, Cambridge (2007)
16. Saito, M.: Exponents and Newton polyhedra of isolated hypersurface singularities. Math. Ann. **281**(3), 411–417 (1988)
17. Steenbrink, J.H.M.: Mixed Hodge structure on the vanishing cohomology. In: Holm, P. (ed.) Real and complex singularities (Proc. Ninth Nordic Summer School/NAVF Sympos. Math., Oslo, 1976), pp 525–563. Sijthoff and Noordhoff, Alphen aan den Rijn (1977)
18. Varchenko, A.N.: Zeta-function of Monodromy and Newton's Diagram. Invent. Math. **37**, 253–262 (1976)

Non-Abelian Resonance: Product and Coproduct Formulas

Ştefan Papadima and Alexander I. Suciu

Abstract We investigate the resonance varieties attached to a commutative differential graded algebra and to a representation of a Lie algebra, with emphasis on how these varieties behave under finite products and coproducts.

Keywords Resonance variety • Differential graded algebra • Lie algebra • product • Coproduct

1 Introduction

Resonance varieties emerged as a distinctive object of study in the late 1990s, from the theory of hyperplane arrangements. Their usefulness became apparent in the past decade, when a slew of applications in geometry, topology, group theory, and combinatorics appeared.

The idea consists of turning the cohomology ring of a space X into a family of cochain complexes, parametrized by the first cohomology group $H^1(X,\mathbb{C})$, and extracting certain varieties $\mathcal{R}^i_m(X,\mathbb{C})$ from these data, as the loci where the cohomology of those cochain complexes jumps. Part of the importance of these resonance varieties is their close connection with a different kind of jumping loci: the characteristic varieties of X, which record the jumps in homology with coefficients in rank 1 local systems.

In recent years, various generalizations of these notions have been introduced in the literature, for instance in [2, 3, 5, 7]. The basic idea now is to replace the

Ş. Papadima
Simion Stoilow Institute of Mathematics, P.O. Box 1-764, RO-014700 Bucharest, Romania

Partially supported by the Romanian Ministry of National Education, CNCS-UEFISCDI, grant PNII-ID-PCE-2012-4-0156
e-mail: Stefan.Papadima@imar.ro

A.I. Suciu (✉)
Department of Mathematics, Northeastern University, Boston, MA 02115, USA

Partially supported by NSF grant DMS–1010298 and NSA grant H98230-13-1-0225
e-mail: a.suciu@neu.edu

D. Ibadula, W. Veys (eds.), *Bridging Algebra, Geometry, and Topology*,
Springer Proceedings in Mathematics & Statistics 96,
DOI 10.1007/978-3-319-09186-0_17

cohomology ring of a space by an algebraic analogue, to wit, a commutative, differential graded algebra (A, d), and to replace the coefficient group $\mathbb{C}$ by a finite-dimensional vector space V, endowed with a representation $\theta\colon \mathfrak{g} \to \mathfrak{gl}(V)$, for some finite-dimensional Lie algebra $\mathfrak{g}$. In this setting, the parameter space for the higher-rank resonance varieties, $\mathcal{R}^i_m(A, \theta)$, is no longer $H^1(A)$, but rather, the space of flat, $\mathfrak{g}$-valued connections on A, which, according to the results of Goldman and Millson from [4], is the natural replacement for the variety of rank 1 local systems on X.

In a previous paper, [6, §13], we established some basic product and coproduct formulas for the classical resonance varieties. In this note, we extend those results to the non-abelian case, using some of the machinery developed in [5]. In Theorem 1, we give a general upper bound on the varieties $\mathcal{R}^i_1(A \otimes \bar{A}, \theta)$ in terms of the resonance varieties of the factors and the space of $\mathfrak{g}$-flat connections on the tensor product of the two cdga's. In Theorem 2, we improve this bound to an equality of a similar flavor, in the case when the respective cdga's have zero differentials, and $\mathfrak{g}$ is either $\mathfrak{sl}_2$ or $\mathfrak{sol}_2$. Finally, in Corollary 5 and Theorem 3 we give precise formulas for the varieties $\mathcal{R}^i_1(A \vee \bar{A}, \theta)$ associated with the wedge sum of two cdga's.

2 Flat Connections and Holonomy Lie Algebras

We start by introducing some basic notions (cdga's, flat connections, holonomy Lie algebras), following in rough outline the exposition from [5].

2.1 *Differential Graded Algebras and Lie Algebras*

Let $A = (A^\bullet, d)$ be a commutative, differential graded algebra (cdga) over the field of complex numbers, that is, a positively graded $\mathbb{C}$-vector space $A = \bigoplus_{i\geq 0} A^i$, endowed with a graded-commutative multiplication map $\cdot\colon A^i \otimes A^j \to A^{i+j}$ and a differential $d\colon A^i \to A^{i+1}$ satisfying $d(a \cdot b) = da \cdot b + (-1)^i a \cdot db$, for every $a \in A^i$ and $b \in A^j$.

We will assume throughout that A is connected, i.e., $A^0 = \mathbb{C}$, and of finite q-type, for some $q \geq 1$, i.e., A^i is finite-dimensional, for all $i \leq q$. Let $Z^i(A) = \ker(d\colon A^i \to A^{i+1})$, $B^i(A) = \operatorname{im}(d\colon A^{i-1} \to A^i)$, and $H^i(A) = Z^i(A)/B^i(A)$. For each $i \leq q$, the dimension of this vector space, $b_i(A) = \dim_{\mathbb{C}} H^i(A)$, is finite.

Now let $\mathfrak{g}$ be a Lie algebra over $\mathbb{C}$. On the vector space $A \otimes \mathfrak{g}$, we may define a bracket by $[a \otimes x, b \otimes y] = ab \otimes [x, y]$ and a differential given by $\partial(a \otimes x) = da \otimes x$, for $a, b \in A$ and $x, y \in \mathfrak{g}$. This construction produces a differential graded Lie algebra (dgla), $A \otimes \mathfrak{g} = (A^\bullet \otimes \mathfrak{g}, \partial)$. It is readily verified that the assignment $(A, \mathfrak{g}) \rightsquigarrow A \otimes \mathfrak{g}$ is functorial in both arguments.

2.2 Flat, $\mathfrak{g}$-Valued Connections

Definition 1. An element $\omega \in A^1 \otimes \mathfrak{g}$ is called an *infinitesimal,* $\mathfrak{g}$*-valued flat connection* on (A, d) if ω satisfies the Maurer–Cartan equation,

$$\partial\omega + [\omega, \omega]/2 = 0. \tag{1}$$

We will denote by $\mathcal{F}(A, \mathfrak{g})$ the subset of $A^1 \otimes \mathfrak{g}$ consisting of all flat connections. A typical element in $A^1 \otimes \mathfrak{g}$ is of the form $\omega = \sum_j \eta_j \otimes x_j$, with $\eta_j \in A^1$ and $x_j \in \mathfrak{g}$; the flatness condition amounts to

$$\sum_j d\eta_j \otimes x_j + \sum_{j<k} \eta_j \eta_k \otimes [x_j, x_k] = 0. \tag{2}$$

In the rank one case, i.e., the case when $\mathfrak{g} = \mathbb{C}$, the space $\mathcal{F}(A, \mathbb{C})$ may be identified with the vector space $H^1(A) = \{\omega \in A^1 \mid d\omega = 0\}$. In particular, if $d = 0$, then $\mathcal{F}(A, \mathbb{C}) = A^1$.

The bilinear map $P\colon A^1 \times \mathfrak{g} \to A^1 \otimes \mathfrak{g}$, $(\eta, g) \mapsto \eta \otimes g$ induces a map $P\colon H^1(A) \times \mathfrak{g} \to \mathcal{F}(A, \mathfrak{g})$. The *essentially rank one* part of the set of flat $\mathfrak{g}$-connections on A is the image of this map:

$$\mathcal{F}^1(A, \mathfrak{g}) = P(H^1(A) \times \mathfrak{g}). \tag{3}$$

2.3 Holonomy Lie Algebra

An alternate view of the parameter space of flat connections is as follows. Let $A_i = \operatorname{Hom}(A^i, \mathbb{C})$ be the dual vector space. Let $\nabla\colon A_2 \to A_1 \wedge A_1$ be the dual to the multiplication map $A^1 \wedge A^1 \to A^2$, and let $d_1\colon A_2 \to A_1$ be the dual of the differential $d^1\colon A^1 \to A^2$.

Definition 2 ([5]). The *holonomy Lie algebra* of a cdga $A = (A^\bullet, d)$ is the quotient of the free Lie algebra on the $\mathbb{C}$-vector space A_1 by the ideal generated by the image of $\partial_A = d_1 + \nabla$:

$$\mathfrak{h}(A) = \operatorname{Lie}(A_1)/(\operatorname{im}(\partial_A)). \tag{4}$$

Remark 1. In the case when $d = 0$, the above definition coincides with the classical holonomy Lie algebra $\mathfrak{h}(A) = \operatorname{Lie}(A_1)/(\operatorname{im}(\nabla))$ of K.T. Chen [1]. In this situation, $\mathfrak{h}(A)$ inherits a natural grading from the free Lie algebra, compatible with the Lie bracket. Consequently, $\mathfrak{h}(A)$ is a finitely presented, graded Lie algebra, with generators in degree 1 and relations in degree 2.

In general, though, the ideal generated by $\mathrm{im}(\partial_A)$ is not homogeneous, and the Lie algebra $\mathfrak{h}(A)$ is not graded. Here is a concrete example, extracted from [5].

Example 1. Let A be the exterior algebra on generators x, y in degree 1, endowed with the differential given by $dx = 0$ and $dy = y \wedge x$, and let $\mathfrak{sol}_2$ be the Borel subalgebra of $\mathfrak{sl}_2$. Then $\mathfrak{h}(A) \cong \mathfrak{sol}_2$, as (ungraded) Lie algebras.

The next lemma (see [5, §4] for details) identifies the set of flat, $\mathfrak{g}$-valued connections on a cdga (A, d) with the set of Lie algebra morphisms from the holonomy Lie algebra of (A, d) to $\mathfrak{g}$.

Lemma 1. *The canonical isomorphism* $A^1 \otimes \mathfrak{g} \cong \mathrm{Hom}(A_1, \mathfrak{g})$ *restricts to isomorphisms* $\mathcal{F}(A, \mathfrak{g}) \cong \mathrm{Hom}_{\mathrm{Lie}}(\mathfrak{h}(A), \mathfrak{g})$ *and* $\mathcal{F}^1(A, \mathfrak{g}) \cong \mathrm{Hom}^1_{\mathrm{Lie}}(\mathfrak{h}(A), \mathfrak{g})$.

Here, $\mathrm{Hom}^1_{\mathrm{Lie}}(\mathfrak{h}(A), \mathfrak{g})$ denotes the subset of Lie algebra morphisms with at most one-dimensional image.

3 Resonance Varieties

In this section, we recall the definition of the Aomoto complexes associated with a cdga (A, d) and a representation of a Lie algebra $\mathfrak{g}$, as well as the resonance varieties associated with these data, following the approach from [2, 3, 5].

3.1 *Twisted Differentials*

Let $\theta \colon \mathfrak{g} \to \mathfrak{gl}(V)$ be a representation of our Lie algebra $\mathfrak{g}$ in a finite-dimensional, nonzero $\mathbb{C}$-vector space V. For each flat connection $\omega \in \mathcal{F}(A, \mathfrak{g})$, we make $A \otimes V$ into a cochain complex,

$$(A \otimes V, d_\omega)\colon\ A^0 \otimes V \xrightarrow{d_\omega} A^1 \otimes V \xrightarrow{d_\omega} A^2 \otimes V \xrightarrow{d_\omega} \cdots, \tag{5}$$

using as differential the covariant derivative

$$d_\omega = d \otimes \mathrm{id}_V + \mathrm{ad}_\omega, \tag{6}$$

where ad_ω is defined via the Lie semi-direct product $V \rtimes_\theta \mathfrak{g}$. The flatness condition insures that $d_\omega^2 = 0$. In coordinates, if $\omega = \sum_j \eta_j \otimes x_j$, then

$$d_\omega(\alpha \otimes v) = d\alpha \otimes v + \sum_j \eta_j \alpha \otimes \theta(x_j)(v), \tag{7}$$

for all $\alpha \in A$ and $v \in V$.

It is readily seen that the multiplication map

$$\mu\colon (A,d)\otimes(A\otimes V,d_\omega)\to(A\otimes V,d_\omega),\quad a\otimes(b\otimes v)\mapsto ab\otimes v \tag{8}$$

defines the structure of a differential $A^\bullet$-module on the Aomoto complex $(A^\bullet\otimes V, d_\omega)$. In particular, the graded vector space $H^\bullet(A\otimes V,d_\omega)$ is, in fact, a graded module over the ring $H^\bullet(A)$.

3.2 *Resonance Varieties of a* cdga

Associated with the above data are the resonance varieties

$$\mathcal{R}^i_m(A,\theta)=\{\omega\in\mathcal{F}(A,\mathfrak{g})\mid \dim_{\mathbb{C}} H^i(A\otimes V,d_\omega)\geq m\}. \tag{9}$$

If $\mathfrak{g}$ is finite-dimensional, the sets $\mathcal{R}^i_m(A,\theta)$ are Zariski closed subsets of $\mathcal{F}(A,\mathfrak{g})$, for all $i\leq q$ and $m\geq 0$. In the case when $\mathfrak{g}=\mathbb{C}$ and $\theta=\mathrm{id}_{\mathbb{C}}$, we will simply write $\mathcal{R}^i_m(A)$ for these varieties, viewed as algebraic subsets of $H^1(A)$. Clearly,

$$0\in\mathcal{R}^i_1(A,\theta)\Leftrightarrow 0\in\mathcal{R}^i_1(A)\Leftrightarrow H^i(A)\neq 0. \tag{10}$$

When $d=0$, the varieties $\mathcal{R}^i_m(A)$ are homogeneous subsets of A^1. This happens in the classical case when X is a path-connected space, and $A=H^\bullet(X,\mathbb{C})$ is its cohomology algebra, endowed with the zero differential.

In general, though, the resonance varieties of a cdga are not homogeneous sets, even in the rank 1 case.

Example 2. Let (A,d) be the cdga from Example 1. Then $H^1(A)=\mathbb{C}$, while $\mathcal{R}^1_1(A)=\{0,1\}$.

Lemma 2. *Let* $\omega=\eta\otimes g\in\mathcal{F}^1(A,\mathfrak{g})$.

1. *If* $\omega\in\mathcal{R}^i_1(A,\theta)$, *then* $A^i\neq 0$.
2. *Suppose* $A^i\neq 0$ *and* $d=0$. *Then* $\omega\in\mathcal{R}^i_1(A,\theta)$ *if and only if either* $\eta\in\mathcal{R}^i_1(A)$ *or* $\det(\theta(g))=0$.

Proof. The first claim is clear. When $d=0$, recall that the rank one resonance variety $\mathcal{R}^i_1(A)$ is homogeneous. The second claim then follows from [5, Corollary 3.6].

3.3 *Resonance Varieties of a Lie Algebra*

Let $\mathfrak{h}$ be a finitely generated Lie algebra, and let $\theta\colon\mathfrak{g}\to\mathfrak{gl}(V)$ be a representation of another Lie algebra. Associated with these data are the resonance varieties

$$\mathcal{R}^i_m(\mathfrak{h},\theta)=\{\varphi\in\mathrm{Hom}_{\mathrm{Lie}}(\mathfrak{h},\mathfrak{g})\mid \dim_{\mathbb{C}} H^i(\mathfrak{h},V_{\theta\circ\varphi})\geq m\}, \tag{11}$$

where $V_{\theta\circ\varphi}$ denotes the $\mathbb{C}$-vector space V, viewed as a module over the enveloping algebra $U(\mathfrak{h})$ via the representation $\theta \circ \varphi : \mathfrak{h} \to \mathfrak{gl}(V)$.

Now suppose $\mathfrak{g}$ is finite-dimensional. Then the resonance varieties $\mathcal{R}^i_m(\mathfrak{h}, \theta)$ are Zariski-closed subsets of $\mathrm{Hom}_{\mathrm{Lie}}(\mathfrak{h}, \mathfrak{g})$, for all $i \le 1$ and $m \ge 0$.

Lemma 3 ([5]). *For each $i \le 1$ and $m \ge 0$, the canonical isomorphism $\mathcal{F}(A, \mathfrak{g}) \cong \mathrm{Hom}_{\mathrm{Lie}}(\mathfrak{h}(A), \mathfrak{g})$ restricts to an isomorphism*

$$\mathcal{R}^i_m(A, \theta) \cong \mathcal{R}^i_m(\mathfrak{h}(A), \theta). \tag{12}$$

Example 3. Let $x_1, \dots, x_n$ be a basis for A_1. Using Lemma 3 and [5, Lemma 2.3], we find that

$$\mathcal{R}^0_1(\mathfrak{h}(A), \theta) = \Big\{ \varphi \in \mathrm{Hom}_{\mathrm{Lie}}(\mathfrak{h}(A), \mathfrak{g}) \;\Big|\; \bigcap_{i=1}^{n} \ker(\theta \circ \varphi(x_i)) \neq 0 \Big\}. \tag{13}$$

4 Products

In this section, we study the way the various constructions outlined so far behave under (finite) product operations.

4.1 Holonomy Lie Algebra and Products

Let (A, d) and $(\bar{A}, \bar{d})$ be two cdga's. The tensor product of these two $\mathbb{C}$-vector spaces, $A \otimes \bar{A}$, is again a cdga, with grading $(A \otimes \bar{A})^q = \bigoplus_{i+j=q} A^i \otimes \bar{A}^j$, multiplication $(a \otimes \bar{a}) \cdot (b \otimes \bar{b}) = (-1)^{|\bar{a}||b|}(ab \otimes \bar{a}\bar{b})$, and differential D given on homogeneous elements by $D(a \otimes \bar{a}) = da \otimes \bar{a} + (-1)^{|a|} a \otimes \bar{d}\bar{a}$.

The definition is motivated by the cartesian product of spaces, in which case the Künneth formula gives an isomorphism

$$(H^{\bullet}(X \times \overline{X}, \mathbb{C}), D = 0) \cong (H^{\bullet}(X, \mathbb{C}), d = 0) \otimes (H^{\bullet}(\overline{X}, \mathbb{C}), \overline{d} = 0). \tag{14}$$

In [3, §9], we gave a product formula for holonomy Lie algebras in the 1-formal case. We now extend this formula to cdga's with nonzero differential.

Proposition 1. *Let A and $\bar{A}$ be two connected cdga's. Then the Lie algebra $\mathfrak{h}(A \otimes \bar{A})$ is generated by $A_1 \oplus \bar{A}_1$, subject to the relations $\partial_A(A_2) = 0$, $\partial_{\bar{A}}(\bar{A}_2) = 0$, and $[A_1, \bar{A}_1] = 0$.*

Proof. By construction, $(A \otimes \bar{A})^1 = A^1 \oplus \bar{A}^1$ and $(A \otimes \bar{A})^2 = A^2 \oplus \bar{A}^2 \oplus (A^1 \otimes \bar{A}^1)$. Plainly, D^1 restricts to d^1 on A^1 and to $\bar{d}^1$ on $\bar{A}^1$. It is readily seen that the multiplication map on $A \otimes \bar{A}$ restricts to the multiplication maps on $A^1 \wedge A^1$ on $\bar{A}^1 \wedge \bar{A}^1$, respectively, and to the identity map on $A^1 \otimes \bar{A}^1$. By taking duals, we conclude that $\mathfrak{h}(A \otimes \bar{A})$ has the asserted presentation.

Corollary 1. *The holonomy Lie algebra of a tensor product of* cdga*'s is isomorphic to the (categorical) product of the respective holonomy Lie algebras,*

$$\mathfrak{h}(A \otimes \bar{A}) \cong \mathfrak{h}(A) \times \mathfrak{h}(\bar{A}).$$

4.2 *Flat Connections and Products*

Proposition 1 also yields a formula for the representation variety of a tensor product of cdga's.

Corollary 2. *For any Lie algebra* $\mathfrak{g}$,

$$\begin{aligned}\mathrm{Hom}_{\mathrm{Lie}}(\mathfrak{h}(A \otimes \bar{A}), \mathfrak{g}) = \{(\varphi, \bar{\varphi}) \in \mathrm{Hom}_{\mathrm{Lie}}(\mathfrak{h}(A), \mathfrak{g}) \times \\ \mathrm{Hom}_{\mathrm{Lie}}(\mathfrak{h}(\bar{A}), \mathfrak{g}) \mid [\varphi(x), \bar{\varphi}(\bar{x})] = 0, \ \forall (x, \bar{x}) \in A_1 \oplus \bar{A}_1\}.\end{aligned}$$

Furthermore, if $\mathfrak{g}$ *is abelian, then*

$$\mathrm{Hom}_{\mathrm{Lie}}(\mathfrak{h}(A \otimes \bar{A}), \mathfrak{g}) = \mathrm{Hom}_{\mathrm{Lie}}(\mathfrak{h}(A), \mathfrak{g}) \times \mathrm{Hom}_{\mathrm{Lie}}(\mathfrak{h}(\bar{A}), \mathfrak{g}).$$

For the simple Lie algebra $\mathfrak{sl}_2$ and its Borel subalgebra $\mathfrak{sol}_2$, the above corollary can be made more explicit.

Corollary 3. *If* $\mathfrak{g} = \mathfrak{sl}_2$ *or* $\mathfrak{sol}_2$, *then*

$$\begin{aligned}\mathrm{Hom}_{\mathrm{Lie}}(\mathfrak{h}(A \otimes \bar{A}), \mathfrak{g}) = \{0\} \times \mathrm{Hom}_{\mathrm{Lie}}(\mathfrak{h}(\bar{A}), \mathfrak{g}) \cup \\ \mathrm{Hom}_{\mathrm{Lie}}(\mathfrak{h}(A), \mathfrak{g}) \times \{0\} \cup \mathrm{Hom}^1_{\mathrm{Lie}}(\mathfrak{h}(A \otimes \bar{A}), \mathfrak{g}).\end{aligned}$$

Proof. The inclusion $\supseteq$ is clear. To prove the reverse inclusion, fix bases $\{x_1, \dots, x_n\}$ and $\{\bar{x}_1, \dots, \bar{x}_m\}$ for A_1 and $\bar{A}_1$. Let $\varphi\colon \mathfrak{h}(A \otimes \bar{A}) \to \mathfrak{g}$ be a morphism of Lie algebras, and suppose there are indices i and j such that $\varphi(x_i) \neq 0$ and $\varphi(\bar{x}_j) \neq 0$. We need to prove that the family $\{\varphi(x_1), \dots, \varphi(x_n), \varphi(\bar{x}_1), \dots, \varphi(\bar{x}_m)\}$ has rank 1.

We know from Corollary 2 that $[\varphi(x_k), \varphi(\bar{x}_l)] = 0$, for all $k \in [n]$ and $l \in [m]$. Now note that for any $0 \neq g, h \in \mathfrak{g}$, the following holds: $[g, h] = 0$ if and only if $g = \lambda h$, for some $\lambda \in \mathbb{C}^\times$. The desired conclusion is now immediate.

4.3 Resonance and Products

We now turn to the jump loci of a tensor product of cdga's. We start with a general upper bound for the depth 1 resonance varieties.

Theorem 1. *For any representation* $\theta\colon \mathfrak{g} \to \mathfrak{gl}(V)$,

$$\mathcal{R}_1^q(A \otimes \bar{A}, \theta) \subseteq \Big(\Big(\bigcup_{i \le q} \mathcal{R}_1^i(A,\theta)\Big) \times \Big(\bigcup_{j \le q} \mathcal{R}_1^j(\bar{A},\theta)\Big)\Big) \cap \mathcal{F}(A \otimes \bar{A}, \mathfrak{g}).$$

Proof. By Lemma 1 and Corollary 2, every element $\Omega \in \mathcal{F}(A \otimes \bar{A}, \mathfrak{g})$ can be written as $\Omega = \omega + \bar{\omega}$, for some $\omega \in \mathcal{F}(A, \mathfrak{g})$ and $\bar{\omega} \in \mathcal{F}(\bar{A}, \mathfrak{g})$. Setting up a first-quadrant double complex with $E_0^{i,j} = A^i \otimes \bar{A}^j \otimes V$, horizontal differential $d_\omega\colon E_0^{i,j} \to E_0^{i+1,j}$, and vertical differential $d_{\bar{\omega}}\colon E_0^{i,j} \to E_0^{i,j+1}$, we obtain spectral sequences starting at

$$_{\mathrm{hor}}E_1^{i,j} = H^i(A \otimes V, d_\omega) \otimes \bar{A}^j \ \text{ and } \ _{\mathrm{vert}}E_1^{i,j} = A^i \otimes H^j(\bar{A} \otimes V, d_{\bar{\omega}}), \tag{15}$$

respectively, and converging to $H^{i+j}(A \otimes \bar{A} \otimes V, d_\Omega)$; see (7).

Consequently, if either $H^{\le q}(A \otimes V, d_\omega)$ or $H^{\le q}(\bar{A} \otimes V, d_{\bar{\omega}})$ vanishes, then $H^q(A \otimes \bar{A} \otimes V, d_\Omega) = 0$. In view of definition (9), this completes the proof.

In general, the inclusion from Theorem 1 is strict. We illustrate this phenomenon with a simple example.

Example 4. Let A be the exterior algebra on a single generator in degree 1, let $\mathfrak{g} = \mathfrak{gl}_2$, and let $\theta = \mathrm{id}_{\mathfrak{g}}$. Using Example 3 and Corollary 2, we see that $\mathcal{R}_1^0(A,\theta) = \{g \in \mathfrak{gl}_2 \mid \det(g) = 0\}$, yet

$$\mathcal{R}_1^0(A \otimes A, \theta) = \{(g,h) \in \mathfrak{gl}_2 \times \mathfrak{gl}_2 \mid [g,h] = 0,\ \mathrm{rank}(g \mid h) < 2\},$$

which is a proper subset of $(\mathcal{R}_1^0(A,\theta) \times \mathcal{R}_1^0(A,\theta)) \cap \mathcal{F}(A \otimes A, \mathfrak{g})$.

4.4 Product Formulas for Resonance

Under certain additional hypotheses, the upper bound from Theorem 1 may be improved to an equality. First, as shown in [6, Proposition 13.1], such an equality holds in the formal, rank 1 case.

Proposition 2 ([6]). *Assume both A and $\bar{A}$ have zero differential. Then*

$$\mathcal{R}_1^q(A \otimes \bar{A}) = \bigcup_{i+j=q} \mathcal{R}_1^i(A) \times \mathcal{R}_1^j(\bar{A}).$$

Using this result, we now show that an analogous resonance formula holds for the non-abelian Lie algebras $\mathfrak{sl}_2$ and $\mathfrak{sol}_2$.

Theorem 2. *Assume both A and $\bar{A}$ have zero differential, and $\mathfrak{g} = \mathfrak{sl}_2$ or $\mathfrak{sol}_2$. Then, for any representation $\theta\colon \mathfrak{g} \to \mathfrak{gl}(V)$,*

$$\mathcal{R}_1^q(A \otimes \bar{A}, \theta) = \Big(\bigcup_{i+j=q} \mathcal{R}_1^i(A,\theta) \times \mathcal{R}_1^j(\bar{A},\theta)\Big) \cap \mathcal{F}(A \otimes \bar{A}, \mathfrak{g}).$$

Proof. Proof of inclusion $\subseteq$. Let $\Omega \in \mathcal{R}_1^q(A \otimes \bar{A}, \theta)$. In view of Lemma 2 and Corollary 3, there are two cases to consider: either $\Omega = \omega \in \mathcal{F}(A, \mathfrak{g})$ (the case $\Omega = \bar{\omega} \in \mathcal{F}(\bar{A}, \mathfrak{g})$ being similar), or $\Omega = (\eta + \bar{\eta}) \otimes g$.

In the first case, $H^q(A \otimes \bar{A} \otimes V, d_\Omega) = \bigoplus_{i+j=q} H^i(A \otimes V, d_\omega) \otimes \bar{A}^j$. Hence, $\Omega \in \mathcal{R}_1^q(A \otimes \bar{A}, \theta)$ if and only if there exist indices i and j with $i + j = q$ such that $\omega \in \mathcal{R}_1^i(A, \theta)$ and $\bar{A}^j \neq 0$. Therefore, $\Omega = \omega + 0 \in \mathcal{R}_1^i(A, \theta) \times \mathcal{R}_1^j(\bar{A}, \theta)$, as asserted; see (10).

In the second case, suppose $\Omega \in \mathcal{R}_1^q(A \otimes \bar{A}, \theta)$. There exist then indices i and j with $i + j = q$ such that A^i and $\bar{A}^j$ are nonzero. If $\det \theta(g) = 0$, then we must have $\eta \otimes g \in \mathcal{R}_1^i(A, \theta)$ and $\bar{\eta} \otimes g \in \mathcal{R}_1^j(\bar{A}, \theta)$, and so we are done. Otherwise, Proposition 2 implies that $\eta + \bar{\eta} \in \bigcup_{i+j=q} \mathcal{R}_1^i(A) \times \mathcal{R}_1^j(\bar{A})$. Therefore, $\eta \otimes g \in \mathcal{R}_1^i(A, \theta)$ and $\bar{\eta} \otimes g \in \mathcal{R}_1^j(\bar{A}, \theta)$, for some i and j with $i + j = q$. This completes the first half of the proof.

Proof of inclusion $\supseteq$. Again, we have to analyze the two cases from Corollary 3. When $\Omega = \omega + 0$ with $\omega \in \mathcal{F}(A, \mathfrak{g})$, we know that $\omega \in \mathcal{R}_1^i(A, \theta)$ and $\bar{A}^j \neq 0$, for some i and j with $i + j = q$. As before, we infer that $H^q(A \otimes \bar{A} \otimes V, d_\Omega) \supseteq H^i(A \otimes V, d_\omega) \otimes \bar{A}^j \neq 0$. Hence, $\Omega \in \mathcal{R}_1^q(A \otimes \bar{A}, \theta)$, as claimed.

Finally, assume that $\Omega = (\eta + \bar{\eta}) \otimes g$, where $\eta \otimes g \in \mathcal{R}_1^i(A, \theta)$ and $\bar{\eta} \otimes g \in \mathcal{R}_1^j(\bar{A}, \theta)$, for some i and j with $i + j = q$. When $\det(\theta(g)) \neq 0$, we deduce that $\eta \in \mathcal{R}_1^i(A)$ and $\bar{\eta} \in \mathcal{R}_1^j(\bar{A})$, hence $\eta + \bar{\eta} \in \mathcal{R}_1^q(A \otimes \bar{A})$, by Proposition 2. Consequently, $\Omega \in \mathcal{R}_1^q(A \otimes \bar{A}, \theta)$, as asserted. If $\det(\theta(g)) = 0$, the fact that A^i and $\bar{A}^j$ are both nonzero forces $(A \otimes \bar{A})^q \neq 0$. Hence, once again, $\Omega \in \mathcal{R}_1^q(A \otimes \bar{A}, \theta)$, and we are done.

5 Coproducts

In this final section, we study the way our various constructions behave under (finite) coproducts.

5.1 Holonomy Lie Algebras and Coproducts

Let $A = (A^{\bullet}, d)$ and $\bar{A} = (\bar{A}^{\bullet}, \bar{d})$ be two connected cdga's. Their wedge sum, $A \vee \bar{A}$, is a new connected cdga, whose underlying graded vector space in positive degrees is $A^{+} \oplus \bar{A}^{+}$, with multiplication $(a, \bar{a}) \cdot (b, \bar{b}) = (ab, \bar{a}\bar{b})$, and differential $D = d + \bar{d}$.

The definition is motivated by the wedge operation on pointed spaces, in which case we have a well-known isomorphism

$$(H^{\bullet}(X \vee \overline{X}), D = 0) \cong (H^{\bullet}(X), d = 0) \vee (H^{\bullet}(\overline{X}), \overline{d} = 0). \tag{16}$$

We now extend the coproduct formula for 1-formal spaces from [3, §9], as follows.

Proposition 3. *The holonomy Lie algebra $\mathfrak{h}(A \vee \bar{A})$ is generated by $A_1 \oplus \bar{A}_1$, with relations $\partial_A(A_2) = 0$ and $\partial_{\bar{A}}(\bar{A}_2) = 0$.*

Proof. By construction, $(A \vee \bar{A})^1 = A^1 \oplus \bar{A}^1$, $(A \vee \bar{A})^2 = A^2 \oplus \bar{A}^2$, and $D^1 = d^1 \oplus \bar{d}^1$. Moreover, the multiplication map on $A \vee \bar{A}$ restricts to the multiplication maps on $A^1 \wedge A^1$ and $\bar{A}^1 \wedge \bar{A}^1$, respectively, and is zero when restricted to $A^1 \otimes \bar{A}^1$. The conclusion follows at once.

Corollary 4. *The holonomy Lie algebra of a wedge sum of cdga's is isomorphic to the (categorical) coproduct of the respective holonomy Lie algebras,*

$$\mathfrak{h}(A \vee \bar{A}) \cong \mathfrak{h}(A) \coprod \mathfrak{h}(\bar{A}).$$

5.2 Resonance and Coproducts

As shown in [6, Proposition 13.3], the classical resonance varieties behave nicely with respect to wedges of spaces. Let us recall this result, in a form adapted to our purposes.

Proposition 4 ([6]). *Assume both A and $\bar{A}$ have zero differential. Then, for all $i > 1$,*

$$\mathcal{R}^i_1(A \vee \bar{A}) = \mathcal{R}^i_1(A) \times H^1(\bar{A}) \cup H^1(A) \times \mathcal{R}^i_1(\bar{A}).$$

If, moreover, $b_1(A) > 0$ and $b_1(\bar{A}) > 0$, then

$$\mathcal{R}^1_1(A \vee \bar{A}) = H^1(A) \times H^1(\bar{A}).$$

Our goal for the rest of this section will be to extend the above proposition to the non-abelian setting, for cdga's with nonzero differential. To that end, let $\mathfrak{g}$ be a Lie algebra, and let $\omega \in A^1 \otimes \mathfrak{g}$ and $\bar{\omega} \in \bar{A}^1 \otimes \mathfrak{g}$. Set $\Omega = \omega + \bar{\omega} \in (A \vee \bar{A})^1 \otimes \mathfrak{g}$.

Lemma 4. *Ω is a flat connection if and only if both ω and $\bar{\omega}$ are flat.*

Proof. By definition of multiplication in $A \vee \bar{A}$, we have that $a \cdot \bar{a} = 0$ for every $a \in A^+$ and $\bar{a} \in \bar{A}^+$. Hence, $[\omega, \bar{\omega}] = 0$, and the conclusion follows.

Now let $\theta\colon \mathfrak{g} \to \mathfrak{gl}(V)$ be a representation. Given an element $\omega \in \mathcal{F}(A, \mathfrak{g})$, we write $Z^i_\omega = \ker(d_\omega\colon A^i \otimes V \to A^{i+1} \otimes V)$ and $B^i_\omega = \operatorname{im}(d_\omega\colon A^{i-1} \otimes V \to A^i \otimes V)$, and set $H^i_\omega = Z^i_\omega / B^i_\omega$.

Lemma 5. *For $i > 0$,*

$$d^i_\Omega = d^i_\omega \oplus d^i_{\bar{\omega}}\colon (A^i \otimes V) \oplus (\bar{A}^i \otimes V) \longrightarrow (A^{i+1} \otimes V) \oplus (\bar{A}^{i+1} \otimes V),$$

while for $i = 0$

$$d^0_\Omega = (d^0_\omega, d^0_{\bar{\omega}})\colon (A \vee \bar{A})^0 \otimes V \cong V \longrightarrow (A^1 \otimes V) \oplus (\bar{A}^1 \otimes V).$$

Proof. Both claims follow from (7) and the construction of $A \vee \bar{A}$, by straightforward direct computation.

Corollary 5. *For each $i > 1$ and for any representation $\theta\colon \mathfrak{g} \to \mathfrak{gl}(V)$,*

$$\mathcal{R}^i_1(A \vee \bar{A}, \theta) = \mathcal{R}^i_1(A, \theta) \times \mathcal{F}(\bar{A}, \mathfrak{g}) \cup \mathcal{F}(A, \mathfrak{g}) \times \mathcal{R}^i_1(\bar{A}, \theta).$$

Proof. By Lemma 5, $H^i_\Omega \cong H^i_\omega \oplus H^i_{\bar{\omega}}$. Using this isomorphism, the desired conclusion follows from Lemma 4.

5.3 A Coproduct Formula for Degree 1 Resonance

To conclude, we compute the degree 1 resonance variety of a wedge sum, $\mathcal{R}^1_1(A \vee \bar{A}, \theta)$. We start with two lemmas.

Lemma 6. *There is a surjective homomorphism*

$$H^1((A \vee \bar{A}) \otimes V, d_\Omega) \xrightarrow{\ \Phi\ } H^1(A \otimes V, d_\omega) \oplus H^1(\bar{A} \otimes V, d_{\bar{\omega}}),$$

whose kernel is isomorphic to $(B^1_\omega \oplus B^1_{\bar{\omega}})/\operatorname{im}((d^0_\omega \oplus d^0_{\bar{\omega}}) \circ \Delta)$, where $\Delta\colon V \to V \oplus V$ is the diagonal map.

Proof. Follows from Lemma 5.

Lemma 7. *The homomorphism Φ is injective if and only if $V = Z^0_\omega + Z^0_{\bar{\omega}}$.*

Proof. Start by noting that $V \oplus V = \mathrm{im}(\Delta) \oplus (V \times 0)$. A standard linear algebra argument, then, finishes the proof.

Theorem 3. *Suppose both $b_1(A)$ and $b_1(\bar{A})$ are positive, and at least one of them is greater than 1. Then, for any representation $\theta\colon \mathfrak{g} \to \mathfrak{gl}(V)$,*

$$\mathcal{R}_1^1(A \vee \bar{A}, \theta) = \mathcal{F}(A \vee \bar{A}, \mathfrak{g}).$$

Proof. Set $r = \dim V$. Using our hypothesis, we may assume that $b_1(A) > 1$ and $b_1(\bar{A}) \geq 1$. Supposing $H^1_{\Omega} = 0$ for some $\Omega = \omega + \bar{\omega} \in \mathcal{F}(A \vee \bar{A}, \mathfrak{g})$, we derive a contradiction, as follows.

Lemma 6 implies that $Z^1_{\omega} = B^1_{\omega}$ and $Z^1_{\bar{\omega}} = B^1_{\bar{\omega}}$. Furthermore, the discussion from Sect. 3.1 shows that $Z^1(A) \otimes Z^0_{\omega} \subseteq Z^1_{\omega}$ and $Z^1(\bar{A}) \otimes Z^0_{\bar{\omega}} \subseteq Z^1_{\bar{\omega}}$. Hence,

$$r - \dim Z^0_{\omega} = \dim B^1_{\omega} = \dim Z^1_{\omega} \geq b_1(A) \cdot \dim Z^0_{\omega},$$

and so $\dim Z^0_{\omega} \leq r/(b_1(A) + 1) < r/2$. Similarly, $\dim Z^0_{\bar{\omega}} \leq r/2$.

Using again Lemma 6, we deduce that Φ must be injective. By Lemma 7,

$$r = \dim(Z^0_{\omega} + Z^0_{\bar{\omega}}) \leq \dim Z^0_{\omega} + \dim Z^0_{\bar{\omega}} < r,$$

a contradiction.

Acknowledgements This work was started while the two authors visited the Max Planck Institute for Mathematics in Bonn in April–May 2012. The work was pursued while the second author visited the Institute of Mathematics of the Romanian Academy in June 2012 and June 2013, and MPIM Bonn in September–October 2013. Thanks are due to both institutions for their hospitality, support, and excellent research atmosphere.

References

1. Chen, K.-T.: Extension of C^{∞} function algebra by integrals and Malcev completion of π_1. Adv. Math. **23**, 181–210 (1977)
2. Dimca, A., Papadima, S.: Nonabelian cohomology jump loci from an analytic viewpoint. Comm. Contemp. Math. **16**(4), 1350025 (2014)
3. Dimca, A., Papadima, S., Suciu, A.: Topology and geometry of cohomology jump loci. Duke Math. J. **148**(3), 405–457 (2009)
4. Goldman, W., Millson, J.: The deformation theory of representations of fundamental groups of compact Kähler manifolds. Inst. Hautes Études Sci. Publ. Math. **67**, 43–96 (1988)
5. Măcinic, A., Papadima, S., Popescu, R., Suciu, A.: Flat connections and resonance varieties: from rank one to higher ranks. Preprint `arXiv:1312.1439v2`
6. Papadima, S., Suciu, A.: Bieri–Neumann–Strebel–Renz invariants and homology jumping loci. Proc. Lond. Math. Soc. **100**(3), 795–834 (2010).
7. Papadima, S., Suciu, A.: Jump loci in the equivariant spectral sequence. Math. Res. Lett. **21**(4), arXiv:1302.4075v3 (2014)

Complements of Hypersurfaces, Variation Maps, and Minimal Models of Arrangements

Mihai Tibăr

To Alexandru Dimca and Ştefan Papadima, on the occasion of a great anniversary

Abstract We prove the minimality of the CW-complex structure for complements of hyperplane arrangements in $\mathbf{C}^n$ by using the theory of Lefschetz pencils and results on the variation maps within a pencil of hyperplanes. This also provides a method to compute the Betti numbers of complements of arrangements via global polar invariants.

Keywords Complements of arrangements • Vanishing cycles • Second Lefschetz theorem • Isolated singularities of functions on stratified spaces • Monodromy

1 Introduction

To study the topology of the complement $\mathbf{C}^n \setminus V$ of an affine hypersurface $V \subset \mathbf{C}^n$ one employs Morse theory, see for instance Randell [14], or the Lefschetz method of scanning by pencils of hyperplanes, as done e.g. by Dimca and Papadima in [2]. Both methods yield in particular a CW-complex model of the complement $\mathbf{C}^n \setminus V$. It was proved in the above two papers that whenever V is a union of hyperplanes, then there exists a CW-complex model which is *minimal*, in the sense that the number of q-cells equals the Betti number $b_q(\mathbf{C}^n \setminus V)$, for any q. This notion of minimality was introduced by Papadima and Suciu in [13] for studying the higher homotopy groups of complements of hyperplane arrangements. We give here a new proof of the minimality by using another method. We first prove the following result:

Theorem 1. *Let $\mathscr{A}$ be an affine arrangement of hyperplanes, not necessarily central. Let $V_{\mathscr{A}} \subset \mathbf{C}^n$ denote the union of hyperplanes in $\mathscr{A}$ and let $\mathscr{H}$ be*

M. Tibăr (✉)
Mathématiques, UMR 8524 CNRS, Université Lille 1
59655 Villeneuve d'Ascq, France
e-mail: tibar@math.univ-lille1.fr

D. Ibadula, W. Veys (eds.), *Bridging Algebra, Geometry, and Topology*, Springer Proceedings in Mathematics & Statistics 96,
DOI 10.1007/978-3-319-09186-0_18

a generic hyperplane with respect to $\mathscr{A}$. *Then, one has the isomorphisms of* $\mathbf{Z}$*-modules:* $H_j(\mathbf{C}^n \setminus V_{\mathscr{A}}) \simeq H_j(\mathscr{H} \setminus V_{\mathscr{A}} \cap \mathscr{H})$, *for* $j \leq n-1$ *and* $H_n(\mathbf{C}^n \setminus V_{\mathscr{A}}) \simeq H_n(\mathbf{C}^n \setminus V_{\mathscr{A}}, \mathscr{H} \setminus V_{\mathscr{A}} \cap \mathscr{H})$.

Moreover, the complement $\mathbf{C}^n \setminus V_{\mathscr{A}}$ *has a minimal model.*

Our alternate proof uses the behavior of the *variation maps* within a pencil of hyperplanes on $\mathbf{C}^n \setminus V_{\mathscr{A}}$. It is based on a particular case (see Theorem 2) of a general result on vanishing cycles of pencils, which involves variation maps, proved in [21, 24]. We discuss in Sect. 2 some aspects of the topology of pencils on complements of affine hypersurfaces, extracted from a general theory of *non-generic Lefschetz pencils* of hypersurfaces, which we have developed in a series of papers [21–25]. In this context, we also give a method to compute inductively the betti numbers of complements of arrangements by using global polar invariants [20].

This question was brought to our attention by Ştefan Papadima in spring 2000 in connection with [13] (a preprint at that time) and with the earlier paper [20] in which we construct CW-complex models for affine hypersurfaces by using pencils and global polar curves (see Sect. 3.3). This note was essentially written in 2003 but not published ever since. However, we think that it might be still of current interest also because of the recent proof by J. Huh [6] of a conjecture about the *polar degree* stated by Dimca and Papadima [2], of which one of the main ingredients is the non-generic Lefschetz pencil theory which we also use here.

2 Complements of Hypersurfaces

Let $V = \{f = 0\}$ be a hypersurface in $\mathbf{C}^n$. The complement $\mathbf{C}^n \setminus V$ is a Stein manifold, since it can be viewed as the hypersurface $\{tf(x) = 1\}$ in $\mathbf{C}^{n+1}$. It therefore has the homotopy type of a CW-complex of dimension $\leq n$, by Hamm's result [3]. For a generic hyperplane $\mathscr{H} \in \mathbf{C}^n$ we have that the pair $(\mathbf{C}^n \setminus V, \mathscr{H} \setminus V \cap \mathscr{H})$ is $(n-1)$-connected, by Lefschetz type results [3, 4], see also [23, Thm.4.1]. One has the following well-known consequence:

Proposition 1. *The space* $\mathbf{C}^n \setminus V$ *is obtained, up to homotopy type, by attaching to the slice* $\mathscr{H} \setminus V \cap \mathscr{H}$ *a certain number of* n*-cells.* □

In general, hypersurface complements do not have minimal models: examples are given in [13], one of the simplest being the case of the plane cusp $V = \{x^2 - y^3 = 0\}$.

It has been observed that the topology of the complement $\mathbf{C}^n \setminus V$ depends on the singularities of V and also on their *position*, see [9, 10]. Moreover, if V is not (stratified) transversal to the hyperplane at infinity, then the non-transversality points may influence the topology, see [9, 11].

A new viewpoint appeared more recently [11,22]: consider a polynomial function $f : \mathbf{C}^n \to \mathbf{C}$ of which V is a fiber, and relate the topology of the complement to the singularities of f. It is shown in [11] that one has two situations: either V is a general fiber of f or a special one. For some fixed f, special (or "atypical") fibers are finitely many and have either singularities in $\mathbf{C}^n$ or have, in some sense, singularities "at infinity" (see e.g. [16]). We have:

Proposition 2 ([11]). *Let V be a general fiber of some polynomial $f : \mathbf{C}^n \to \mathbf{C}$. Then $\mathbf{C}^n \setminus V$ is homotopy equivalent to the wedge $S^1 \vee S(V)$, where $S(V)$ denotes the suspension over V. The cup-product in the cohomology ring of $\mathbf{C}^n \setminus V$ is trivial.*

Even if in the above statement V is non-singular, the complexity of the singularities at infinity of the polynomial f influences the topology of V (see [11, 16] for examples). In certain situations, the general fiber of a polynomial function may be a bouquet of spheres of dimension $n-1$. It is the case when f has isolated singularities at infinity. We send to [23] for a survey and more bibliography on singularities at infinity of polynomials.

When V is an atypical fiber of a polynomial, we have the following result.

Proposition 3 ([11]). *Let $V = f^{-1}(0)$ be an atypical fiber of the polynomial function $f : \mathbf{C}^{n+1} \to \mathbf{C}$. If the general fiber of f is s-connected, $s \geq 2$, then $\pi_i(\mathbf{C}^{n+1} \setminus V) = 0$, for $1 < i \leq s$, and $\pi_1(\mathbf{C}^{n+1} \setminus V) = \mathbf{Z}$.* □

In particular, if f has "isolated singularities at infinity," then the above discussion yields that $\pi_i(\mathbf{C}^{n+1} \setminus V) = 0$ for $1 < i \leq n-1$.

3 Variation Maps of Pencils of Affine Hypersurfaces

3.1 Pencils with Isolated Singularities

The two methods of investigating the topology of complements, by Morse functions or by Lefschetz pencils, are actually close in spirit. The latter allows one to use the full power of complex geometry and we shall stick to it in this paper.

In several recent papers we have introduced and used a general concept of *non-generic pencils of hypersurfaces* (e.g., [19,22,23]), which may have singularities in the axis. Here we only use pencils of hyperplanes and with "no singularities in the axis," as we describe in the following.

We consider our complement $\mathbf{C}^n \setminus V$ as embedded into the projective space $\mathbf{P}^n$, and we identify it to $\mathbf{P}^n \setminus (\bar{V} \cup H^\infty)$, where $\bar{V}$ denotes the projective closure of the affine hypersurface $V = f^{-1}(0)$ and H^∞ is the hyperplane at infinity of $\mathbf{P}^n$. Then consider the following pencil of hyperplanes:

$$l(x) - tx_0 = 0, \tag{1}$$

where $t \in \mathbf{C}$, $l : \mathbf{C}^n \to \mathbf{C}$ is a linear function and x_0 is the coordinate at infinity of $\mathbf{P}^n$. This pencil defines a holomorphic function $t := l/x_0$ on $\mathbf{C}^n = \mathbf{P}^n \setminus H^\infty$, where H^∞ denotes the hyperplane at infinity. Such a pencil is not generic with respect to the divisor $\bar{V} \cup H^\infty$ since the axis $A := \{l = x_0 = 0\}$ is included into H^∞ and hence A is not transversal to any Whitney stratification of the pair $(\mathbf{P}^n, \bar{V} \cup H^\infty)$. Nevertheless, we show that this pencil is *without singularities in the axis*, in the sense of [22, Definitions 2.2, 2.3].

The projective hypersurface $\bar{V} \cup H^\infty$ has a canonical minimal Whitney stratification, which we denote by $\mathscr{W}$. In particular, the intersection $\bar{V} \cap H^\infty$ is a union of strata. Then we consider the product stratification $\mathscr{W} \times \mathbf{C}$ in the product space $\mathbf{P}^n \times \mathbf{C}$.

By a Bertini type result, there is a Zariski-open dense set $\Omega \subset \check{\mathbf{P}}^{n-1}$ of linear forms $l : \mathbf{C}^n \to \mathbf{C}$ such that for any $l \in \Omega$, the projective hyperplane $\{l = 0\} \subset H^\infty \simeq \mathbf{P}^{n-1}$ is transversal within H^∞ to all strata included into $\bar{V} \cap H^\infty$. In particular $\{l = 0\}$ avoids all point-strata inside H^∞.

For such $l \in \Omega$, the hyperplane $\mathbf{H} \subset \mathbf{P}^n \times \mathbf{C}$ defined by the equation (1) is transversal within $\mathbf{P}^n \times \mathbf{C}$ to all product-strata included into $H^\infty \times \mathbf{C}$. Then the stratification $\mathscr{W} \times \mathbf{C}$ induces a stratification on $\mathbf{H}$, call it $\mathscr{S}$, which is also Whitney, by the transversality of the intersection.

Moreover, $\mathscr{S}$ has the property that all its strata which are included into $H^\infty \times \mathbf{C}$ have a product structure, by the line $\mathbf{C}$. It then follows that each member of the pencil (i.e., for fixed $t \in \mathbf{C}$) is transversal to all strata of $\mathscr{S}$ included into $H^\infty \times \mathbf{C}$. Equivalently, the projection to $\mathbf{C}$ has no stratified singularities in the neighborhood of $\mathbf{H} \cap (H^\infty \times \mathbf{C})$. In such a case we say that the pencil (1) has no singularities in the axis. It follows that this pencil can have singularities only outside the axis and that they are isolated. Namely, there are finitely many points on V where the projection to the second factor $p : \mathbf{H} \to \mathbf{C}$ has a stratified singularity, with respect to the stratification $\mathscr{S}$. The set of these points will be denoted by $\mathrm{Sing}_{\mathscr{S}}\, p$.

3.2 Variation Maps

We recall from [21, 24, 25] and adapt to our case the construction of the global variation maps associated with a pencil. Let us fix some notation. Let $X := \mathbf{C}^n \setminus V$ and note that X can be identified to $\mathbf{H} \cap ((\mathbf{C}^n \setminus V) \times \mathbf{C})$.

For any $M \subset \mathbf{C}$, we denote $\mathbf{H}_M := p^{-1}(M)$ and $X_M := \mathbf{H}_M \cap ((\mathbf{C}^n \setminus V) \times \mathbf{C})$. Let $\mathrm{Sing}_{\mathscr{S}}\, p = \cup_{i,j}\{a_{ij}\}$, where $\Lambda := p(\mathrm{Sing}_{\mathscr{S}}\, p) = \{a_1, \ldots, a_p\}$ and a_{ij} denotes some point of $\mathrm{Sing}_{\mathscr{S}}\, p \cap p^{-1}(a_i)$.

For $c \in \mathbf{C} \setminus \Lambda$ we say that $\mathbf{H}_c$, resp. X_c, is a *general fiber* of $p : \mathbf{H} \to \mathbf{C}$, resp. of $p_| : \mathbf{H} \cap ((\mathbf{C}^n \setminus V) \times \mathbf{C}) \to \mathbf{C}$. Indeed, $p_|$ can be identified to $l_| : \mathbf{C}^n \setminus V \to \mathbf{C}$ and X_c is just $l^{-1}(c) \cap X$.

At some singularity $a_{ij} \in V$, we choose a ball B_{ij} centered at a_{ij}. For a small enough radius of B_{ij}, this is a "Milnor ball" of the holomorphic function p at a_{ij}. Next we may take a small enough disc $D_i \subset \mathbf{C}$ at $a_i \in \mathbf{C}$, so that (B_{ij}, D_i) is Milnor data for p at a_{ij}. Moreover, we may do this for all (finitely many) singularities in the fiber $\mathbf{H}_{a_i}$, keeping the same disc D_i, provided it is small enough.

Now the restriction of p to $\mathbf{H}_{D_i} \setminus \cup_j B_{ij}$ is a trivial fibration over D_i. One may construct a stratified vector field which trivializes this fibration and such that this vector field is tangent to the boundaries of the balls $\mathbf{H}_{D_i} \cap \partial \bar{B}_{ij}$. Using this, we may also construct a geometric monodromy of the fibration $p_| : \mathbf{H}_{\partial \bar{D}_i} \to \partial \bar{D}_i$ over the circle $\bar{D}_i$, such that this monodromy is the identity on the complement of the balls, $\mathbf{H}_{\partial \bar{D}_i} \setminus \cup_j B_{ij}$. The same is then true, when replacing $\mathbf{H}_{\partial \bar{D}_i}$ by $X_{\partial \bar{D}_i}$.

Fix some point $c_i \in \partial \bar{D}_i$. We have the geometric monodromy representation:

$$\rho_i : \pi_1(\partial \bar{D}_i, c_i) \to \mathrm{Iso}(X_{c_i}, X_{c_i} \setminus \cup_j B_{ij}),$$

where Iso(.,.) denotes the group of relative isotopy classes of stratified homeomorphisms (which are C^∞ along each stratum). It follows that the geometric monodromy restricted to $X_{c_i} \setminus \cup_j B_{ij}$ is the identity.

As shown above, we may identify the fiber $X_{c_i} \setminus \cup_j B_{ij}$ to the fiber $X_{a_i} \setminus \cup_j B_{ij}$ in the trivial fibration over D_i. Furthermore, in local coordinates at a_{ij}, X_{a_i} is a germ of a complex analytic space; hence, for a small enough ball B_{ij}, the set $B_{ij} \cap X_{a_i}$ retracts to $\partial \bar{B}_{ij} \cap X_{a_i}$, by the local conical structure of analytic sets [1]. Therefore X_{a_i} is homotopy equivalent, by retraction, to $X_{a_i} \setminus \cup_j B_{ij}$.

Notation. *Due to the above homotopy equivalences, we shall freely use* $X^*_{a_i}$ *as notation for* $X_{c_i} \setminus \cup_j B_{ij}$ *whenever we consider the pair* (X_{c_i}, X_{a_i}).

It then follows that the geometric monodromy induces an algebraic monodromy, in any dimension q:

$$\nu_i : H_q(X_{c_i}, X^*_{a_i}; \mathbf{Z}) \to H_q(X_{c_i}, X^*_{a_i}; \mathbf{Z}),$$

such that the restriction $\nu_i : H_q(X^*_{a_i}) \to H_q(X^*_{a_i})$ is the identity.

Consequently, any relative cycle $\delta \in H_q(X_{c_i}, X^*_{a_i}; \mathbf{Z})$ is sent by the morphism $\nu_i - \mathrm{id}$ to an absolute cycle. In this way we define a *variation map*, for any $q \geq 0$:

$$\mathrm{var}_i : H_q(X_{c_i}, X^*_{a_i}; \mathbf{Z}) \to H_q(X_{c_i}; \mathbf{Z}). \tag{2}$$

Variation morphisms are basic ingredients in the description of the behavior of vanishing cycles of global and local fibrations at singular fibers of holomorphic functions, see e.g. [7,12,15], [19, 4.4]. Zariski already used $\nu_i - \mathrm{id}$ in dimension 2, in his well-known theorem for the fundamental group. We shall use [24, Theorem 4.4] in the following form adapted to our particular case.

Theorem 2 ([21,24]). *Let* $V \subset \mathbf{C}^n$, $l \in \Omega$ *and let* X_c *be a general member of the pencil, as above. Then* $H_q(X, X_c) = 0$ *for* $q \leq n-1$ *and the kernel of the surjection* $H_{n-1}(X_c) \to H_{n-1}(X)$ *is generated by the images of the variation maps* var_i*, for* $i = \overline{1, p}$.

The first claim is also a consequence of the connectivity result stated in Proposition 1. The second claim is highly nontrivial and is proved in [24]. All the assumptions made in [24, Theorem 4.4] are clearly verified, except of one, which we still need to verify: $H_q(X_c, X^*_{a_l}) = 0$ for $q \leq n - 2$. This is indeed true by the following reason. In [24, 3.7, 3.9] it is shown that the named condition is satisfied whenever $H_q(X_D, X_c) = 0$ for $q \leq n-1$, where D is a small enough disc centered at some value $a \in \Lambda$. But the later condition is fulfilled by our [19, Corollary 2.7], which is based on Hamm and Lê's results in [5].

3.3 Number of Cells and Polar Invariants

Vanishing cycles in a pencil of hypersurfaces have been investigated in large generality, for example in [16, 19, 20, 23]. If the hypersurface $V \subset \mathbf{C}^n$ is given by $f = 0$ then, for some linear function l, one defines the global polar variety:

$$\Gamma(l, f) := \text{closure}\{\text{Sing}(l, f) \setminus \text{Sing} f\} \subset \mathbf{C}^n.$$

By the *global polar curve lemma* [19, Lemma 2.4], it follows that $\Gamma(l, f)$ is either empty or it is a curve, provided that l is general enough. This means that l can be taken out of a Zariski-open set $\tilde{\Omega} \subset \Omega \subset \check{\mathbf{P}}^{n-1}$, see *loc.cit.*) Global polar curves appeared for the first time in [19] in the study of the topology at infinity of polynomial functions. Local polar varieties have been introduced by Lê D.T. and B. Teissier and are currently used in the literature. We refer the reader to [17, 20, 23] for different aspects of global polar curves.

By [19, Theorem 4.6] and especially [23, Corollary 4.3] we have that the Betti number $b_n(X, X_c)$ is equal to $\lambda := \sum_{i=1}^{p} \lambda_{a_i}$, where λ_{a_i} is the polar number at the atypical value of the pencil a_i. According to [20, Definition 3.5], λ_{a_i} is a non-negative integer equal to the following difference of intersection multiplicities:

$$\lambda_{a_i} = \text{int}(\Gamma(l, f), X_{c_i}) - \text{int}(\Gamma(l, f), X_{a_i}),$$

where c_i is a nearby typical value of the pencil.

The difference of intersection numbers appears as follows. First observe that $\Gamma(l, f)$ does not intersect some small neighborhood of $\bar{V} \cap H^\infty$. Next, the curve $\Gamma(l, f)$ is algebraic, therefore it intersects V at a finite number of points. It is a general fact proved by Lê D.T [8] that these points are among the stratified singularities of the restriction of the function l to V. On the other hand, these singularities are isolated since l is general. In the particular case of arrangements of hyperplanes, the stratified singularities of the restriction $l_{|V}$ are precisely the zero-dimensional strata of the canonical Whitney stratification of V. So the polar curve is eventually nonempty in the neighborhood of these points. As c_i tends to a_i, the points of intersection of $\Gamma(l, f)$ with X_{c_i}, in some neighborhood of some zero-dimensional stratum of V which is also on $\bar{X}_{a_i}$, tend to this point-stratum. Consequently, there is loss of intersection multiplicity from $\text{int}(\Gamma(l, f), X_{c_i})$ to $\text{int}(\Gamma(l, f), X_{a_i})$ and this loss is localized near the point-strata of V.

Moreover, the space X is obtained from the slice X_c by attaching cells of dimension n only, by [18, 23], see also [25, Theorem 9.3.1]. We thus have a geometric interpretation of the topological quotient space X/X_c as a bouquet of $\lambda = b_n(X, X_c)$ n-spheres. By repeated slicing we get similar formulas in lower dimensions. In case of complements of hyperplane arrangements, we shall see in the next section that the relative betti number $b_n(X, X_c)$ equals the absolute betti number $b_n(X)$.

4 Proof of Theorem 1

We proceed by induction on the dimension. Our arrangement of hyperplanes $\mathscr{A}$ defines a natural Whitney stratification $\mathscr{W}' = \{W_B\}_{B\subset\mathscr{A}}$ on $V_{\mathscr{A}}$ which is also the coarsest one. More explicitly, the strata are defined as follows. Let V_B denote the intersection of all hyperplanes corresponding to the indices of some subset $B \subset \mathscr{A}$. Then $W_B := V_B \setminus \cup_{C\subsetneq B} V_C$. This stratification is Whitney since along any stratum W_B, by some analytic local change of coordinates, the space $V_{\mathscr{A}}$ has the product structure {transversal slice} $\times W_B$.

Since the hyperplane at infinity $H^\infty \subset \mathbf{P}^n$ is transversal to all the strata, the induced natural stratification on $\bar{V}_{\mathscr{A}} \cap H^\infty$ is Whitney and it is the coarsest one. This is what we have denoted by $\mathscr{W}$ in Sect. 3.1.

Let $l \in \Omega$ define a generic pencil of hyperplanes in $\mathbf{C}^n$, as in Sect. 3.1. We have seen before that the genericity of the pencil amounts to the condition that the axis of the pencil $A = \{l = 0\} \cap H^\infty$ is transversal to all the strata of $\mathscr{W}$.

Let $\mathscr{H}$ denote a generic member of the pencil. By Proposition 1, we get that the long exact sequence of the pair $(\mathbf{C}^n\setminus V_{\mathscr{A}}, \mathscr{H}\setminus V_{\mathscr{A}}\cap\mathscr{H})$ splits into the isomorphisms $H_j(\mathbf{C}^n\setminus V_{\mathscr{A}}) \simeq H_j(\mathscr{H}\setminus V_{\mathscr{A}}\cap\mathscr{H})$, for $j \le n-1$, and the following exact sequence:

$$\begin{aligned} 0 \to H_n(\mathbf{C}^n \setminus V_{\mathscr{A}}) \to H_n(\mathbf{C}^n \setminus V_{\mathscr{A}}, \mathscr{H} \setminus V_{\mathscr{A}} \cap \mathscr{H}) \to \\ \to H_{n-1}(\mathscr{H} \setminus V_{\mathscr{A}} \cap \mathscr{H}) \xrightarrow{\iota_*} H_{n-1}(\mathbf{C}^n \setminus V_{\mathscr{A}}) \to 0. \end{aligned} \tag{3}$$

We claim that ι_* is injective. By Theorem 2 we have that $\ker \iota_* = \sum_{i=1}^{p} \mathrm{im}(\mathrm{var}_i)$. In our case, we may show that var_i is *trivial*, for any i. Our pencil has no singularities in the axis, it is a pencil of hyperplanes and $V_{\mathscr{A}}$ is a union of hyperplanes too. It follows that the singularities of the pencil are exactly the point-strata of the canonical stratification $\mathscr{W}$ of $V_{\mathscr{A}}$. Then the atypical members of the pencil are those which pass through such points. The pencil can be chosen generic enough such that each member of it contains at most one such point-stratum.

Let us focus on some atypical value a_i. We may assume, without affecting the generality, that the singularity of $\bar{X}_{a_i}$ is the origin of $\mathbf{C}^n$. Consider the map germ $(\mathbf{C}^n, 0) \to (\mathbf{C}^n, 0)$ such that $x \mapsto x\exp(2i\pi t)$ for any coordinate x. Taking t as parameter, this defines a family of diffeomorphisms which preserve the arrangement $V_{\mathscr{A}}$ and its complement $\mathbf{C}^n\setminus V_{\mathscr{A}}$, and moves the hyperplane X_{c_i} of our pencil into the hyperplane $X_{\exp(2i\pi t)c_i}$, over the circle $\partial \bar{D}_i \subset \mathbf{C}$. For $t = 1$, this yields a geometric monodromy of X_{c_i} around the value a_i, at the origin of $\mathbf{C}^n$.

By its definition, this geometric monodromy is the identity on the hyperplane X_{c_i} and therefore also on $X^*_{a_i} \subset X_{c_i}$ (see the definition of the notation $X^*_{a_i}$ at Sect. 3.2). It then follows (from the definition of the variation map, see Sect. 3.2) that the variation of this monodromy is trivial, i.e., $\mathrm{im}(\mathrm{var}_i) = 0$. We have proved in this way that $\ker \iota_* = 0$, which also means that the above exact sequence (3) splits in the middle. This proves the second part of our first statement.

By the attaching result discussed at the end of Sect. 3 (see also [25, Theorem 9.3.1]) we get that the number of the n-cells attached to $\mathcal{H} \setminus V_{\mathcal{A}} \cap \mathcal{H}$ in order to obtain $\mathbf{C}^n \setminus V_{\mathcal{A}}$ is equal to $b_n(\mathbf{C}^n \setminus V_{\mathcal{A}})$. The minimal model claim follows then by iterated slicing.

References

1. Burghelea, D., Verona, A.: Local homological properties of analytic sets. Manuscripta Math. **7**, 55–66 (1972)
2. Dimca, A., Papadima, S.: Hypersurface complements, Milnor fibers and higher homotopy groups of arrangments. Ann. Math. 158(2), 473–507 (2003)
3. Hamm, H.: Zum Homotopietyp Steinischer Räume. J. Reine Angew. Math. 339, 121–135 (1983)
4. Hamm, H.: Lefschetz theorems for singular varieties. Arcata Singularities Conference. Proc. Symp. Pure Math. **40**(I), 547–557 (1983)
5. Hamm, H.A., Lê, D.T.: Rectified homotopical depth and Grothendieck conjectures. The Grothendieck Festschrift, Collect. Artic. in Honor of the 60th Birthday of A. Grothendieck. Vol. II. Prog. Math. **87**, 311–351 (1990)
6. Huh, J.: Milnor numbers of projective hypersurfaces with isolated singularities. Duke Math. J. **163**(8), 1525–1548 (2014)
7. Lamotke, K.: The topology of complex projective varieties after S. Lefschetz. Topology **20**, 15–51 (1981)
8. Lê, D.T.: Calcul du nombre de cycles évanouissants d'une hypersurface complexe. Ann. Inst. Fourier (Grenoble) 23(4), 261–270 (1973)
9. Libgober, A.: Homotopy groups of the complements to singular hypersurfaces, II. Ann. Math. **139**, 117–144 (1994)
10. Libgober, A.: Position of singularities of hypersurfaces and the topology of their complements. J. Math. Sci. New York **82**(1), 3194–3210 (1996)
11. Libgober, A., Tibăr, M.: Homotopy groups of complements and nonisolated singularities. Int. Math. Res. Not. **17**, 871–888 (2002)
12. Milnor, J.: Singular points of complex hypersurfaces. Annals of Mathematics Studies, No. 61, Princeton University Press, Princeton, N.J.; University of Tokyo Press, Tokyo pp iii+122 (1968)
13. Papadima, S., Suciu, A.: Higher homotopy groups of complements of complex hyperplane arrangements. Adv. Math. **165**(1), 71–100 (2002)
14. Randell, R.: Morse theory, Milnor fibers and minimality of hyperplane arrangements. Proc. Am. Math. Soc. **130**(9), 2737–2743 (2002)
15. Siersma, D.: Variation mappings on singularities with a 1-dimensional critical locus. Topology **30**(3), 445–469 (1991)

16. Siersma, D., Tibăr, M.: Singularities at infinity and their vanishing cycles. Duke Math. J. **80**(3), 771–783 (1995)
17. Siersma, D., Tibăr, M.: Singularities at infinity and their vanishing cycles, II. Monodromy. Publ. Res. Inst. Math. Sci. **36**(6), 659–679 (2000)
18. Siersma, D., Tibăr, M.: On the vanishing cycles of a meromorphic function on the complement of its poles. Real and Complex Singularities, pp. 277–289. Contemp. Math., vol. 354. American Mathematical Society, Providence (2004)
19. Tibăr, M.: Topology at infinity of polynomial maps and Thom regularity condition. Compositio Math. **111**(1), 89–109 (1998)
20. Tibăr, M.: Asymptotic equisingularity and topology of complex hypersurfaces. Int. Math. Res. Not. **18**, 979–990 (1998)
21. Tibăr, M.: Topology of Lefschetz fibrations in complex and symplectic geometry. Newton Institute preprint NI01029 (2001)
22. Tibăr, M.: Connectivity via nongeneric pencils. Int. J. Math. **13**(2), 111–123 (2002)
23. Tibăr, M.: Singularities and topology of meromorphic functions. Trends in Singularities, pp. 223–246. Trends Math., Birkhäuser, Basel (2002)
24. Tibăr, M.: Vanishing cycles of pencils of hypersurfaces. Topology **43**(3), 619–633 (2004)
25. Tibăr, M.: Polynomials and vanishing cycles. Cambridge Tracts in Mathematics, vol. 170. Cambridge University Press, Cambridge (2007)

MIX
Papier aus verantwortungsvollen Quellen
Paper from responsible sources
FSC® C105338

If you have any concerns about our products,
you can contact us on
ProductSafety@springernature.com

In case Publisher is established outside the EU,
the EU authorized representative is:
Springer Nature Customer Service Center GmbH
Europaplatz 3, 69115 Heidelberg, Germany

Printed by Libri Plureos GmbH
in Hamburg, Germany